AF358749

Optical Systems for Astronomy

Optical Systems for Astronomy

Guest Editors

Eduard Muslimov
Gennady G. Valyavin

Basel • Beijing • Wuhan • Barcelona • Belgrade • Novi Sad • Cluj • Manchester

Guest Editors

Eduard Muslimov
NOVA Optical IR Instrumentation Group
ASTRON
Dwingeloo
The Netherlands

Gennady G. Valyavin
Special Astrophysical Observatory of the
Russian Academy of Sciences
Nizhnij Arkhyz
Russia

Editorial Office
MDPI AG
Grosspeteranlage 5
4052 Basel, Switzerland

This is a reprint of articles from the Special Issue published online in the open access journal *Photonics* (ISSN 2304-6732) (available at: www.mdpi.com/journal/photonics/special_issues/optial_systems_for_astronomy).

For citation purposes, cite each article independently as indicated on the article page online and using the guide below:

Lastname, A.A.; Lastname, B.B. Article Title. *Journal Name* **Year**, *Volume Number*, Page Range.

ISBN 978-3-7258-1946-1 (Hbk)
ISBN 978-3-7258-1945-4 (PDF)
https://doi.org/10.3390/books978-3-7258-1945-4

Cover image courtesy of Gennady G. Valyavin
Special Astrophysical Observatory of RAS, Russia

Contents

Preface

Astronomical instrumentation is at the doorstep of a new era. In the domain of space astronomy, we have obtained the first scientific outcome of the recently launched James Webb Space Telescope and are experiencing rapid growth in the field of small missions, including CubeSats, a newly available platform. In the domain of ground-based astronomy, we are approaching the commissioning of extremely large telescopes and their first-light instruments, but we are also witnessing a number of small missions, which have become possible with new image sensors, robotic mounts, and other technologies. The present Special Issue of *Photonics* is focused on novel developments for optical systems in astronomy. It could be of interest for engineers developing scientific instrumentation and for astronomers.

Eduard Muslimov and Gennady G. Valyavin
Guest Editors

Review

Overview on Space-Based Optical Orbit Determination Method Employed for Space Situational Awareness: From Theory to Application

Zhe Zhang [1,2], Gaopeng Zhang [1,2,*], Jianzhong Cao [1,2,*], Cheng Li [1,2], Weining Chen [1,2], Xin Ning [3] and Zheng Wang [4]

1   Xi'an Institute of Optics and Precision Mechanics of CAS, Xi'an 710119, China; zhangzhe@opt.ac.cn (Z.Z.); licheng@opt.ac.cn (C.L.); cwn@opt.ac.cn (W.C.)
2   Xi'an Key Laboratory of Spacecraft Optical Imaging and Measurement Technology, Xi'an 710119, China
3   School of Astronautics, Northwestern Polytechnical University, Xi'an 710119, China; ningxin@nwpu.edu.cn
4   Research Center for Unmanned System Strategy Developments, Northwestern Polytechnical University, Xi'an 710119, China; wz_rcussd@nwpu.edu.cn
*   Correspondence: zhanggaopeng@opt.ac.cn (G.Z.); cjz@opt.ac.cn (J.C.)

Abstract: Leveraging space-based optical platforms for space debris and defunct spacecraft detection presents several advantages, including a wide detection range, immunity to cloud cover, and the ability to maintain continuous surveillance on space targets. As a result, it has become an essential approach for accomplishing tasks related to space situational awareness. However, the prediction of the orbits of space objects is crucial for the success of such missions, and current technologies face challenges related to accuracy, reliability, and practical efficiency. These challenges limit the performance of space-based optical space situational awareness systems. To drive progress in this field and establish a more effective and reliable space situational awareness system based on space optical platforms, this paper conducts a retrospective overview of research advancements in this area. It explores the research landscape of orbit determination methods, encompassing orbit association methods, initial orbit determination methods, and precise orbit determination methods, providing insights from international perspectives. The article concludes by highlighting key research areas, challenges, and future trends in current space situational awareness systems and orbit determination methods.

Keywords: space situational awareness; orbit association methods; initial orbit determination; precise orbit determination; angle-only measurement

Citation: Zhang, Z.; Zhang, G.; Cao, J.; Li, C.; Chen, W.; Ning, X.; Wang, Z. Overview on Space-Based Optical Orbit Determination Method Employed for Space Situational Awareness: From Theory to Application. *Photonics* **2024**, *11*, 610. https://doi.org/10.3390/photonics11070610

Received: 19 December 2023
Revised: 8 June 2024
Accepted: 10 June 2024
Published: 27 June 2024

1. Introduction

Since the launch of the first satellite on 4 October 1957, approximately 15,302 artificial spacecraft have been deployed through 6346 launch missions worldwide by the end of 2022 (as shown in Figures 1 and 2) [1]. The fragments generated from those rocket launches and disabled spacecraft have resulted in millions of pieces of space debris, posing a significant threat to operational satellites in space [2,3]. In 1983, the space shuttle Challenger experienced the first ever significant collision with space debris in human history. This incident resulted in damage to Challenger's window and the premature termination of the mission. Subsequently, several severe instances of space debris colliding with operational satellites in orbit have been observed. In each case, the impact caused substantial damage or even the complete disintegration of the affected spacecraft [4]. With the space environment becoming increasingly crowded, it is crucial to ensure the safety of space assets in orbit and prevent potential damage from space debris collisions. This urgency highlights the need for timely and continuous sensing and measurement of the operational orbits of objects in space.

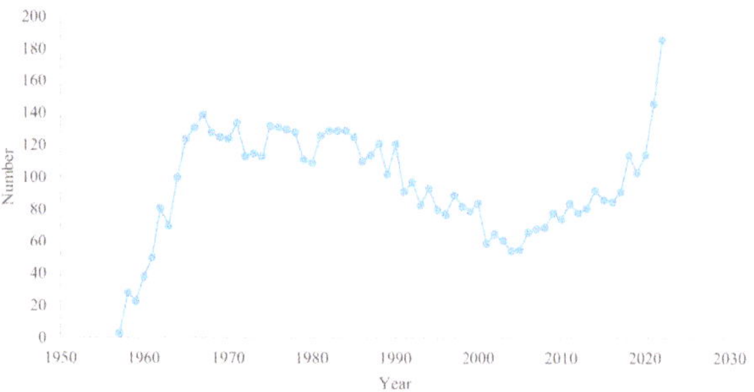

Figure 1. The number of launch missions worldwide from 1957 to 2022.

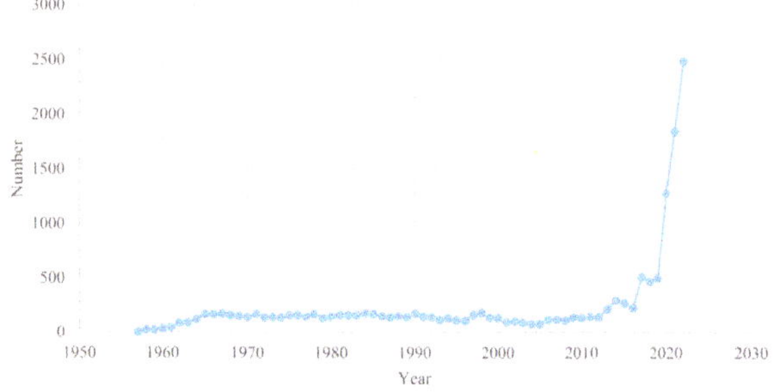

Figure 2. The number of spacecraft launched worldwide from 1957 to 2022.

Recognizing the imperative of protecting on-orbit spacecraft from collisions with space debris, major spacefaring nations worldwide have universally developed their own space situational awareness systems. In the mid-20th century, the United States and the Soviet Union pioneered the development of ground-based space situational awareness systems. As we entered the 21st century, other spacefaring nations, including Europe, Japan, India, and China, significantly enhanced their capabilities in space situational awareness, aiming to address the challenges posed by the increasingly congested space environment and ensure the security of their respective space assets.

In recent years, significant attention has been directed towards space-based optical systems for space situational awareness. Unlike ground-based systems, those dependent on space-based optical platforms provide unique advantages, such as a wide observational field, extended daily monitoring time, and immunity to atmospheric or cloud cover. These features facilitate the timely and continuous surveillance of both space regions and targets. Furthermore, through orbital networking or orbital maneuvers, space-based optical space situational awareness systems can attain persistent monitoring and conduct close-range reconnaissance of space targets.

In the technical framework of space situational awareness utilizing optical monitoring systems in space, the determination of the orbits of space targets constitutes its core and foundation. After acquiring detection data, space-based situational awareness platforms employ orbit determination methods to convert sensor data into precise parameters defining the orbital state of the targets. This process serves as the essential bridge for transforming sensor detection information into assessments of the orbits of space targets and alert information. This foundational procedure is integral for estimating the on-orbit operational status of space targets, cataloging space objects, and conducting analyses related to space contact and collisions.

In this paper, a comprehensive overview of the development history of space situational awareness systems and the present state of research on orbit determination methods employing space-based optical platforms is provided. First, it delineates the status of space situational awareness policies and equipment in major spacefaring nations, analyzing the practical engineering demands in this domain. Subsequently, the existing state of relevant theoretical research in the three stages of orbit determination—short-arc association, initial orbit determination, and precise orbit determination—is introduced. Finally, the paper summarizes key issues in the current theoretical research and underscores research directions that merit particular attention in the future development of this field.

2. The Evolution of Space Situational Awareness Policies

Research on space situational awareness originated in the 1980s and gained significant attention in the 1990s [5]. In August 1998, the U.S. government underscored the critical significance of space situational awareness in its official document [6]. In 2018, space situational awareness was officially recognized as one of the four primary U.S. space missions. The related document explicitly emphasized that a comprehensive understanding of the space environment is foundational to the execution of U.S. space commands and operations in the space domain [7].

In 2019, the United States formally introduced the term Space Domain Awareness (SDA) to supplant space situational awareness. This signifies the official acknowledgment of space as an operational domain, alongside land, sea, and air [8]. The documents titled "Space Capstone Publication, Spacepower" published in 2020 and "Space Doctrine Notes, Operations" published in 2022 by the U.S. government underscored that the Space Domain Awareness mission entails understanding all factors, intentions, capabilities, and current conditions that may impact U.S. space activities. These documents particularly highlighted the importance of space situational awareness in anticipating potential future scenarios in the space domain [9,10]. Important milestones in the development of the U.S. space situational awareness policies are shown in Figure 3.

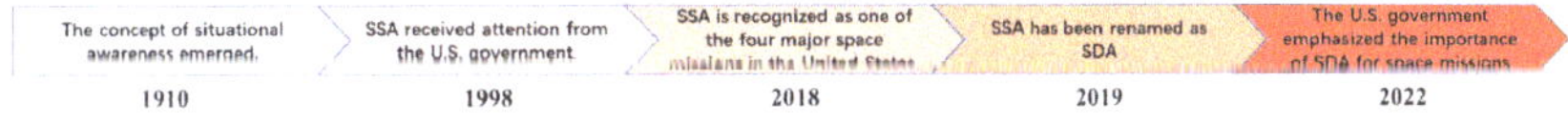

Figure 3. Milestones of the U.S. SSA policies.

In addition to the United States, European countries have embarked on establishing an independent system to detect, assess, and forecast space threats since the last century. The development of a European space situational awareness system was initially led by European spacefaring countries, notably France [11]. In 2008, the Council of the European Union adopted resolutions to advance European space policies, emphasizing the need for Europe to enhance its monitoring capabilities for space facilities and debris to strengthen its global standing in space endeavors [12]. Since 2009, the European Space Agency (ESA) has spearheaded the Space Situational Awareness Program, categorizing it among the six European major space programs alongside the Galileo Global Satellite Navigation System [13].

The primary objectives of the European space situational awareness system include monitoring and tracking space objects, observing and forecasting space environments, and detecting space targets in low Earth orbits (LEOs). These objectives aim to achieve space collision avoidance, re-entry trajectory analysis, and orbit evolution of space targets to ensure the safety of European space facilities [14–16]. In 2010, the European Council formally acknowledged the importance of space situational awareness in both military and civilian areas in its seventh Space Council resolution [17]. Subsequently, the European Council reiterated the significance of space situational awareness in a series of related meetings in 2011 [18,19].

In 2014, the European Parliament adopted Resolution No. 541 concerning the establishment of the Space Surveillance and Tracking (SST) System, systematically outlining concepts, necessity, mission goals, and series of resolutions related to SST [20]. In 2015, the EU established the SST Alliance to coordinate SST-related affairs. In 2016, the European

Space Strategy was promulgated, emphasizing Europe's commitment to strengthening existing space target monitoring and tracking capabilities and evolving them into a more robust space situational awareness system in the future [21]. In 2018, the EU Commission submitted the European SST system development report for the years 2014–2017 to the European Parliament. The report systematically summarized the development of the European space monitoring and space target tracking system during that period, providing prospects for the future development of the system in both technical and managerial aspects, underscoring European high regard for constructing a space target monitoring system [22].

In 2022, European leaders proposed considering space as a strategic zone for Europe and called for the formulation of a European Space Security and Defense Strategy. In response, the European Commission put forth the first European Space Security and Defense Strategy, emphasizing the prominent position of space situational awareness [23,24]. Presently, aligning with the European development plan for its space situational awareness system, Europe has established ground radar systems, such as the Grand Réseau Adapté à la Veille Spatiale (GRAVES), for space situational awareness. Concurrently, the EU is planning to respectively launch the Hera satellite in 2024 and the ClearSpace-1 satellite in 2026 to carry out tasks related to space debris cleanup. Plans are also underway to launch the Vigil satellite in the 2020s to monitor the space environment, gradually advancing toward the goal of constructing the European space situational awareness and space defense system. Figure 4 shows the important milestones in the development of European space situational awareness policies.

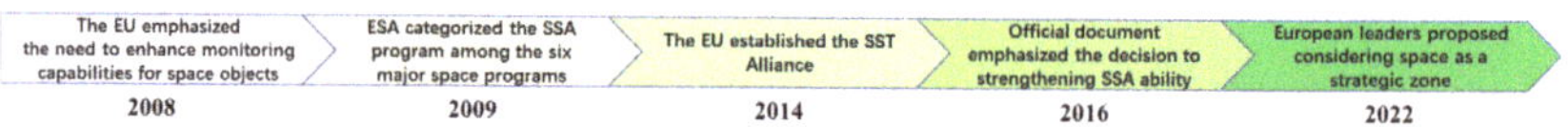

Figure 4. The development of Europe SSA policies.

In addition to the United States and Europe, Russia has developed its space situational awareness system known as the Russian Space Surveillance System (RSSS) [25]. The origins of the Russian space situational awareness system can be traced back to the Soviet era in the 1960s when a world-class network of ground-based radar stations and optical telescopes was established, ensuring the efficient monitoring of space targets [26]. Moreover, owing to historical factors, Russia has maintained a top-tier level of theoretical research in the field of space situational awareness [27].

In recent years, India has demonstrated a growing interest in space situational awareness. India established the Directorate of Space Situational Awareness and Management and, in 2019, initiated the construction of the Space Situational Awareness Control Centre to specifically oversee India's activities and research programs in space situational awareness [28,29]. In 2022, India entered into a cooperation agreement with the United States, covering aspects such as the sharing of space situational awareness information [30]. Additionally, starting from 2022, the Indian Space Research Organisation (ISRO) has annually released a Space Situational Assessment Report on its official website. This report provides information on the global launch scenario and typical space collision risks of the previous year, highlights future trends in space situation, and offers insights into the development in the field of space situational awareness [31,32]. Japan is also in the planning stages for constructing a space-based situational awareness system composed of radars, optical telescopes, and analysis systems. The goal is to monitor space debris and analyze its potential threats to on-orbit satellites and astronauts [33].

In general, the concept of space situational awareness emerged in the 1990s and has garnered widespread attention in the 21st century. With the ongoing advancement of space technology, the concept of space situational awareness has evolved, expanding from its initial focus on the discovery and perception of space objects to encompass the measurement and estimation of the motion states of these objects, continuous monitoring of key space targets and regions, orbit prediction of space objects, and warning of collision threats.

Looking ahead, as space technology continues to advance, more complex space missions will demand higher requirements for the scope and precision of space situational awareness. Existing space situational awareness systems primarily detect and perceive unknown targets one by one within their detection range. The development of large-scale spacecraft constellations urgently calls for perception systems capable of addressing clustered targets.

Furthermore, to ensure the long-term operation of increasingly sophisticated space payloads on orbit, it is crucial to minimize the probability of collisions between these payloads and space debris. This places higher demands on the sensitivity of space situational awareness systems, necessitating the research and development of equipment and systems capable of detecting extremely small space debris.

In order to reduce the detection time of space collision threats, providing more time for spacecraft to generate and execute collision avoidance strategies, space situational awareness systems need to have a longer detection range and higher recognition speed. This enables the early and swift identification of potential close encounters and collision threats.

3. Typical Space-Based Situational Awareness Facilities

The current global systems for detecting and alerting against threats to space assets are broadly categorized into ground-based and space-based surveillance systems based on their spatial distribution [34–37]. Ground-based systems, being the initial development, formed the early space surveillance infrastructure for major spacefaring nations, leveraging their high power and capacity for large detection payloads. However, the Earth's curvature and susceptibility to atmospheric and cloud cover may limit ground-based systems, imposing significant constraints on their detection range and effective operational time. Additionally, their distance from space targets results in lower resolution for detecting targets on HEO.

In contrast, space-based surveillance systems utilize detection payloads on satellites, mitigating issues related to atmosphere and clouds on Earth. They offer a broader field of view compared to ground-based systems through maneuverability. Moreover, these systems, equipped with satellite maneuvering and multi-satellite networking capabilities, can ensure continuous monitoring and close observation of specific space targets or regions. To achieve comprehensive, continuous, and high-precision space situational awareness on a global scale, nations worldwide have increasingly emphasized the development of space-based platforms in recent years [34]. In this section, typical detection payloads for space situational awareness systems shown in Table 1 are presented to provide a more intuitive and comprehensive understanding of current development status of this field.

In the 1990s, the United States launched one of the earliest and most representative space-based situational awareness satellites, the Midcourse Space Experiment (MSX). It carried a suite of optical detection instruments, including an infrared imaging telescope, a space-based visible (SBV) camera, and the ultraviolet and visible imagers and spectrographic imagers (UVISI) [38]. The SBV camera is considered a groundbreaking space-based visible light detection instrument. The optical aperture of SBV is 15 cm, operating in the spectral range of 0.3–0.9 μm. Its focal plane is composed of four low-noise visible light CCD chips with a field of view of $1.4° \times 1.4°$ each, stitched together to form a total field of view of $1.4° \times 5.6°$ [39]. This camera signifies the validation of space-based optical detection technology and provides crucial insights for the development of subsequent space detection systems.

Building upon the success of the MSX satellite, the United States further strengthened its space situational awareness capabilities by establishing the space-based surveillance system (SBSS). The first satellite of the SBSS mission, launched in 2010, carried a TMA visible light camera with 30 cm aperture, $3° \times 3°$ field of view, and 2.4 million pixels. Mounted on a gimbal, TMA demonstrated the capability to flexibly adjust its field of view under a fixed satellite posture. The detection range of TMA covered LEO, MEO, and HEO, with a positioning accuracy of 10 m for low-orbit targets and 500 m for high-orbit targets. According to the plan of the SBSS project, the system was originally intended to establish a constellation of space-based monitoring satellites operating in LEO, thereby enhancing positioning accuracy for targets in HEO to approximately 250 m [40].

To address the issue of too many false collision alerts caused by insufficient ground-based detection resolution, the Space-based Telescopes for the Actionable Refinement of Ephemeris (STARE) satellite was launched into a polar orbit to validate the capability of space-based surveillance systems to update the orbital parameters of on-orbit targets. According to relevant research, this satellite carries a coaxial reflection system payload with an 85 mm aperture, a field-of-view angle of $2.08° \times 1.67°$, and a pixel size of 6.7 μm, comprising 1280×1024 pixels [39,40].

To validate whether satellite maneuvers can enhance target detection capability, Canada launched the smallest space-based telescope to date, the Microvariability and Oscillations of Stars (MOST) satellite [41]. This satellite carries a 15 cm aperture optical telescope with a field-of-view angle of $2° \times 2°$ and features 1024×1024 pixels [42].

After the launch of the MOST satellite in Canada, the United States launched the Wide Area Space Surveillance System (WASS), utilizing a network of multiple LEO satellites to monitor stationary orbit targets on Earth. Each satellite in this system carries multiple wide-angle cameras, with each camera having a field-of-view angle of up to $60° \times 4°$ and an aperture of 28 mm. The cameras are equipped with a long sunshade to achieve a $4°$ sun avoidance angle [41].

The most well-known space situational awareness satellite of Canada is the Sapphire satellite. This satellite employs a three-mirror anastigmat design which is similar to the SBV, with a 15 cm aperture and a field of view of $1.4°$. The detector of this satellite utilizes a trajectory speed mode to achieve a higher signal-to-noise ratio. It is capable of monitoring and tracking objects with a magnitude as faint as 15 [40,42].

The space-based optical telescope (SBO), developed in Europe for space debris monitoring, features a CMOS sensor with an optical aperture of 20 cm, pixel size of 18 μm, and a field-of-view angle of $6°$. It possesses robust capabilities for monitoring space debris targets, enabling observation of 2 cm-sized objects in HEO [39].

In addition, the AsteroidFinder satellite developed by Germany carries an optical payload with a 25 cm aperture. It employs an off-axis three-reflection dispersive imaging catadioptric optical system with a field-of-view angle of $2° \times 2°$, pixel size of 13 μm, and a resolution of $2k \times 2k$ pixels. By incorporating a field stop in the optical system, the camera of this satellite achieves effective stray light suppression, enabling the detection of spacecraft and space debris in LEO. This satellite validates the capability of space-based optical detection systems for detecting centimeter-sized space targets, marking a significant milestone in the development of space situational awareness systems [39,40].

Table 1. Key performance parameters of space situational awareness satellite payloads.

Satellite	Optical Aperture	Pixel Size	Pixel Number	Field of View
MSX	15 cm	Unknown	Unknown	$1.4° \times 5.6°$
SBSS	30 cm	Unknown	2.4 million	$3° \times 3°$
STARE	8.5 cm	6.7 μm	1280×1024	$2.08° \times 1.67°$
MOST	15 cm	Unknown	1024×1024	$2° \times 2°$
Sapphire	15 cm	Unknown	Unknown	$1.4°$
SBO	20 cm	18 μm	Unknown	$6°$
Asteroid	25 cm	13 μm	$2k \times 2k$	$2° \times 2°$

In general, space-based situational awareness facilities are predominantly equipped with an optical detection payload to monitor and assess space objects and the space environment. Compared to radar detection, optical detection is envisioned as the primary means for future space-based target monitoring. Optical detection consumes less energy for long-range detection and can be combined with real-time image processing to enhance resolution. In terms of monitoring capabilities, space-based situational awareness systems predominantly rely on visible light cameras as their effective payloads. However, due to the diversity of space targets, there is a growing trend towards multi-spectral joint detection, such as combining visible light with infrared detection or utilizing multi-band infrared detection.

Regarding the selection of monitoring constellations and the number of satellites, with the advancement of small satellite technology and space satellite formation techniques, space situational awareness payloads are increasingly integrated with small or multiple satellites. Additionally, a single monitoring satellite may carry various sensors to cover a broader monitoring range and multiple spectral bands.

In the choice of space camera, off-axis three-mirror optical systems and even more complex optical systems are becoming more widely used to address the growing demand for wide-field space-based surveillance.

4. Orbit Association Methods

Space-based optical space situational awareness systems offer several advantages, including a short detection wavelength, a large frame information capacity, and the ability to monitor multiple space targets. Currently, this kind of system has emerged as the mainstream solution for monitoring space target and has entered the experimental and application phases.

In the development of space-based space situational awareness systems, one of the crucial technologies is predicting the orbits of identified space targets based on optical imaging results. This process, often referred to as orbit determination, serves as the core and foundation of space-based space situational awareness tasks [35]. Orbit determination can be categorized into batch processing and sequential processing, depending on specific processing methods [36]. It can also be classified into real-time tracking and post-event tracking based on processing timeliness.

According to the orbit model used in the orbit determination process, orbit determination tasks can be classified into dynamic orbit determination, kinematic orbit determination, and reduced-dynamic orbit determination. In terms of processing steps, the orbit determination process involves orbit association, initial orbit determination (IOD), and orbit refinement (or precise orbit determination) [37].

In scenarios where optical observations are the sole source of information, obtaining direct measurements of the distance and velocity of space targets in relation to the observing platform is unattainable. Only information about the relative angles and angular velocities of the target's motion can be gathered. Before commencing the initial orbit determination, it is crucial to confirm, through orbit association, whether multiple observation results pertain to the same target [43].

The admissible region method is a classical approach for correlating diverse observation outcomes [44]. This method utilizes the mechanical principles governing the motion of space targets as constraint conditions, restricting the feasible range of distances and its change rate (equivalent to the orbital elements) from observation platforms to the target during angle-only measurement process. Subsequently, the orbital element of the target will be determined within the feasible range.

Expanding upon the admissible region method, as shown in Table 2, Tommei et al. introduced an approach that integrates the gravity as a constraint on the the motion of the targets. Simultaneously they utilized the orbital altitudes of the targets as a constraint to construct the feasible domain of their motion. This method, grounded in the admissible region framework, successfully achieves the association of the observed short arcs of the space targets [45].

Following the principles laid out by Tommei et al., Maruskin et al. proposed a recursive intersection method to address the problem of orbit association. This approach employs orbital mechanical energy, distance from Earth, and permissible ranges for perigee and apogee as constraints for feasible target orbital solutions. It calculates intersections of different orbital segments corresponding to multiple observations, and associates different observed orbit segments according to the existence of intersections associated with various observation results [46]. The recursive intersection method simplifies the orbital association process but requires manual screening and processing of results during implementation, making it unsuitable for associating large-scale orbital observation data.

Fujimoto et al. mapped the feasible domain where the targets might exist to a 6D Poincaré space, which is divided into multiple hypercubes. In the Poincaré space, the density of virtual particles, representing the probability of the existence of the observed target in space, is distributed among different hypercubes. Bayesian theorem is then employed to achieve the association between different observation segments [47–49]. The approach, characterized by linearizing the mapping process and discretizing the feasible domain, effectively mitigates computational costs and maintains robustness even in the presence of observation errors. Nevertheless, it does not apply to the GEO association problems [50].

Siminski et al. reframed the orbit association problem into an optimization process by employing a loss function and corresponding optimization methods. Depending on whether the optimization problem used in the trajectory association process is an initial value problem (IVP) or a boundary value problem (BVP), the associated methods are categorized into two classes [51,52]. Among them, the BVP-based approach utilizes distance assumptions and Lambert problem solvers to determine feasible orbits for observed targets, exhibiting better completeness and efficiency. On the other hand, the IVP-based methods determine the orbits of the targets through the admissible region method, offering advantages in terms of clarity.

Table 2. Comparison of the orbit association methods mentioned in this section.

Developers	Characteristics	Inputs	Outputs	Main Equations
Tommei et al. [45]	Regard the gravity as a constraint on the motion of targets and utilized the orbital altitudes as a constraint to construct the feasible domain	Arc of observations and the corresponding optical attributable	Admissible region of the orbit of target	$\mathcal{C} = \mathcal{C}_1 \cap \mathcal{C}_2$
Maruskin et al. [46]	Simplified the association process and further limited the size of admissible region in the range range-rate plane	Arc of observations and the corresponding optical attribute	Admissible region of the orbit of target	$\mathcal{C} = \overset{4}{\underset{i=1}{\cap}} \mathcal{C}_i$
Fujimoto et al. [47–49]	Mapped the feasible domain where the target might exist to a 6D Poincaré space	Multiple observations and the sensor position	Correction of observations	$X \approx X^* + \Phi[\delta\rho, \delta\dot{\rho}]^T$
Siminski et al. [51,52]	Reframed the association problem into an optimization process	Range and range-rate hypothesis	Improved range and range-rate hypothesis	$L(p) = (a_2 - \hat{a}_2)^T (C_{\hat{a}_2} + C_{a_2})(a_2 - \hat{a}_2);$ $L(p,k) = (\dot{z} - \hat{\dot{z}})^T (C_{\hat{\dot{z}}} + C_{\dot{z}})(\dot{z} - \hat{\dot{z}})$

It is notable that a significant challenge in existing orbit association methods lies in the high rate of erroneous orbit associations for distinct targets within the same satellite constellation. To enhance the accuracy of orbit associations for different targets within the same constellation, Cai et al. introduced a novel approach known as the "Common Ellipse Method". This method addresses the orbit association problem by examining the proximity of the hypothetical ellipses representing the target trajectories in the observational data to a common ellipse. Simulation results demonstrate a notable improvement in the accuracy of orbit associations using this method [43].

5. Initial Orbit Determination Methods

In the initial stages of orbit determination, challenges arise due to the limited field of view of monitoring camera and the high relative velocity between space targets and monitoring platforms. A single monitoring spacecraft often struggles to maintain continuous tracking of the target motion. Consequently, the estimation of a space target orbit is usually based on limited observational data, a technique referred to as short-arc orbit determination [53]. Representing the orbit of a space target uniquely requires six orbital elements. However, a single optical observation can only provide information about two angles of the target and their corresponding times. Hence, achieving short-arc orbit determination during the initial phase typically requires obtaining six sets of observations, including angles and times from three separate instances as depicted in Figure 5.

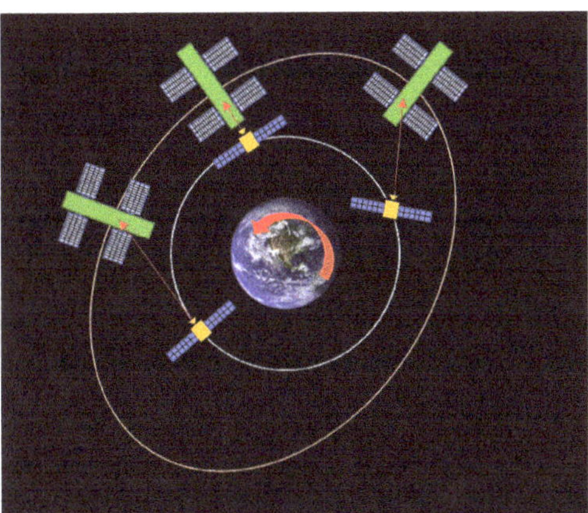

Figure 5. Schematic diagram of angle-only orbit determination based on three observations.

The most classic algorithms for initial orbit determination are the Gauss method and the Laplace method shown in Table 3 [54,55]. Among them, the Gauss method initiates by formulating a system of equations for target positioning, involving 18 unknowns. This process leverages the geometric relationship between the observation platform and the target, along with the spatial target position vector expression based on Lagrange coefficients and the conservation theorem of angular momentum. Finally, the Gauss method utilizes multiple observations, incorporating target angle information and the position of the observation platform, to solve these equations [56]. However, the accuracy of the Gauss method diminishes when the curvature of the observed arc is low since the curvature of the observed trajectory is used as a divisor during the computation, and it may encounter singularity and instability problems during its usage [50]. Considering the impact of observational errors, results obtained by the Gauss method may face challenges in subsequent differential corrections for least squares fitting, potentially resulting in divergence in the orbit determination process [57].

On the other hand, the Laplace method solves the geocentric distance of the target iteratively by utilizing the azimuth of the observed vector, the azimuthal velocity, and the angular acceleration derived differentially from the observed data. Building upon this, the method utilizes observational geometry to ascertain the position and acceleration of the target, enabling the determination of the target's six orbital elements. Theoretical research and application instances show that the Laplace method proves to be a more straightforward and practical method for orbit determination missions, particularly when dealing with orbit determination problems with very limited observational data [58]. However, it is crucial to acknowledge the inherent difficulty in accurately determining the geocentric distance of space targets due to errors in determining the target's geocentric distance relative to angle information from the observation platform. This challenge becomes particularly pronounced in scenarios with sparse observational data during short-arc orbit determination. The inherent limitations of the Laplace method make it difficult to precisely determine the geocentric distance of observed targets, potentially leading to an ill-condition during the matrix inversion process when calculating the position vector of the target, sometimes even resulting in orbit determination failure [43,59,60].

Distinct from the Gauss method and the Laplace method, the Gooding method and the double r method are other two widely recognized techniques for initial orbit determination [61,62]. Schaeperkoetter systematically compared the Gauss method, the Laplace method, the double r method and the Gooding method in terms of accuracy, convergence, robustness and suitability for space-based platforms. The results consistently showed that the Gooding method exhibited optimal performance in initial orbit determination problems in the majority of cases [63]. When estimating the direction of the target, the accuracy of the Gooding method typically surpassed other methods by several orders of magnitude. In terms of the shape estimation of orbits, the precision of the Gooding method and double r method was generally comparable. In scenarios involving near-polar orbits and issues where the initial distance estimation for the target orbit is unknown, the Laplace method often performs the best.

To further enhance the precision and speed for solving the initial orbit determination problems, Schmidt et al. investigated the observability of the target relative to the monitoring platform. They proposed an algorithm for determining the shape of the target orbit under the condition of angle-only measurements [64]. Newman and others addressed observability issues in the angle-only initial orbit determination problems through nonlinear second-order Volterra sequences [65,66]. Subsequently, Shubham et al. optimized the algorithms proposed by Newman and achieved higher computational efficiency [67]. Geller et al. developed another orbit determination method by setting the camera offset from the vehicle [68,69]. Gong et al. employed neural networks to map the observation vectors to space orbits, effectively addressing the nonlinear challenges introduced by complex dynamic models [70]. While machine learning methods exhibit commendable performance in orbit determination, they concurrently escalate the computational demands for space situational awareness tasks, significantly hampering the operational efficiency of associated algorithms. Within the highly constrained computational hardware of onboard computers, these substantial increases in computational workload may even lead to a considerable escalation in overall computation. According to the research of Moniruzzaman et al. [71], GPUs can extend the computing speed of CPU-based algorithms, providing the possibility of real-time computing. Dai et al. [72] pointed out that GPU can improve the computational efficiency by concurrently executing the matrix operation process that decouples each other in the calculation of orbit dynamics problems, thereby greatly reducing the computational time complexity. Based on this principle and in order to reduce the on-board computation workload, Lim et al. utilized GPU to handle the increased computational demands caused by artificial intelligence algorithms in space situational awareness, and they achieved significant acceleration effects in addressing these concerns [42].

With the development of spacecraft control and cluster techniques, observing space objects through multiple observation stations, as shown in Figure 6, has become practical and attracted widespread attention [61,73]. In the late 20th century, Jia et al. proposed the reference vector method which utilizes statistical characteristics of the observation data to achieve an approximate optimal estimation of the target state [74]. However, this method has low accuracy and cannot be used in practice [75]. Subsequently, Lu et al. introduced a unit vector method to iteratively approximate the initial state of space targets by constructing two coordinate systems. This method has a wide application range and converges fast, but it may be inaccurate or even diverge when the observed arcs are too long [76,77]. Building upon the unit vector method, Yang et al. proposed a space target initial orbit determination method for multi-camera arraies. This method first establishes equations for solving the initial state of the targets, then incorporates measurements from different time into the equation to form a system of equations. By solving the equations using the least squares method, the dynamic state of the target will be ultimately obtained. By comprehensively utilizing the computational results from multiple measurement devices, this approach leverages the advantages of camera arrays, achieving higher computational accuracy compared to traditional methods such as the Laplace method and the Gauss method [78].

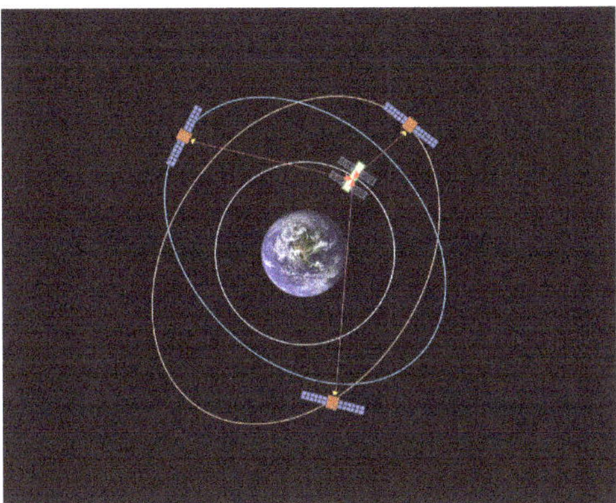

Figure 6. Schematic diagram of angle-only orbit determination based on multiple observation devices.

Table 3. Comparison of the mentioned initial orbit determination methods in this section.

Methods	Characteristics	Inputs	Outputs	Main Equations
Gauss method	The most classic method for initial orbit determination	The direction of three observations	Geocentric distance of the target	$\mathbf{r}_2 = c_1\mathbf{r}_1 + c_3\mathbf{r}_3$
Laplace method	Straightforward and practical	The direction of three observations	Geocentric distance of the target	$\mathbf{r} + \mathbf{R} = \rho\hat{\rho}$
Proposed by Schmidt et al. [64]	Proposed method for determining the shape of orbit	relative angle measurements	values of several of the relative orbit elements	$[i_x, i_y, i_z]^T = ([x_d - a_e cos(\beta_0)/2]\hat{i} + [y_{d0} + a_e sin(\beta_0)]\hat{j} + [z_{max} sin(\psi_0)]\hat{k}) / sqrt(a_e^2(1 - 3cos^2(\beta_0)/4) + a_e(2y_{d0}sin(\beta_0) - x_d cos(\beta_0)) + x_d^2 + y_{d0}^2 + z_{max}^2 sin^2(\psi_0))$
Proposed by Geller et al. [68,69]	Developed novel orbit determination method by setting the camera offset from the vehicle	three line-of-sight observations for relative motion coasting trajectory to the center of mass of an object	initial position and velocity of target	$k_i \boldsymbol{i}_{los}(i) = \phi_{rr}(i)\{k_0 \boldsymbol{i}_{los}(0) - \boldsymbol{d}(0)\} + \phi_{rv}(i) \{\phi_{rv}^{-1}(1)[k_1 \boldsymbol{i}_{los}(1) - \phi_{rr}(1)\{k_0 \boldsymbol{i}_{los}(0) - \boldsymbol{d}(0)\} - \boldsymbol{d}(1)]\} + \boldsymbol{d}(i)$
Proposed by Gong et al. [70]	Effectively addressed the nonlinear challenges in orbit determination by using machine learning methods	three sets of bearing angle and the absolute orbit state of the observer	the initial relative orbit state of the target	$L = \dfrac{1}{m} \sum_{i=1}^{m} (y_i - \hat{y}_i)^2 + \dfrac{\lambda}{2n} \Sigma_\omega\, \omega^2$ (Loss Function)

6. Precise Orbit Determination Methods

In the initial orbit determination phase, the perturbation factors are usually not taken into consideration [58]. To augment the accuracy of the predicted orbit of the target, it is imperative to iteratively refine the initial orbit based on subsequent observational data and in conjunction with a precise orbit dynamic model.

Distinguished from the initial orbit determination, precise orbit determination involves optimizing the estimation of the initial states of space targets and parameters of the orbit dynamic model using a series of observations with associated errors. Essentially, it is a optimal state estimation problem [58].

The methods employed for solving precise orbit determination problems generally revolve around the concept of least squares estimation [79]. Building upon this concept, a series of estimation methods have been developed, including unbiased least squares estimation, standard least squares estimation, weighted least squares estimation, and least squares estimation. Typically, the methods grounded in the least squares framework for precise orbit determination begin with the initial orbit determination process. Then, the estimates are incorporated into the precise orbit determination model to update the calculation results iteratively to improve the accuracy.

As shown in Table 4, Kozai proposed the first analytical prediction algorithm for precisely determining the orbital elements of LEO objects [80]. Building upon that, Hilton et al. simplified the gravity model as well as the drag model and introduced the Simplified General Perturbations method (SGP) [81]. Lane et al. further refined the SGP, leading to the development of the first practical orbit precise determination method, the SGP4, used by the North American Aerospace Defense Command (NORAD) [82,83]. Hujsak extended SGP4 and introduced the Simplified Deep-space Perturbation 4 method (SDP4) to precisely determine the orbits of deep-space targets [81]. Han et al. conducted an accuracy test on SDP4 and SGP4 using typical orbits. The results showed that SGP4/SDP4 models exhibit high computational speeds and can achieve the required accuracy for orbit prediction when dealing with near-circular orbits in MEO and GEO. The maximum deviation from the determined orbit to the nominal orbit was less than 3 km. However, for near-circular orbits with altitudes less than 50 km or highly eccentric orbits with eccentricity greater than 0.6 and perigee altitude less than 50 km, the computational results showed large discrepancies, and more precise computational models are needed to assist in updating the precise orbit determination of the targets [84,85].

The convergence speed and accuracy are core indicators for evaluating the performance of precise orbit determination methods and can be influenced by various factors such as the performance of the orbit prediction model, distribution of observation arcs, accuracy of the observation data, data density, and more [86]. A systematic study conducted by the U.S. National Research Council on the impact of different factors on orbit prediction models revealed that factors including the accuracy of atmospheric drag models, solar radiation pressure models, gravity field models, the performance of orbit correlation and orbit propagation methods during initial orbit determination, the algorithm for approximating the uncertainty of the orbit determination results, sensor measurement errors, and nonlinear estimation and filtering methods for handling uncertainties of measurement result may influence the orbit prediction results. Simultaneously, the research emphasized that with the rapid increase in the number of space objects, traditional orbit determination methods are no longer sufficient to meet current needs in determining the orbit of space objects. There is a need for research in orbit determination algorithms with superior performance to achieve higher demand in space situational awareness [87].

Against the backdrop of rapid development in space satellite networking technology, precise orbit determination techniques for multi-satellite platforms have emerged as a new research hotspot. Wang et al. proposed a dual-satellite optical orbit determination method for GEO satellites by deploying monitoring satellites on both sides of the targets. Experimental studies have shown that this method can improve the observability of targets, and the orbit determination accuracy of the target can be improved by three orders of magnitude compared to single-satellite orbit determination [88]. Liu et al. conducted research on the orbit determination effects for targets in near-Earth orbits under different conditions, deploying two, four and six observation satellites. The results revealed a positive correlation between orbit determination accuracy and the number of observation satellites. However, the enhancement effect diminishes gradually as the number of observing satellites increases. Therefore, adopting a dual-satellite evenly distributed space-based target orbit determination scheme is deemed to have the highest efficiency-to-cost ratio [89]. Shao et al. conducted a study on the monitoring effects of targets in GEO when deploying multiple surveillance satellites in a inclination LEO orbits. The findings suggest that increasing the number of observation satellites from two to six can significantly enhance the observation accuracy from the kilometer level to the decimeter level [90].

Due to the rapid increase in the number of space debris, precise orbit determination and prediction for space debris cluster have become another hot spot [91]. Building upon the investigation of factors influencing the growth of space debris, Toshiya et al. proposed a method to quantitatively estimate the density and quantity of space debris. However, it does not provide a method to estimate the orbits of space debris [92]. Hu et al. conducted a dedicated study on the selection of dynamic models in the orbit prediction of space debris. They systematically examined the influence of disturbance factors, including gravity field models, atmospheric drag, three-body gravity fields, and solar radiation, on the orbit prediction accuracy of space debris. The research offers quantitative insights into the selection of dynamic models in the precise orbit prediction processes for space debris in different orbital regimes. However, it primarily relied on established initial value problem integrators for recursive calculations and did not specifically delve into the investigation of the orbit estimation model [93]. In addressing the challenges of low computational efficiency with numerical methods and low accuracy with analytical methods in the precise orbit determination process, Li Bin proposed an analytical method for solving the orbit propagation problems with both rapidity and high precision. This method employs numerical integration method with large step size to compute the mean orbital elements of space targets, utilizes an analytical algorithm to reconstruct the short-period terms of the targets, and finally recombines the mean orbital elements and short-period terms to determine orbits fast and accurately [94]. Zhang et al. presented a method for orbit prediction in large-scale space debris clusters. They categorized the debris cluster into nominal and correlated debris. Numerical integration was employed for a small number of nominal debris to ensure accuracy in the prediction. For a larger quantity of correlated debris, a semi-analytical method based on Taylor expansion was utilized to enhance the computational speed while maintaining prediction accuracy. This approach achieved a balanced consideration between computational efficiency and accuracy [95].

Table 4. Comparison of the mentioned precise orbit determination methods in this section.

Developers	Characteristics	Inputs	Outputs	Main Equations
Kozai [80]	Established the first analytical precise orbit determination method	Initial states of a close earth satellite	Prediction of satellite orbit	$R_1 = GMA_2(I/3 - Isin^2 i/2)(I - e^2)^{-3/2}/a^3$
Wang et al. [88]	Proposed a dual-satellite optical orbit determination method for GEO satellites	Initial state of the target, priori of estimate and related covariance matrix	Optimal estimation of the orbits of space targets	$(H^T R^{-1} H + \bar{P}_0^{-1})\hat{x}_0 = H^T R^{-1} y + \bar{P}_0^{-1} X_0$
Li Bin [94]	Proposed an analytical method for solving the orbit propagation to determine orbit fast and accurately	Initial states of a close earth satellite	Prediction of satellite orbit	$\hat{r} = f(\hat{r}, t)_{3\times 3}$
Zhang et al. [95]	Presented a method for orbit prediction in large-scale space debris clusters	Initial states of nominal space debris	Predicted orbits of space debris cluster	$[x^i] = x_0 + \Delta x_o^i$

7. Challenges and Development Trends in Orbit Determination Methods for Space-Based Optical Platforms

The rapid increase in the quantity of space objects and space debris has heightened the need for improved speed and precision in orbit determination for space targets. Effectively addressing the challenge of enhancing the efficiency and accuracy of existing orbit determination methods within space situational awareness systems is of paramount importance. Furthermore, the development of strategies to swiftly and accurately predict the orbits of satellite constellations or extensive space debris resulting from space collisions has become a central concern for contemporary space situational awareness systems. Given the current state of space situational awareness systems and the anticipated future demands for orbit prediction, essential research in this field can be focused on the following key areas.

7.1. High-Precision Orbit Prediction Methods for Space Targets at Long-Distance

With the rapid escalation in the quantity of spacecrafts and space debris, the responsibilities of space situational awareness systems are undergoing a substantial surge in complexity. A critical focal point for advancing these systems lies in the development of algorithms for orbit determination that are not only more robust but also more efficient and accurate. This becomes especially challenging given the constraints posed by the limited number of space situational awareness platforms and their detection capabilities. Tackling these challenges is indispensable for realizing the precise and prompt tracking of distant space targets, thereby forming the linchpin for future endeavors in space collision prevention.

7.2. High-Performance Orbit Determination Methods for Large-Scale Satellite Cluster Targets

Due to the characteristics of space missions, large-scale constellations of satellites are commonly positioned in LEO, where they encounter specific perturbations from atmospheric drag. Compounding this, since these constellations tend to share similar orbital altitudes, the perturbations affecting their orbits may display similarities. The paramount challenge is to adeptly and precisely forecast the orbits of these extensive satellite constellations using their characteristics while operating within the constraints of limited on-orbit computational resources. Addressing this issue is imperative to fulfill the stringent requirements for avoiding space collisions arising from on-orbit activities of large constellations, elevating it to a pressing matter that demands attention.

7.3. Orbit Prediction Methods for Large-Scale Space Debris

Space collision will generate a large number of space fragments, posing a significant threat to orbiting space objects and creating substantial computational loads. Therefore, establishing a precise, low-computational-cost model for the evolution of fragment orbit after a space collision is of great significance for assessing the consequences of space collisions and defending against space debris.

7.4. Orbit Design for Space-Based Situational Awareness Systems

The operational orbits of a space-based surveillance systems have a crucial impact on the observability of space targets relative to the surveillance platforms and the precision of target trajectory measurements. Designing adaptive operational orbits tailored to specific space surveillance platforms and monitoring tasks to achieve the continuous and accurate monitoring of targets is a key research focus of space situational awareness.

8. Conclusions

This article provides a comprehensive overview of space situational awareness systems and orbit determination methods. The escalating presence of space debris has underscored the critical importance of space situational awareness systems, prompting major spacefaring nations to invest in diverse space target surveillance systems. To enhance space situational awareness capabilities, these nations are increasingly directing their efforts towards the development of space-based optical platforms.

Addressing the core challenge of orbit determination for space-based optical space situational awareness platforms, scholars have proposed theoretical frameworks, establishing a three-stage system comprising orbit association, initial orbit determination and precise orbit determination. However, the evolution of huge space constellations and the rapid proliferation of space debris impose heightened requirements on the construction of space situational awareness systems and the efficacy of associated methods. Numerous theoretical and technical challenges persist in achieving precise orbit determination for space targets.

Anticipating future demands for enhanced space situational awareness, further in-depth research into space target orbit determination methods is imperative and will serve as the theoretical bedrock for ensuring the secure operation of spacecraft.

Author Contributions: Conceptualization, resources and writing—original draft preparation, Z.Z.; writing—review and editing, G.Z.; resources and writing—original draft preparation, C.L.; supervision, J.C. and W.C.; revision, X.N. and Z.W. All authors have read and agreed to the published version of the manuscript.

Funding: This research was funded by Youth Innovation Promotion Association of CAS (2022410), Photon Plan of XIOPM (E45542Z1), The National Natural Science Foundation of China (62303378) and The Foundation of Shanghai Astronautics Science and Technology Innovation (SAST2022-114).

Institutional Review Board Statement: Not applicable.

Data Availability Statement: The data presented in this study are available on request from the corresponding author.

Conflicts of Interest: The authors declare no conflicts of interest.

References

1. Chronology of Space Launches. Available online: https://space.skyrocket.de/directories/chronology.htm (accessed on 24 July 2023).
2. Shan, M.; Guo, J.; Gill, E. Review and comparison of active space debris capturing and removal methods. *Prog. Aeosp. Sci.* **2016**, *80*, 18–32. [CrossRef]
3. Xu, T.; Yang, X.; Fu, Z.; Wu, M.; Gao, S. A staring tracking measurement method of resident space objects based on the star tacker. *Photonics* **2023**, *10*, 288. [CrossRef]
4. Geng, W.; Du, X.; Li, Z.; Ma, Z.; Wu, Y.; Geng, G. *Introduction to Space Situational Awareness*; National Defense Industry Press: Beijing, China, 2015; pp. 5–8.
5. Endsley, M.R. Toward a theory of situation awareness in dynamic systems. *J. Hum. Factors Ergon. Soc.* **1995**, *37*, 32–64. [CrossRef]
6. U.S. Air Force. Air Force Doctrine Document 2-2 Space Operations. Available online: https://www.globalsecurity.org/military/library/policy/usaf/afdd/2-2/afdd2-2.pdf (accessed on 1 July 2023).
7. U.S. Air Force. Air Force Doctrine Publication 3–14 Counterspace Operations. Available online: https://www.doctrine.af.mil/Doctrine-Publications/AFDP-3-14-Counterspace-Ops/ (accessed on 26 July 2023).
8. Air Force: SSA Is No More It's 'Space Domain Awareness'. Available online: https://spacenews.com/air-force-ssa-is-no-more-its-space-domain-awareness/ (accessed on 15 August 2023).
9. U.S. Space Force. Space Capstone Publication, SPACE POWER, Doctrine for Space Forces. Available online: https://apps.dtic.mil/sti/pdfs/AD1129735.pdf (accessed on 26 July 2023).
10. U.S. Space Force. Space Doctrine Notes, Operations, Space Doctrine Notes. Available online: https://media.defense.gov/2022/Feb/02/2002931717/-1/-1/0/SDN$%$20OPERATIONS$%$2025$%$20JANUARY$%$202022.PDF (accessed on 26 July 2023).
11. Polkowska, M. Space situational awareness (SSA) for providing safety and security in outer space: implementation challenges for Europe. *Space Policy* **2020**, *51*, 101347. [CrossRef]
12. Council of the European Union. Council Resolution Taking Forward the European Space Policy. Available online: https://data.consilium.europa.eu/doc/document/ST-13569-2008-INIT/en/pdf (accessed on 2 August 2023).
13. The EU Space Programme. Available online: https://www.euspa.europa.eu/pressroom/press-resources/eu-space-programme-overview (accessed on 1 September 2023).
14. Plans for the Future. Available online: https://data.consilium.europa.eu/doc/document/ST-14455-2010-INIT/en/pdf (accessed on 27 August 2023).
15. Space Situational Awareness. Available online: https://www.euspa.europa.eu/european-space/space-situational-awareness (accessed on 30 August 2023).
16. EUSPA Takes on the Space Surveillance and Tracking Helpdesk as of 2023. Available online: https://www.euspa.europa.eu/newsroom/news/euspa-takes-space-surveillance-and-tracking-helpdesk-2023 (accessed on 1 September 2023).
17. Council of the European Union. 7th Space Council Resolution: "Global Challenges: Taking Full Benefit of European Space Systems". Available online: https://www.consilium.europa.eu/uedocs/cms_data/docs/pressdata/en/intm/118012.pdf (accessed on 2 September 2023).
18. Council of the European Union. Towards a Space Strategy for the European Union That Benefits Its Citizens. Available online: https://link.springer.com/chapter/10.1007/978-3-7091-1363-9_5 (accessed on 2 September 2023).
19. Council of the European Union. Orientations Concerning Added Value and Benefits of Space for the Security of European Citizens. Available online: https://eur-lex.europa.eu/legal-content/EN/TXT/PDF/?uri=CELEX:32011G1223(01) (accessed on 5 September 2023).

20. The European Parliament and the Council of the European Union. Establishing a Framework for Space Surveillance and Tracking Support. Available online: https://eur-lex.europa.eu/legal-content/EN/TXT/PDF/?uri=CELEX:32014D0541 (accessed on 5 September 2023).

21. European Commission. Space Strategy for Europe. Available online: https://ec.europa.eu/commission/presscorner/api/files/document/print/en/ac_16_3888/AC_16_3888_EN.pdf (accessed on 8 September 2023).

22. European Commission. On the Implementation of the Space Surveillance and Tracking (SST) Support Framework (2014–2017). Available online: https://www.europarl.europa.eu/RegData/docs\$_\$autres\$_\$institutions/commission\$_\$europeenne/com/2018/0256/COM\$_\$COM(2018)0256\$_\$EN.pdf (accessed on 10 September 2023).

23. European Commission. EU Space Strategy for Security and Defence. Available online: https://defence-industry-space.ec.europa.eu/eu-space-strategy-security-and-defence-stronger-and-more-resilient-eu-2023-03-10_en (accessed on 10 September 2023).

24. European Commission. DG Defence Industry and Space. European Union Space Strategy for Security and Defence. Available online: https://www.europeansources.info/record/european-union-space-strategy-for-security-and-defence/ (accessed on 10 September 2023).

25. Space Situational Awareness Experts Urge Russia to Join Orbital Neighborhood Watch. Available online: https://spacenews.com/space-situational-awareness-experts-urge-russia-to-join-orbital-neighborhood-watch/ (accessed on 10 September 2023).

26. Russian Space Surveillance System(RSSS). Available online: https://www.globalsecurity.org/space/world/russia/space-surveillance.htm (accessed on 12 September 2023).

27. Seidelmann, P.K. Space surveillance: United States, Russia, and China. *J Astronaut Sci.* **2012**, *59*, 265–272. [CrossRef]

28. Explained: Space Situational Awareness (SSA) Control Centre Launched by ISRO. Available online: https://www.jagranjosh.com/general-knowledge/explained-space-situational-awareness-ssa-control-centre-launched-by-isro-1611048676-1 (accessed on 1 October 2023).

29. Russian Space Surveillance System (RSSS). Available online: https://www.isro.gov.in/Foundation\$%\$20stone\$%\$20of\$%\$20Space.html (accessed on 1 October 2023).

30. India Agree to Cooperate on Space Situational Awareness. Available online: https://spacenews.com/us-india-agree-to-cooperate-on-space-situational-awareness/ (accessed on 5 October 2023).

31. Space Situational Assessment 2021. Available online: https://www.isro.gov.in/ISRO\$_\$EN/SSA.html (accessed on 14 October 2023).

32. Indian Space Situational Assessment for the Year 2022—Highlights. Available online: https://www.isro.gov.in/Indian\$_\$Space\$_\$Situational\$_\$Assessment\$_\$2022.html?\$_\$\$_\$cf\$_\$chl\$_\$tk=lV8X9lG9opx0ud73nx2YZtaAnBTr6PfZs8YbJSux1C8-1692020443-0-gaNycGzNCns (accessed on 12 September 2023).

33. Space Situational Awareness (SSA) System. Available online: https://global.jaxa.jp/projects/ssa/index.html (accessed on 14 August 2023).

34. Lu, Z. History and status of US space situational awareness. *J. Ordnance Equip. Eng.* **2016**, *37*, 1–8.

35. Liu, L. Study on the Initial Orbit Determination of Space Target. Ph.D. Thesis, National University of Defense Technology, Changsha, China, 2010.

36. Raol, J.R. On the orbit determination problem. *IEEE Trans. Aerosp. Electron. Syst.* **1985**, *21*, 274–291. [CrossRef]

37. Luo, H.; Liu, L.S.; Li, A.; Zhang, Q. *Technical Dictionary of TTC &C for Missile and Space*; National Defense Industry Press: Beijing, China, 2001; pp. 1–10.

38. Li, Y.; Chen, X.; Rao P. Review of space target positioning algorithms based on spaced-based optical detection. *Chin. Space Sci. Technol.* **2023**, *44*, 1–15.

39. Qi, J. A new large field of view and small distortion optical system. In Proceedings of the 26th National Space Exploration Academic Symposium, Boao, China, 1 October 2013.

40. Xia, C. The Study on Off-Axis Optical System with Wide Field of View for Space-Based Detection Camera. Ph.D. Thesis, University of Chinese Academy of Sciences, Beijing, China, 2023.

41. Xie, D. Optimization of the Indicators of the Space-Based Visible Light Detection System. Ph.D. Thesis, University of Chinese Academy of Sciences, Beijing, China, 2023.

42. Michael, L.; Payam, M; Jelena, S; Huiwen, Y. Onboard Artificial Intelligence for Space Situational Awareness with Low-Power GPUs. In Proceedings of the Advanced Maui Optical and Space Surveillance Technologies Conference, Maui, HI, USA, 15–18 September 2020.

43. Cai, H.; Yang, Y.; Gehly, S.; Wu, S.; Zhang, K. On the orbit determination problem. *Acta Astronaut.* **2018**, *151*, 836–847. [CrossRef]

44. Du, J.; Chen, J.; Li, B.; Sang, J. Tentative design of SBSS constellations for LEO debris catalog maintenance. *Acta Astronaut.* **2019**, *155*, 379–388. [CrossRef]

45. Tommei, G.; Milani, A.; Rossi, A. Orbit Determination of Space Debris: Admissible Regions. *Celest. Mech. Dyn. Astron.* **2007**, *97*, 289–304. [CrossRef]

46. Maruskin, J.M.; Scheeres, D.J.; Alfriend, K.T. Correlation of optical observations of objects in earth orbit. *J. Guid. Control Dyn.* **2009**, *32*, 194–209. [CrossRef]

47. Fujimoto, K.; Scheeres, D.J. Correlation of optical observations of earth-orbiting objects and initial orbit determination. *J. Guid. Control Dyn.* **2012**, *35*, 208–221. [CrossRef]

48. Fujimoto, K.; Scheeres, D.J. Applications of the admissible region to space-based observations. *Adv. Space Res.* **2013**, *52*, 696–704. [CrossRef]
49. Fujimoto, K.; Scheeres, D.J.; Herzog, J.; Schildknecht, T. Association of optical tracklets from a geosynchronous belt survey via the direct Bayesian admissible region approach. *Adv. Space Res.* **2014**, *53*, 295–308. [CrossRef]
50. Gong, B.; Wang, S.; Li, S.; Li, X. Review of space relative navigation based on angles-only measurements. *Astrodynamics* **2023**, *7*, 131–152. [CrossRef]
51. Siminski, J.A.; Montenbruck, O.; Fiedler, H.; Schildknecht, T. Short-arc tracklet association for geostationary objects. *Astrodynamics* **2014**, *53*, 1184–1194. [CrossRef]
52. Siminski, J.A.; Montenbruck, O.; Fiedler, H.; Martin, W. Best hypotheses search on Iso-Energy-Grid for initial orbit determination and track association. *Adv. Astronaut. Sci.* **2013**, *148*, 605–617.
53. Pirovano, L.; Principe, G.; Armellin, R. Data association and uncertainty pruning for tracks determined on short arcs. *Celest. Mech. Dyn. Astron.* **2020**, *132*, 1–23. [CrossRef]
54. Feng, Z. Study on Space-Based Optical Orbit Determination Method for Space Targets. Ph.D. Thesis, Changchun Institute of Optics Fine Mechanics and Physics of Chinese Academy of Sciences, Changchun, China, 2022.
55. Danby, J. *Fundamentals of Celestial Mechanics*; Willman-Bell: Richmond, VA, USA, 1992; pp. 213–234.
56. Curtis, H.D. *Orbital Mechanics for Engineering Students*; Elsevier: Boston, MA, USA, 2009; pp. 631–640.
57. Milani, A.; Geonchi, G.F.; Vitturi, M.D.M.; Knezevic Z. Orbit determination with very short arcs. i admissible regions. *Celest. Mech. Dyn. Astron.* **2004**, *90*, 59–87. [CrossRef]
58. Liu, L. *Orbit Theory of Spacecraft*; National Defense Industry Press: Beijing, China, 2000; pp. 66–75.
59. Li, J. Research on Key Technologies of Space Objects Surveillance and Tracking in Space-Based Optical Surveillance. Ph.D. Thesis, National University of Defense Technology, Changsha, China, 2009.
60. Wu, L.; Jia, P. An analysis of the ill-condition in initial orbit determination. *Acta Astron. Sin.* **1997**, *38*, 288–296.
61. Escobal, P.R. *Methods of Orbit Determination*; John Wiley and Sons: New York, NY, USA, 1965; pp. 25–28.
62. Gooding, R.H. A new procedure for the solution of the classical problem of minimal orbit determination from three lines of sight. *Celest. Mech. Dyn. Astron.* **1996**, *66*, 387–423. [CrossRef]
63. Schaeperkoetter, A.V. A Comprehensive Comparison between Angles-Only Initial Orbit Determination Techniques. Master's Thesis, Texas A&M University, College Station, TX, USA, 2012.
64. Schmidt, J.; Lovell, T.A. Estimating geometric aspects of relative satellite motion using angles-only measurements. In Proceedings of the AIAA/AAS Astrodynamics Specialist Conference and Exhibit, Honolulu, HI, USA, 18–21 August 2008.
65. Newman, B.A.; Pratt, E.; Lovell, A.; Duncan E. Quadratic Hexa-Dimensional Solution for Relative Orbit Determination. In Proceedings of the AIAA/AAS Astrodynamics Specialist Conference and Exhibit, San Diego, CA, USA, 4–7 August 2014.
66. Newman, B.A.; Lovell, A.; Pratt, E.; Duncan E. Hybrid linear-nonlinear initial determination with single iteration refinement for relative motion. In Proceedings of the 25th AAS/AIAA Space Flight Mechanics Meeting, Williamsburg, VA, USA, 11–15 January 2015.
67. Shubham, G.; Sinclair, A. Initial relative orbit determination using second order dynamics and line-of-sight measurements. In Proceedings of the 25th AAS/AIAA Space Flight Mechanics Meeting, Williamsburg, VA, USA, 11–15 January 2015.
68. Gong, B.; Li, W.; Li, S.; Ma, W.; Zheng L. Angles-only initial relative orbit determination algorithm for non-cooperative spacecraft proximity operations. *Astrodynamics* **2018**, *8*, 217–231. [CrossRef]
69. Geller, D.K.; Perez, A. Initial relative orbit determination for close-in proximity operations. *J. Guid. Control Dyn.* **2015**, *38*, 1833–1842. [CrossRef]
70. Gong, B.; Ma, Y.; Zhang, W.; Li, S.; Li X. Deep-neural-network-based anglesonly relative orbit determination for space non-cooperative target. *Acta Astronaut.* **2022**, *204*, 552–567. [CrossRef]
71. Moniruzzaman, M.; Okilly, A.H.; Choi, S.; Baek, J.; Mannan, T.I.; Islam, Z. A Comprehensive Study of Machine Learning Algorithms for GPU based Real-time Monitoring and Lifetime Prediction of IGBTs. In Proceedings of the 2024 IEEE Applied Power Electronics Conference and Exposition, Long Beach, CA, USA, 25–29 February 2024.
72. Dai, H.; Zhang, Z.; Wang, X.; Feng, H.; Wang C. Fast and accurate adaptive collocation iteration method for orbit dynamic problems. *Chin. J. Aeronaut.* **2023**, *36*, 231–242. [CrossRef]
73. Tasif, T.H.; Hippelheuser, J.E.; Elgohary, T.A.; Ma, W.; Zheng L. Analytic continuation extended Kalman filter framework for perturbed orbit estimation using a network of space-based observers with angles-only measurements. *Astrodynamics* **2022**, *8*, 161–187. [CrossRef]
74. Jia, P.; Wu, L. A reference vector algorithm for the initial orbit computation. *Acta Astron. Sin.* **1997**, *38*, 353–358.
75. Chen, L.; Liu, C.; Li, Z.; Kang, Z. Error Analysis of Space Objects Common-View Observation Positioning. *Acta Opt. Sin.* **2022**, *42*, 144–150.
76. Liu, B.; Rong, P.; Wu, J.; Xia, Y.; Zhang, Y.; Zhou, J.; Jiang, X.; Shi, Y.; Mao, Y. The unit vector method for initial orbit determination of artificial satellite. *J. Astronaut.* **1997**, *18*, 2–8.
77. Liu, B.; Ma, J.; Xia, Y.; Zhang, Y. A method of initial orbit determination for long arc. *Acta Astron. Sin.* **2003**, *44*, 369–374.
78. Yang, B.; Li, Y.; Zhang T. Method suitable for initial orbit determination of space targets using camera array. *Acta Opt. Sin.* **2019**, *39*, 57–64.
79. Brouwer, D.; Clemence, G.M. *Methods of Celestial Mechanics*; Academic Press: New York, NY, USA, 1961; pp. 154–196.
80. Kozai, Y. The motion of a close earth satellite. *Astron. J.* **1959**, *64*, 367–377. [CrossRef]

81. Hoots, F.R.; Roehrich R.L. Space Track Report No. 3-Models for Propagation of NORAD Element Sets. Available online: https://celestrak.org/NORAD/documentation/spacetrk.pdf (accessed on 10 November 2023).
82. Lane, M.H.; Hoots, F.R. General Perturbations Theories Derived from the 1965 Lane Drag Theory. Available online: https://www.semanticscholar.org/paper/General-Perturbations-Theories-Derived-from-the-Lane-Hoots/512ca55c1cb795f2ae01c2d00d1105fda62d6da5 (accessed on 10 November 2023).
83. Bolandi, H.; Ashtari, L.M.H.; Sedighy S.H.; Estimation of simplified general perturbations model 4 orbital elements from global positioning system data by invasive weed optimization algorithm. *J. Aerosp. Eng.* **2015**, *229*, 1384–1394. [CrossRef]
84. Han, L.; Chen, L.; Zhou, B. Precision analysis of SGP4/SDP4 implemented in space debris orbit prediction. *Chin. Space Sci. Technol.* **2004**, *24*, 65–71.
85. Wei, D.; Zhao, C. Analysis on the accuracy of the SGP4/SDP4 model. *Acta Astron. Sin.* **2009**, *50*, 332–339.
86. Zhang, Y.; Liu, J. Precise orbit determination method for angle-only observation data of space debris based on angle conversion theory. *J. Beijing Univ. Aeronaut. Astronaut.* **2023**, *49*, 1600–1605.
87. National Research Council. *Continuing Kepler's Quest: Assessing Air Force Space Command's Astrodynamics Standards*; The National Academies Press: Washington, DC, USA, 2012; pp. 1–10.
88. Wang, L.; Song, Y.; Ye, Z.; Zeng, C.; Shao, R. Orbit determination for GEO objects using space-based optical measurement with double GEO platforms. *Sci. Surv. Mapp.* **2022**, *47*, 9–14.
89. Liu, J. Near-Earth Asteroids Orbit Determination by Space-Based Optical Observations. Master's Thesis, ShanghaiTech University, Shanghai, China, 2023.
90. Shao, R.; Song, Y.; Ye, Z.; Zeng, C.; Hu, X. Geostationary satellite orbit determination by LEO networks with small inclination. *Acta Astron. Sin.* **2022**, *63*, 129–140.
91. Stauch, J.; Bessell, T.; Rutten M. Joint probabilistic data association and smoothing applied to multiple space object tracking. *J. Guid. Control Dyn.* **2018**, *41*, 19–33. [CrossRef]
92. Hanada, T.; Yasaka, T. Orbital debris environment model in the geosynchronous region. *J. Spacecr. Rockets* **2002**, *39*, 92–98. [CrossRef]
93. Hu, S.; Tang, J. Reference dynamic models for space debris orbit prediction. *Acta Astron. Sin.* **2023**, *64*, 45–59.
94. Li, B. Researches on Key Technologies of Fast and Accurate Orbit Determination and Prediction of Space Debris. Ph.D. Thesis, Wuhan University, Wuhan, China, 2017.
95. Zhang, Z.; Chen, J.; Sun, C.; Fang, Q.; Zhu, Z. Geostationary satellite orbit determination by LEO networks with small inclination. *Chin. Space Sci. Technol.* **2022**, *42*, 89–98.

 *photonics*

Article

Ghost Fringe Suppression by Modifying the f-Number of the Diverger Lens for the Interferometric Measurement of Catadioptric Telescopes

Yi-Kai Huang [1,2] and Cheng-Huan Chen [1,*]

[1] Department of Photonics, College of Electrical and Computer Engineering, National Yang Ming Chiao Tung University, 1001 University Road, Hsinchu 310403, Taiwan; kai80108@gmail.com
[2] Optical Payload Division, Taiwan Space Agency, 8F, 9 Prosperity 1st Road, Hsinchu Science Park, Hsinchu 310403, Taiwan
* Correspondence: chhuchen@nycu.edu.tw

Abstract: A high-precision catadioptric telescope such as a space-borne telescope is usually tested with interferometer to check the optical quality in assembly. The coarse and fine alignment of the telescope are mainly based on the information from the coordinate measuring machine and the fringe pattern of the interferometer, respectively. In addition, further fine-tuning can be achieved according to the variation in wavefront error and Zernike data. The issue is that the vast majority of the catadioptric telescopes contain plural lens surfaces which could produce unwanted ghost fringes, disturbing the wavefront measurement. Technically, off-axis installation to shift away ghost fringes from central interferogram could be acceptable in some cases. Nevertheless, in this paper, the source of ghost fringe in interferometric measurement for catadioptric telescopes is investigated with light path simulation, and a solution of reducing the f-number of the diverger lens is proposed to eliminate the ghost fringe disturbance. Both simulation and experimental results verify the effectiveness of the proposed concept.

Keywords: catadioptric telescope; interferometry; ghost fringes

Citation: Huang, Y.-K.; Chen, C.-H. Ghost Fringe Suppression by Modifying the f-Number of the Diverger Lens for the Interferometric Measurement of Catadioptric Telescopes. *Photonics* **2024**, *11*, 453. https://doi.org/10.3390/photonics11050453

Received: 26 March 2024
Revised: 3 May 2024
Accepted: 10 May 2024
Published: 11 May 2024

1. Introduction

Optical telescopes such as catadioptric and full-mirror-type systems are often applied in astronomical research, earth observation missions and deep space exploration. The ROCSAT-2 and Pleiades HR both served as successful earth observation satellites with a lifetime of more than 10 years; they featured the high-end catadioptric type and full-mirror-type systems with an aperture over 60 cm, respectively [1,2]. Recently, Canon Electronics Inc. has developed several low-cost micro-satellites equipped with the optical payload consisting of their own digital camera and a 40 cm diameter catadioptric telescope aiming at commercial earth observation applications [3]. The examples above show that the catadioptric telescope plays an important role in high-end and low-cost solutions for space telescopes.

In this paper, a catadioptric telescope was designed to conduct an earth observation mission in a 500–600 km altitude orbit and it could provide multi-spectral images with a 450–890 nm wavelength band. The catadioptric telescope is a circularly symmetric system and features easier fabrication and faster alignment compared to more advanced architectures such as the full-mirror off-axis system which usually replaces the corrector lens with a tertiary mirror. In order to meet the tight schedule of deploying a constellation of remote sensing satellites, the catadioptric telescope could be a decent candidate for the optical payload of the earth observation mission. The optical layout of the telescope is shown in Figure 1; it consists of a 40 cm diameter primary mirror (M1), a 14 cm diameter secondary mirror (M2) and a 4-lens corrector module [4], where the color rays represent the light path of each field.

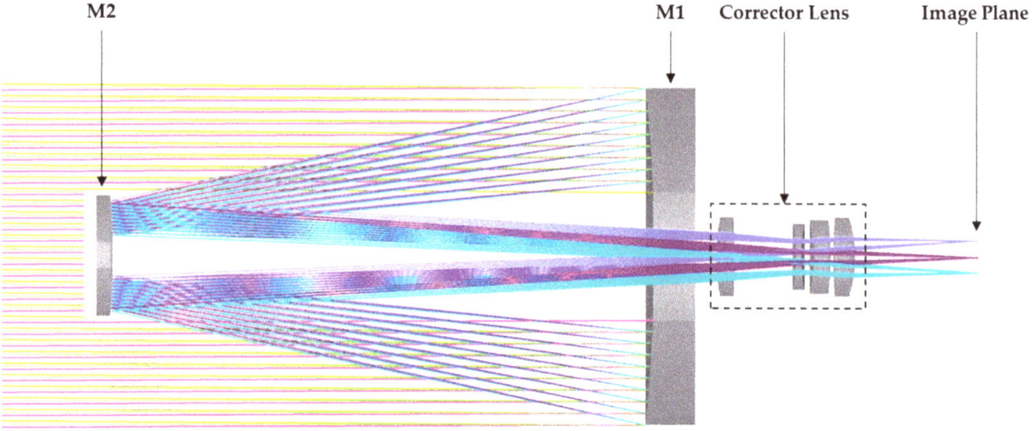

M2   M1   Corrector Lens   Image Plane

Figure 1. Optical layout and light path of catadioptric telescope.

Both mirrors are made of a low coefficient of thermal expansion (CTE) substrate with high-reflectance silver coating on their optical surfaces, and the lenses are made of fused silica with broad-band anti-reflective coating (reflectance less than 2%) on each surface. The alignment and assembly of the corrector lens are accomplished with an allocated maximum range of tilt and decenter error budgets less than ± 60 arcsec and ± 75 μm, respectively.

The optical quality of the entire system is usually measured with an interferometer. Figure 2 shows the full light path of the double-pass interferometric measurement system of the catadioptric telescope with a Twyman–Green interferometer which adopts a spherical diverger lens with an f-number of 7. An auto collimation flat mirror with a clear aperture of 24 inches is employed to construct a double-pass architecture for the interferometric measurement. In addition, a pinhole array, consisting of five pinholes with a diameter of 1 mm, as illustrated in Figure 2, is used to define the measurement fields, including one on-axis field and four off-axis fields, and is placed on the image plane of the telescope.

A converging beam (He-Ne laser 632.8 nm), produced by the interferometer and the diverger lens, focuses onto the pinhole and then passes through the corrector to enter the telescope. After being reflected by M2 and M1 sequentially, it becomes collimatedly incident onto the auto collimation flat mirror. The reflected beam from the auto collimation flat mirror then returns to the telescope and interferometer by following the original light path. This to and fro route becomes the so-called double-pass light path. Finally, an interferogram can be obtained with the superposition of the return beam from the telescope and the reference beam from the reference mirror on the detector of the interferometer.

Before executing the interferometry, a sequence of mechanical alignments with a coordinate measuring machine (CMM) is conducted which will eventually leave an alignment error at the micrometer scale. The interferometry is then used for fine-tuning the position of optical components by monitoring the wavefront error (WFE). The final system wavefront data will be acquired in multiple average based on the standard deviation to eliminate the influence of random disturbances such as vibration and turbulence.

The full-field interferograms and corresponding WFE maps are shown in Figures 3 and 4, respectively. The interferogram shows a coarse fringe from the overall telescope optics and several fine fringes which constitute unwanted disturbance called ghost fringe. Consequently, the WFE map in Figure 4 shows broken patches where the ghost fringes are located, and those broken patches highly disturb the evaluation of the wavefront error. The other four off-axis fields present nominal WFE maps without ghost fringe disturbance.

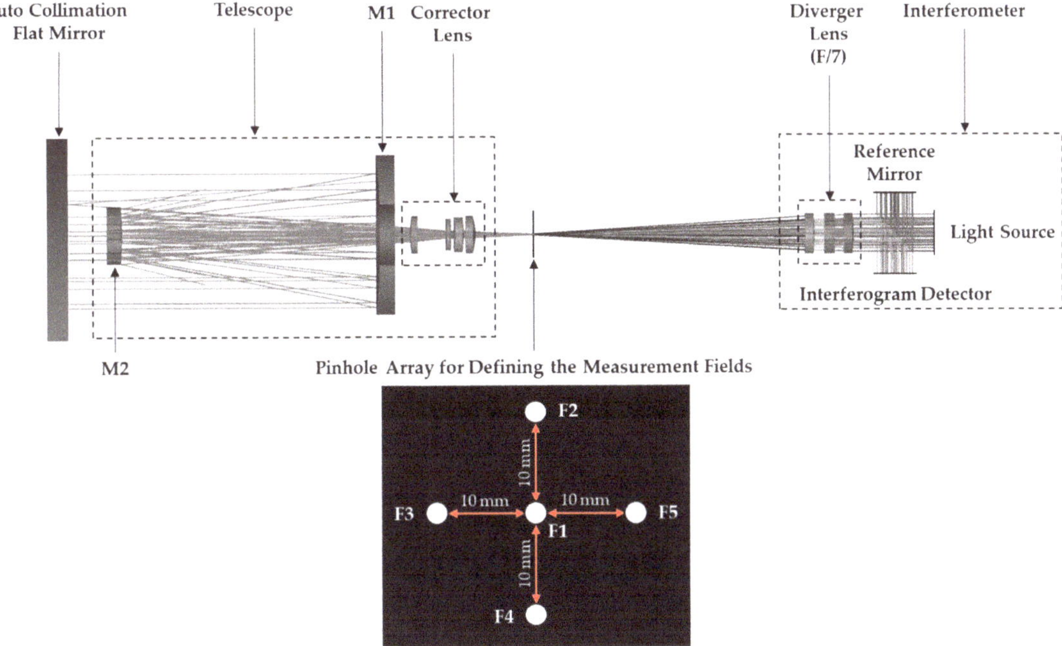

Figure 2. Interferometric measurement setup and light path for catadioptric telescope.

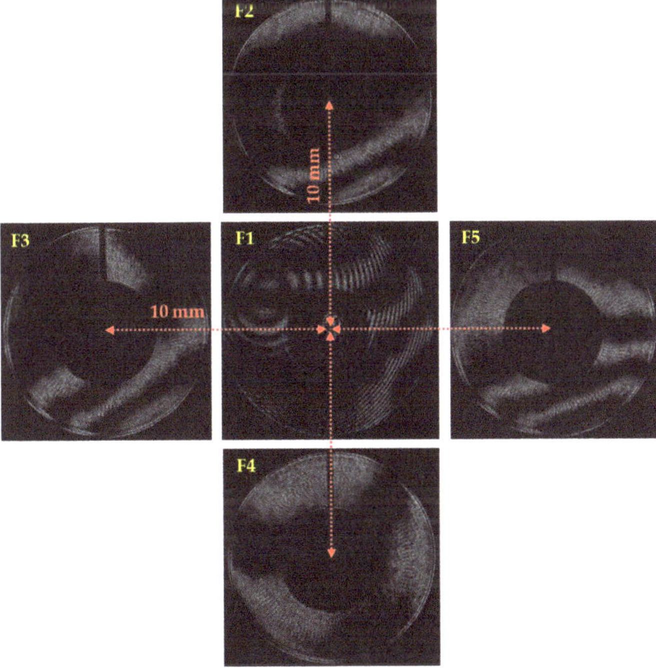

Figure 3. Full-field interferograms of one on-axis field (F1) and four off-axis fields (F2–F5).

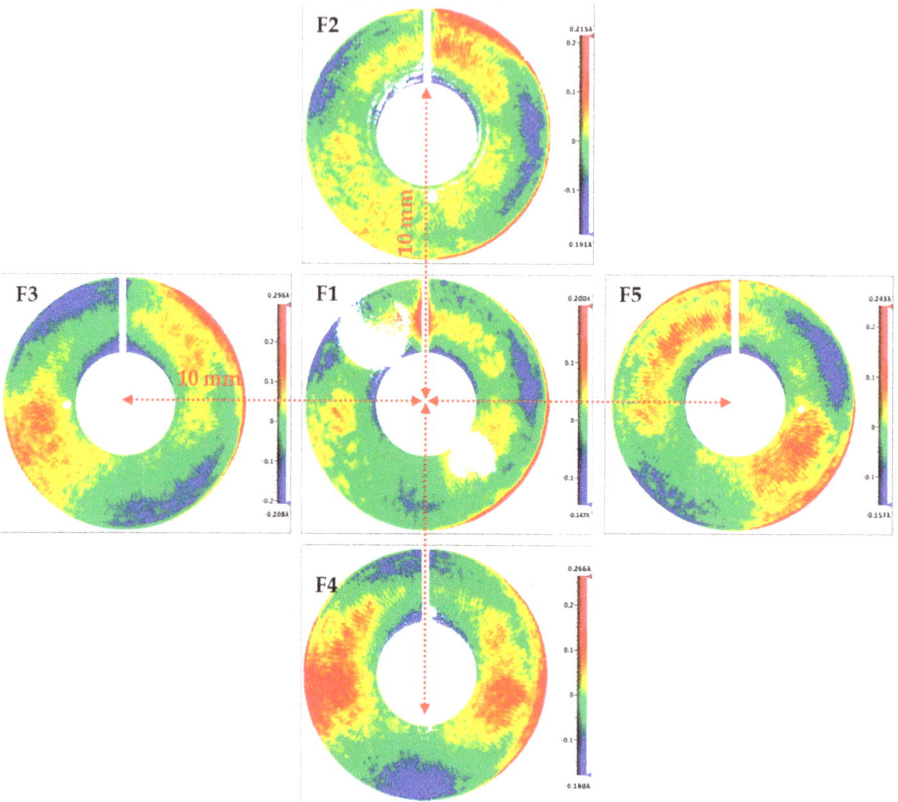

Figure 4. Full-field WFE maps of one on-axis field (F1) and four off-axis fields (F2–F5).

Technically, the ghost fringes can be shifted away from the on-axis interferogram with a slight off-axis installation of the setup shown in Figure 2, and this approach can be acceptable in cases where the full-field WFE results satisfy the requirement with good aberration symmetry, but it is not applicable in general. Ghost images are classified according to the number of reflections encountered in the ghost ray path [5]. In imaging applications, ghost images could cause contrast reduction and veil parts of the image [6]. The light reaching the nominal image plane could also degrade the image quality [7]. Too many concentrated ghosts can disqualify a candidate lens design, equivalent to the effect of aberration [8]. The intensity of a ghost image is proportional to the product of the reflection coefficient of the coating on the refractive interfaces involved in the ghost light path and inversely proportional to the area of the ghost image [9]. In this paper, the source of the ghost fringe for the catadioptric telescope is investigated and the solutions to resolve the issue without off-axis installation are proposed and verified.

2. Source of Ghost Fringe in Catadioptric Telescope

In experiments, when a mask is located between the primary mirror and the corrector, blocking the laser beam from entering the mirrors, the ghost fringes still remain on the interferogram, as shown in Figure 5. This shows that the ghost fringe mainly comes from the corrector lens. When adjusting the position of the corrector, the ghost fringe shifts on the interferogram in accordance.

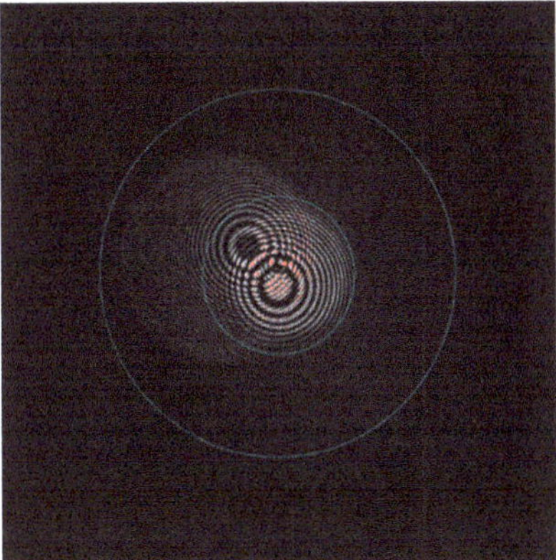

Figure 5. Interferogram of corrector module.

Theoretically, there should be an infinite number of ghost fringes. For practical engineering purposes, only the ones with sufficient intensity to disturb the correct reading of the original system fringe should be considered. Therefore, only the ghost light path with the first reflection from a refractive surface will be examined. As there are four lens elements in the corrector module, there should be a total of eight ghost light paths meeting this condition. However, to form a clear ghost fringe on the interferogram, the ghost light path should have a focus point close to that of the original system interferometric light path, i.e., the position of the pinhole shown in Figure 2. As a consequence, the most likely source of the ghost fringe come from the surfaces which are concave toward the pinhole, i.e., surface 1 of lens 1 (L1S1), L2S2, and L4S1, as illustrated in Figure 6, where the corrector module and the light path through it in Figure 1 are highlighted to show the labels of optical surfaces for the corrector.

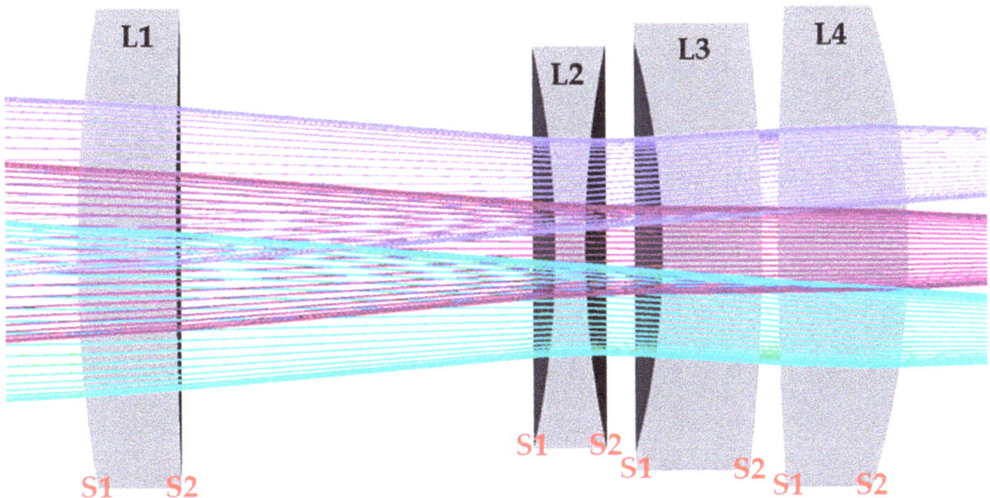

Figure 6. Labels of refractive surfaces for corrector module.

The distances between the center of curvature of those three surfaces and the pinhole are listed in Table 1. According to the table, the center of curvature of L1S1 is the one closest to the pinhole, and the corresponding ghost fringe could be the largest and strongest. Its light path through the pinhole is simulated and illustrated in Figure 7.

Table 1. Distance between pinhole and center of curvature of three surfaces.

Surface	Distance (mm)
Pinhole	0
L1S1	−22.7
L2S2	48.2
L4S1	−99.6

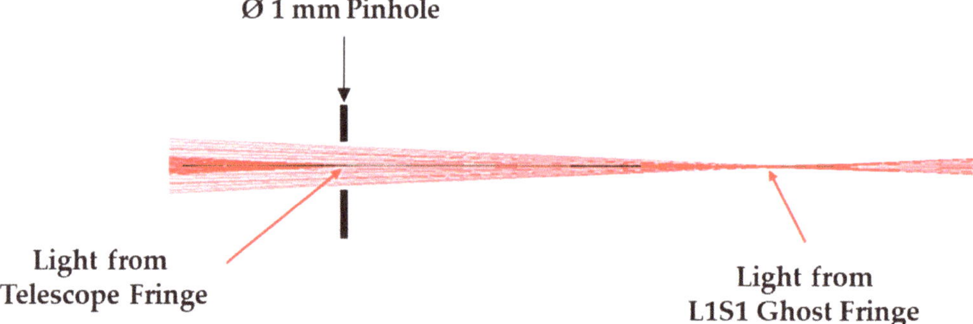

Figure 7. Light path of telescope fringe and L1S1 ghost fringe.

An alternative approach to get around the disturbance of ghost fringes is to tilt the corrector module such that the reflected ghost light is directed out of the path toward the detector or image plane [10]. To further confirm the source of the ghost fringe, the corrector module was tilted with respect to the center of curvature of the three surfaces listed in Table 1. For example, when tilting the corrector with respect to the center of curvature of L1S1, the corresponding ghost fringe from L1S1 will stay at its original position. Based on the abovementioned optical simulation and experimental results, it is confirmed that the largest ghost fringe is from L1S1, whereas the other two smaller ghost fringes are from L2S2 and L4S1.

3. Elimination of the Ghost Fringe

For eliminating the ghost fringes, several solutions have been proposed, including anti-reflection coating, polarization alteration, optical parameter manipulation, etc. The most direct way with minimum modification on the original optical system would be anti-reflection coating at the wavelength of the interferometer on the surfaces causing ghost fringes. Figure 8 shows the spectral reflectivity of the coating for L1S1, where new coating has been designed specially for anti-reflection at the 632.8 nm wavelength of the He-Ne laser, which is used as the light source in the interferometer. The interferogram of the corrector and the WFE of the whole telescope with nominal and new coating on L1S1 are shown in Figure 9, which indicates that the new coating is effective in eliminating the ghost fringe. Nevertheless, the modification is made only for accommodating interferometric measurements in the assembly process, which might not be an optimal design or might even cause side effects in real applications. This means that the approach is not a generic solution.

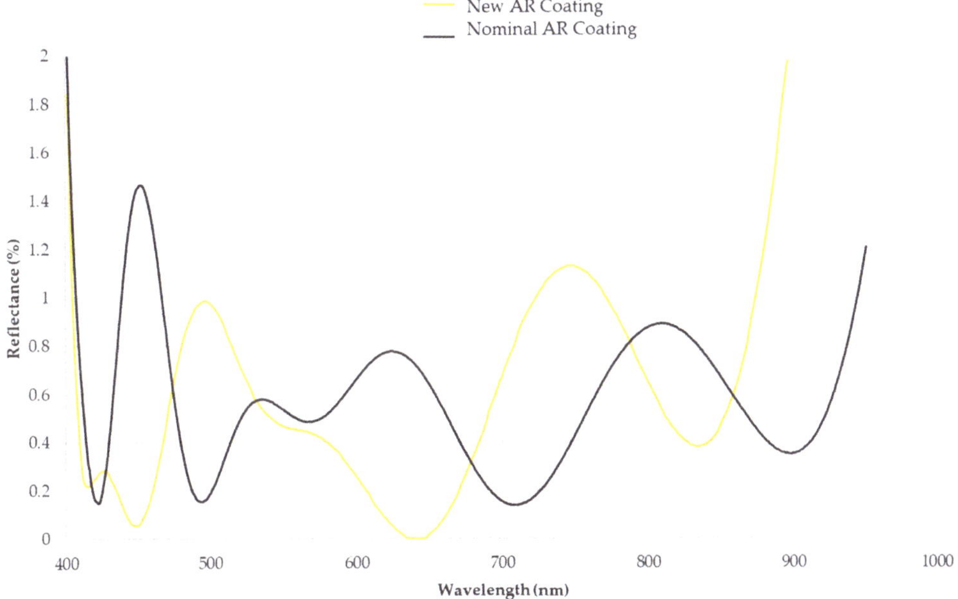

Figure 8. Spectral reflectance of L1S1 with nominal AR coating and new AR coating.

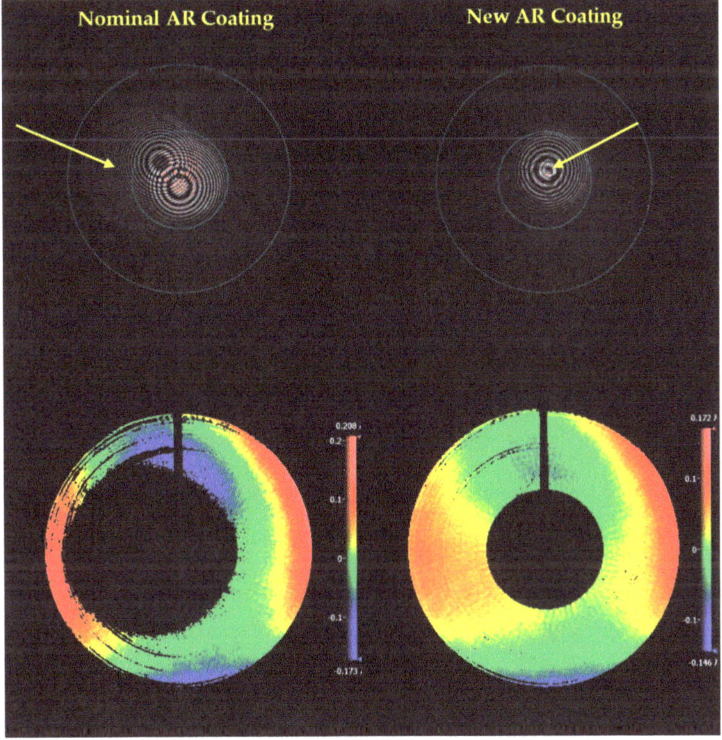

Figure 9. *Cont.*

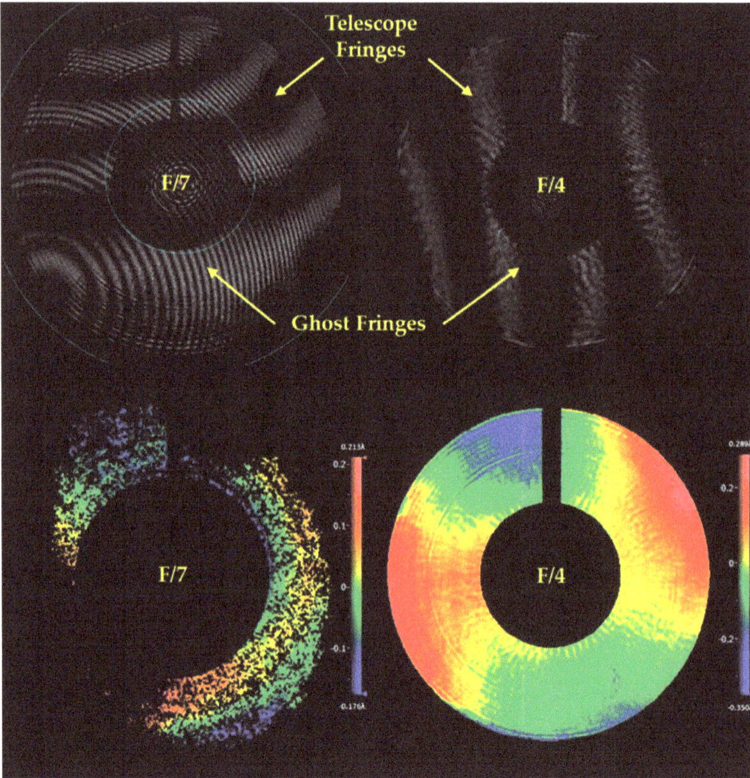

Figure 9. Interferogram of corrector only and WFE map of telescope with nominal AR coating and new AR coating.

By examining the light path shown in Figure 7, reducing the size of the pinhole should be able to block more energy from the ghost light path while allowing through the desired light from the telescope, consequently reducing the relative contrast of the ghost fringe. However, the minimum pinhole size is normally determined by the feasibility of the optical alignment process. An alternative but equivalent modification would be to change the f-number of the diverger lens of the interferometer.

The analysis of the influence of the f-number on the ghost fringe was conducted with the non-sequential mode of Zemax [11,12] by calculating the contrast of coherent irradiance. An aberration-free module, consisting of three lens elements, was designed to act as the diverger lens in the interferometric light path, as shown in Figure 2. A 1 mm diameter pinhole with absorbing coating material was located at the image plane of the telescope. The whole interferometric system consisted of a diverger lens module, a reference mirror, a light source at 632.8 nm with 1 w/cm^2 power, and an interferogram detector with 1000×1000 pixels.

For the first configuration, perfect AR coating with 100% transmittance was applied to all surfaces of the corrector lens module to simulate the interferometric fringe of the whole telescope. The corresponding coherent irradiance on the interferogram and cross-sectional distribution are shown in Figure 10. The maximum value of coherent irradiance was calculated by averaging the top 100 data points and the result was 0.0083 w/cm^2.

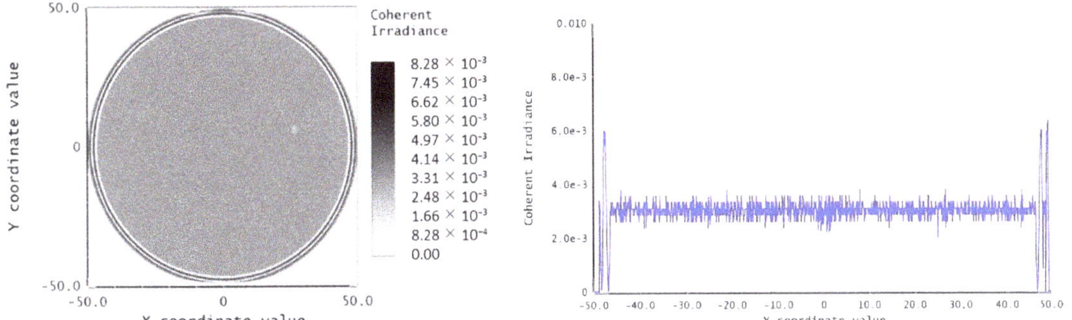

Figure 10. Coherent irradiance of interferogram and cross-sectional distribution for the whole telescope.

Secondly, in order to simulate only the ghost fringe from L1S1, the mirror surface was applied on L1S1, as shown in Figure 11. With this setup, the coherent irradiance purely from L1S1 could be obtained, as shown in Figure 12. The maximum value of coherent irradiance was 0.0106 w/cm^2.

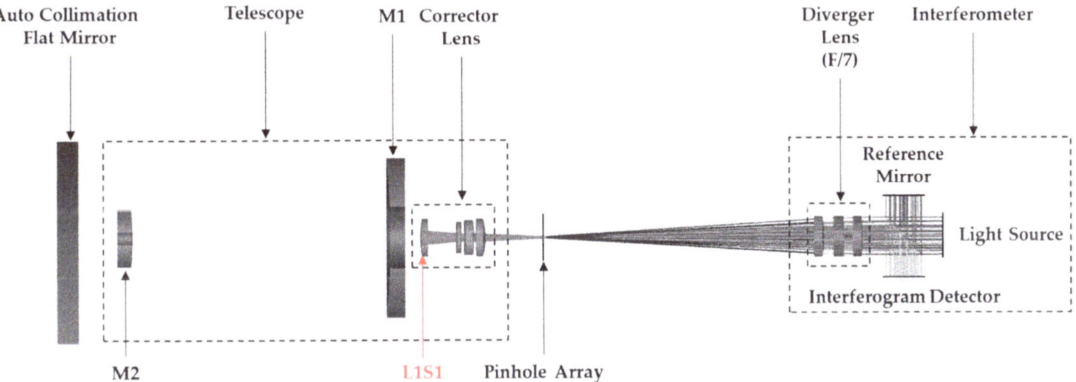

Figure 11. Interferometric light path of ghost fringe from L1S1.

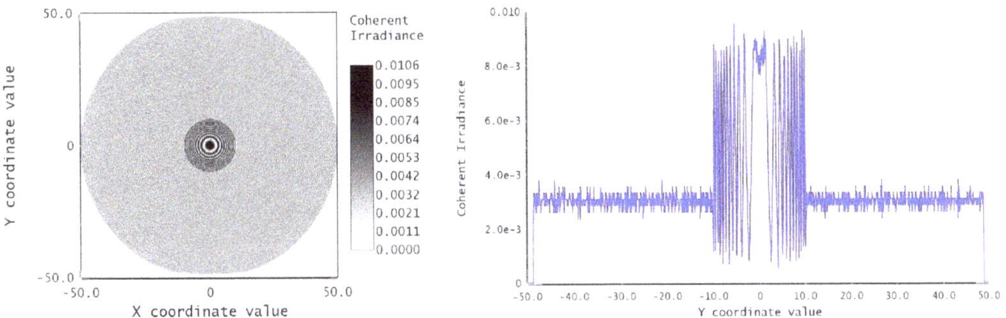

Figure 12. Coherent irradiance of interferogram and cross-sectional distribution for L1S1 ghost fringe only.

Finally, the superposition of the telescope fringe and the ghost fringe was simulated by applying 50% transmittance coating on L1S1. The gray scale graph of coherent irradiance with both the original telescope fringe and the ghost fringe from L1S1 is shown in Figure 13. Based on this diagram, the contrast of the original fringe to the ghost fringe could be evaluated, which was around 60%.

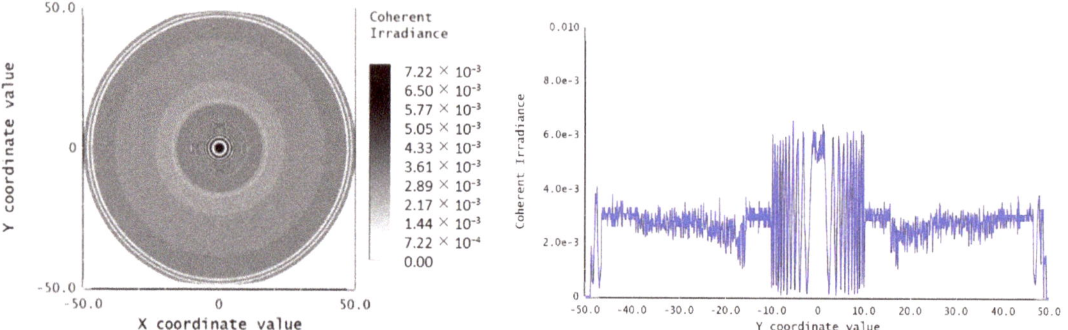

Figure 13. Coherent irradiance of interferogram and cross-sectional distribution for the superposition of telescope fringe and ghost fringe from L1S1.

A similar analysis can be performed for different f-numbers of the diverger lens module. Figure 14 shows the contrast of the telescope fringe to the ghost fringe with different f-numbers of the diverger lens module from F/7 to F/4. It indicates that the contrast increases with the reduction in f-number, and it exceeds 100%, i.e., the intensity of the telescope fringe surpass that of the ghost fringe from L1S1, at F/4 of the diverger lens. The comparison of the superposition fringe for F/7 and F/4 is shown in Table 2.

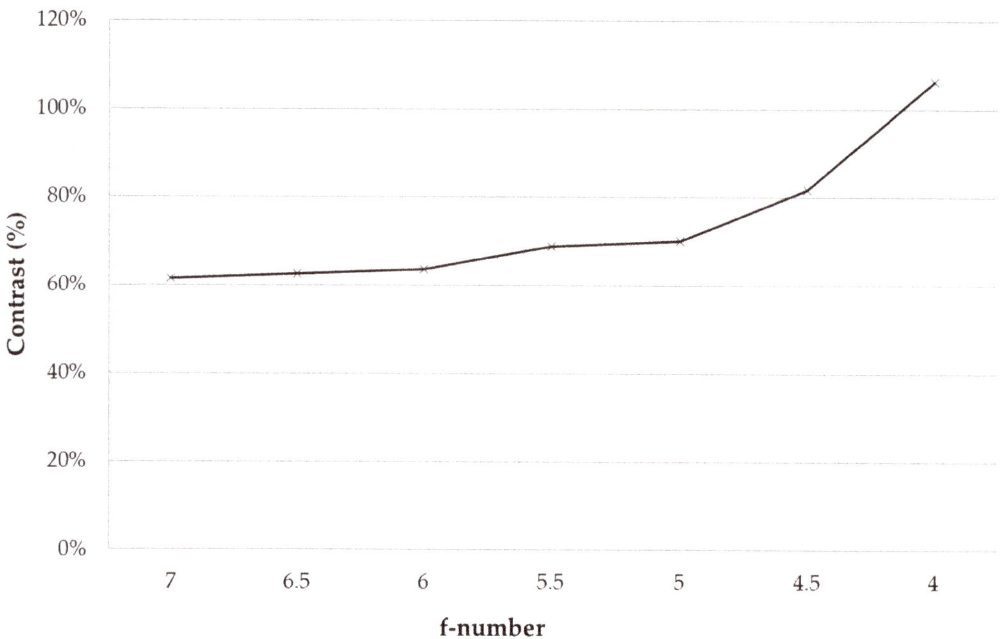

Figure 14. Contrast of telescope fringe to ghost fringe versus f-number of diverger lens module.

Table 2. Comparison of telescope fringe, ghost fringe and superposition fringe for F/7 and F/4.

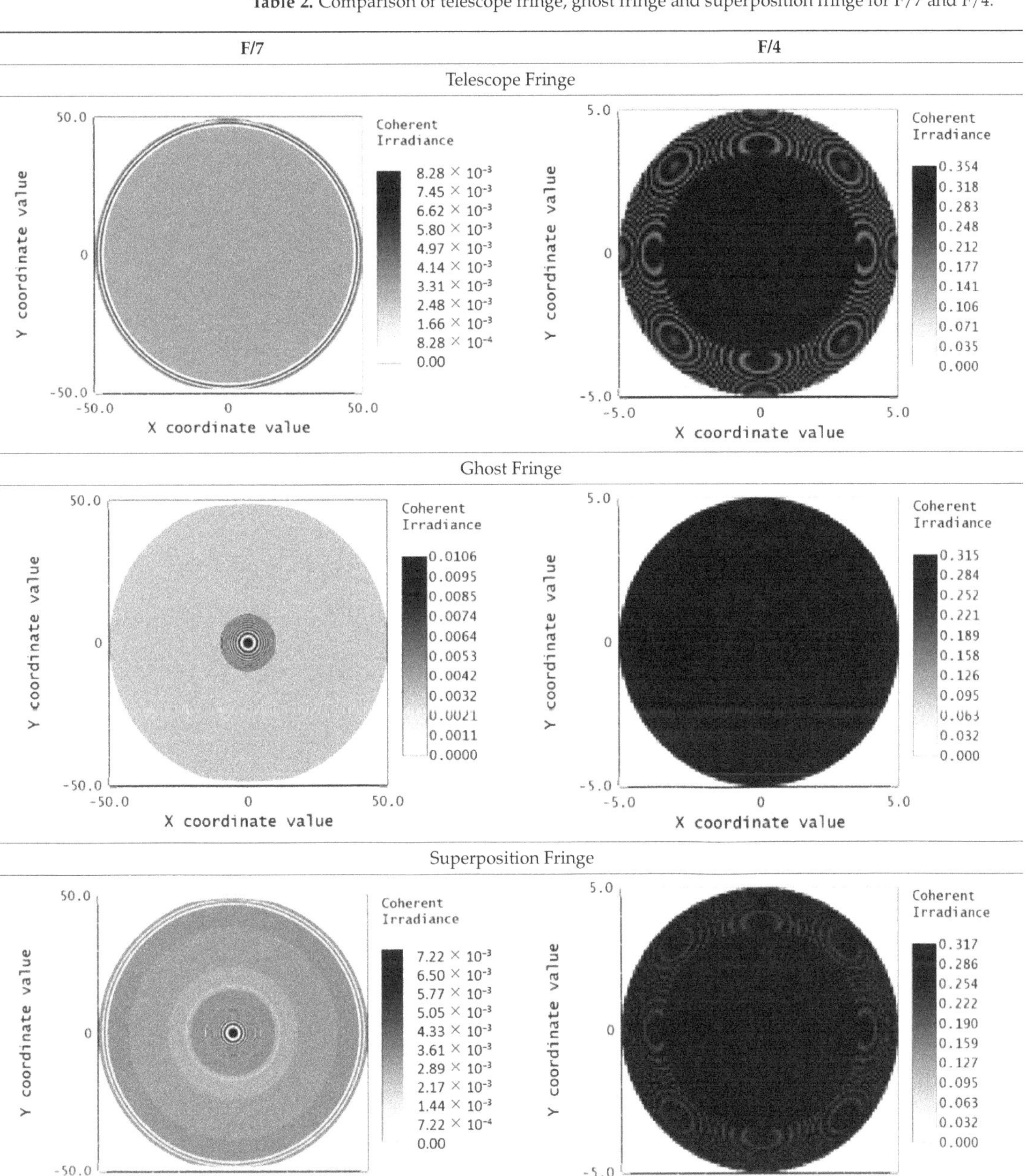

4. Experimental Results

Based on the simulation and analytical results with Zemax, the corresponding experiment was carried out for verification. For the spatial filter (pinhole) obscuration test, the size was changed from the original diameter of 1 mm to 0.3 mm, as shown in Figure 15, and the corresponding test result is shown in Figure 16. The light from the ghost fringe could be partially obscured by a 1 mm diameter pinhole but was largely obscured by a 0.3 mm diameter pinhole. The relative contrast of the telescope fringe to the ghost fringe could be increased, which is beneficial for the reconstruction of the WFE map with less data loss. However, the smaller the pinhole, the more time-consuming the alignment.

Figure 15. Light path through pinhole with different pinhole sizes.

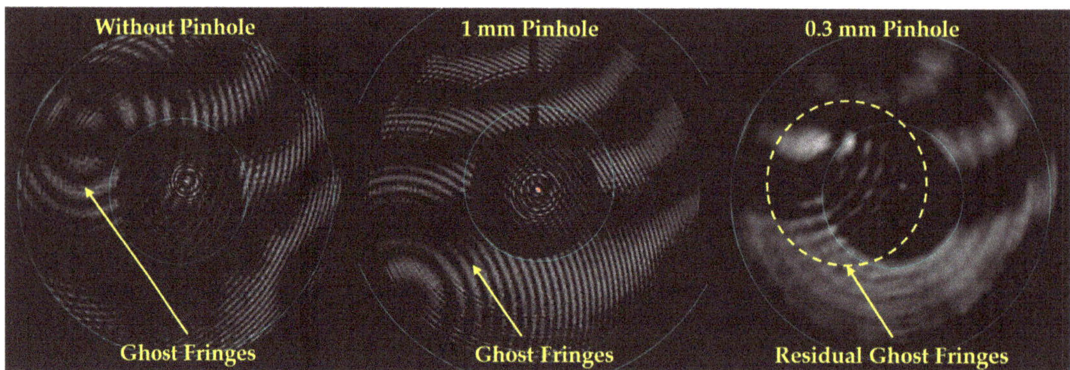

Figure 16. Interferogram of telescope without pinhole and with pinhole sizes of 1 mm and 0.3 mm.

The alternation is to reduce the f-number of the diverger lens while keeping the pinhole size at 1 mm. Figure 17 shows the interferogram and the WFE map of the telescope for the f-number of the diverger lens at F/4 and F/7. It indicates that a nearly perfect WFE map was obtained with F/4, which matched the desired improvement obtained in the simulation.

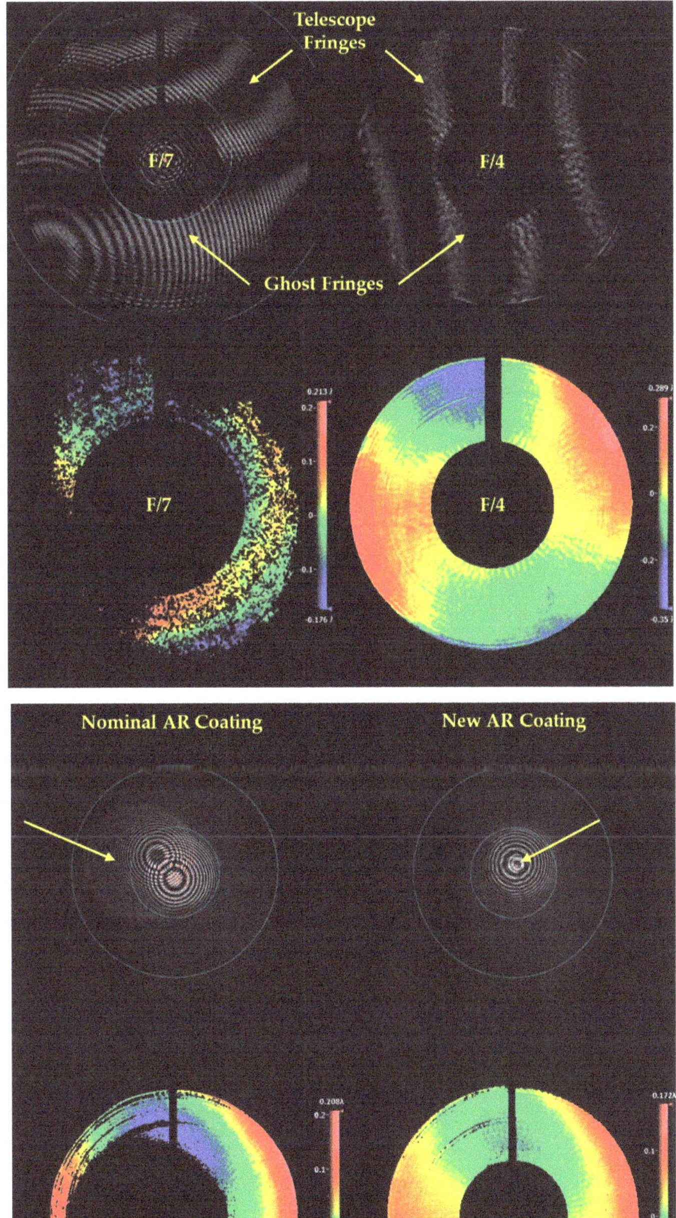

Figure 17. Interferogram and WFE map of telescope with f-number of diverger lens at F/4 and F/7.

5. Conclusions

High-quality catadioptric telescopes have been increasingly used in commercial applications in recent years. The fast assembly of this kind of telescopes, with various specifications to meet the launching schedule of satellites, has become a more important issue. The proposed approach was verified as an effective solution for the alignment of catadioptric telescopes without the disturbance from ghost fringes. Once a suitable f-number of the diverger lens is found, it can be used for the testing of the same batch. For the other batches with different specifications, the only part that needs to be changed in the whole interferometric test system is the diverger lens. Due to the fact that the solution is applied to the interferometer side, a similar concept can be applicable to all the other high-quality optical systems involving transmissive optical elements.

Author Contributions: Conceptualization, Y.-K.H.; methodology, Y.-K.H.; software, Y.-K.H.; validation, Y.-K.H.; formal analysis, Y.-K.H.; investigation, Y.-K.H.; data curation, Y.-K.H.; writing—original draft preparation, Y.-K.H.; writing—review and editing, C.-H.C.; visualization, Y.-K.H.; supervision, C.-H.C. All authors have read and agreed to the published version of the manuscript.

Funding: This research received no external funding.

Institutional Review Board Statement: Not applicable.

Informed Consent Statement: Not applicable.

Data Availability Statement: Data is unavailable due to privacy or ethical restrictions.

Acknowledgments: We would like to express my thanks to Sheng-Feng Lin from Taiwan Space Agency for supporting the preliminary investigations and the discussions of developing the ideas put forward here. Both the success of optical analysis and experiment are attributed to his invaluable advice.

Conflicts of Interest: The authors declare no conflict of interest.

References

1. Uguen, G.; Luquet, P.; Chassat, F. Design and development of the 2m resolution camera for ROCSAT-2. In Proceedings of the 5th International Conference on Space Optics (ICSO 2004), Toulouse, France, 30 March–2 April 2004.
2. Lamard, J.-L.; Gaudin-Delrieu, C.; Valentini, D.; Renard, C.; Tournier, T.; Laherrere, J.-M. Design of the High Resolution Optical Instrument for the Pleiades HR Earth Observation Satellites. In Proceedings of the 5th International Conference on Space Optics (ICSO 2004), Toulouse, France, 30 March–2 April 2004.
3. CE-SAT Series, Remote Sensing Technology Center of Japan. Available online: https://www.restec.or.jp/en/solution/product/ce-sat.html (accessed on 10 March 2024).
4. Lin, S.-F.; Chen, C.-H.C.; Huang, Y.-K.H. Optimal F-Number of Ritchey–Chrétien Telescope Based on Tolerance Analysis of Mirror Components. *Appl. Sci.* **2020**, *10*, 5038. [CrossRef]
5. El-Maksoud, R.H.A.; Sasian, J.M. Paraxial Ghost Image Analysis. In Proceedings of the SPIE Optical Engineering + Applications, San Diego, CA, USA, 2–6 August 2009.
6. El-Maksoud, R.H.A.; Hillenbrand, M.; Sinzinger, S.; Sasian, J.M. Optical Performance of Coherent and Incoherent Imaging Systems in The Presence of Ghost Images. *Appl. Opt.* **2012**, *51*, 7134–7143. [CrossRef]
7. El-Maksoud, R.H.A. Ghost Reflections of Gaussian Beams in Anamorphic Optical Systems with An Application to Michelson Interferometer. *Appl. Opt.* **2016**, *55*, 1302–1309. [CrossRef] [PubMed]
8. El-Maksoud, R.H.A.; Sasian, J.M. Modeling and Analyzing Ghost Images for Incoherent Optical Systems. *Appl. Opt.* **2011**, *50*, 2305–2315. [CrossRef] [PubMed]
9. Grabarnik, S. Optical Design Method for Minimization of Ghost Stray Light Intensity, Optics Section, Directorate of Technical and Quality Management. *Appl. Opt.* **2015**, *54*, 3083–3089. [CrossRef] [PubMed]
10. Rogers, J.D.; Tkaczyk, T.S.; Descour, M.R. Removal of Ghost Images by Using Tilted Element Optical Systems with Polynomial Surfaces for Aberration Compensation. *Opt. Lett.* **2006**, *31*, 504–506. [CrossRef] [PubMed]
11. Younis, R.B.; Khorsheed, S.M.; Al-Dahan, Z.T. Modified Michelson Interferometer Design for Ear Measurement. *Int. J. Sci. Eng. Res.* **2018**, *6*, 36–41.
12. Hitzenberger, C.K. Low-coherence interferometry. In *Handbook of Visual Optics*; CRC Press: Boca Raton, FL, USA, 2017.

Article

Control of the Optical Wavefront in Phase and Amplitude by a Single LC-SLM in a Stellar Coronagraph Aiming for Direct Exoplanet Imaging

Andrey Yudaev *, Alla Venkstern, Irina Shulgina, Alexander Kiselev, Alexander Tavrov and Oleg Korablev

IKI RAS, Space Research Institute of Russian Academy of Sciences, 117997 Moscow, Russia
* Correspondence: yudaev@phystech.edu

Abstract: This article presents a novel approach to actively compensate wavefront errors in both phase and amplitude using a Liquid Crystal Spatial Light Modulator (LC-SLM) for direct exoplanet imaging. This method involves controlling the wavefront to address challenges posed by stellar coronagraphy. Experimental results demonstrate successful wavefront error compensation in both phase and amplitude components. This technique shows promise for direct exoplanet imaging and may be applied onboard orbital telescopes in the future.

Keywords: stellar coronagraphy; direct imaging; LC-SLM; rotational shear interferometer; wavefront sensing; wavefront control and correction; exoplanets

Citation: Yudaev, A.; Venkstern, A.; Shulgina, I.; Kiselev, A.; Tavrov, A.; Korablev, O. Control of the Optical Wavefront in Phase and Amplitude by a Single LC-SLM in a Stellar Coronagraph Aiming for Direct Exoplanet Imaging. *Photonics* **2024**, *11*, 300. https://doi.org/10.3390/photonics11040300

Received: 20 February 2024
Revised: 14 March 2024
Accepted: 21 March 2024
Published: 26 March 2024

1. Introduction

In photonics, one of the most intriguing tasks is to visualize a point-like image of an Earth-type planet in the vicinity of a neighboring star (of Solar type) at a distance on the order of 10 parsecs. Formally, the spatial resolution of a meter class telescope in the optical wavelength range is sufficient by Rayleigh criterion to detect the outstanding peak of an exoplanet point spread function (PSF) at a *stellocentric* is unnecessary distance of $1...10\,\lambda/D$.

However, the luminosity ratio of star/planet is about $10^9 \ldots 10^{10}$. The planet has insufficient contrast as the faint light source relative to the host star light source. We shall consider a two-meter class telescope because the effect of exozodi is surrounding and therefore masking a stellar vicinity [1,2]. If the telescope has an ideal optical quality and has a stellar coronagraph mounted after, such a coronagraph instrument can show a planet on an attenuated diffraction background of the host star or, more precisely, on the attenuated background of the stellar PSF wings.

Modern space optics offers diffraction-limited resolution; however, optics is not free from residual aberrations and micro-roughness, which result in a visible *halo* effect around the main maximum of the PSF. Certainly that masks the image of the faint planet light source. To move from a theoretical to a practical scope, one has to use precise adaptive optics (AO) in order to eliminate the wavefront (WF) error (WFE). The adaptive optics system aims to measure the WFE and then to compensate for it. One fundamentally analyses the diffraction problem resumes wavefront error in terms of complex amplitude [3]. One has to compensate not only for the phase component of the WFE, but also for its amplitude component.

The amplitude component of the WFE is caused by some zonal inhomogeneity in transmittance or/and reflection; as well, it is due to Fresnel diffraction on the micro-roughnesses, residual aberrations, and aperture boundaries of every optical element (or optical surface) mounted in the plane not being conjugated with the pupil of the optical system, e.g., a secondary telescope mirror, downstream coronagraph optics, a spider, apertures etc. Moreover, a fringe field effect can be caused in the LC layer that could distort the phase and amplitude wavefront profiles in outgoing light from the LC-SLM itself [4].

A portrait of the Fresnel diffraction can be primitively illustrated via the Talbot effect for a periodic structure or via sharp boundary oscillating. To compensate for the WFE in the Fresnel zone (not in the Fraunhofer zone), we require an increasing number of controllable pixels or actuators if deformable mirrors (DM) are used.

In the present communication, we show new technique and demonstrate the start-up laboratory experiments aiming at the phase–amplitude correction for the WFE in a stellar interference coronagraph.

Omitting details, we shall address the reader to the published papers on the characteristics and operations of the stellar coronagraph, as constructed by a classical Lyot schematic with and without apodization schemes [5–7], and, in particular, by a coronagraph based on destructive interference process [8,9].

The working principle of an interfero coronagraph [10] is to superimpose with antiphase two pupil images being shifted [8], reversed, or rotated [11], or differently superposed [12]. The listed interferometers result in a destructive interference process. Therefore, an interferometer eliminates the starlight being received from the on-axis direction. Planetary (or companion) light does not interfere destructively because of the off-axis tilt (or shift or shear in a pupil). The image component from a planet becomes split into two copies; by superposition, they are spatially separated and do not interfere destructively. Interfero-coronagraphy is advantageous in terms of its broad spectral band achromaticity and, more generally, because of its small inner working angle (IWA). The IWA characteristic can be considered as the spatial resolution of a coronagraph instrument. The achromatic interfero-coronagraph (AIC) is known to have one of the smallest possible IWAs ($0.38 \lambda/D$), such as it was initially referred to in [10] with the fixed angle of a 180-deg. rotational shear. Later, it was redesigned into a modified Sagnac scheme, implementing the common path (or cyclic path or cavity) (CP-AIC) [13] aiming to relax the mechanic instability.

The severe functional disadvantage of an interfero coronagraph is known as the stellar leakage effect. Due to this effect, the starlight cannot be completely suppressed because the apparent size of the star is not physically infinitesimal [5]. The observed size of a star is far beyond the spatial resolution of a *single dish* optical telescope, but physically, the star size causes a non-fully coherent point-like source; considering spatial coherence, it causes an extended source. Numerous methods to reduce the stellar leakage effect, due to the infinitesimal apparent size of the star, have been proposed from the 180-deg. rotational shift to smaller angles [11] or to apply an apodization, e.g., a Sonine type [14].

In the present communication, the proposed method is applicable to the listed possible efforts to reduce the star leakage effect.

2. Method

Precise wavefront correction in phase and amplitude is ultimately required as pre-optics schematics for the coronagraph for its functionality. AO compensates for optical aberrations including optical defects given by the optical surfaces and apertures. Light radiation is collected by a telescope under or above the telluric atmosphere, and it is analyzed after a coronagraph by a field camera, spectroscopy, or different instrument. It is important to note that the measurement of the wavefront has to be organized after the coronagraph. Otherwise, a non-common-path wavefront error (NCP WFE) is caused by a different or incompletely similar optical path to the wavefront sensor (WFS). This generates additional phase and amplitude errors that become magnified [15] by the coronagraph.

Aiming active adaptive optics to control the WFE in phase and amplitude, we used a phase-only liquid crystal spatial light modulator (LC-SLM) mounted in a specific polarization schematic as shown in Figure 1 inside the dashed-line box I. In our work, we used the reflective type LC-SLM. In Figure 1, the LC-SLM is shown in transmittance mode for the purpose of simplification. It is better to mount the LC-SLM by an incident angle less than $15°$ to minimize depolarization effects, which can reduce the SLM modulation contrast [16]. In our setup the incident angle was about $4°$.

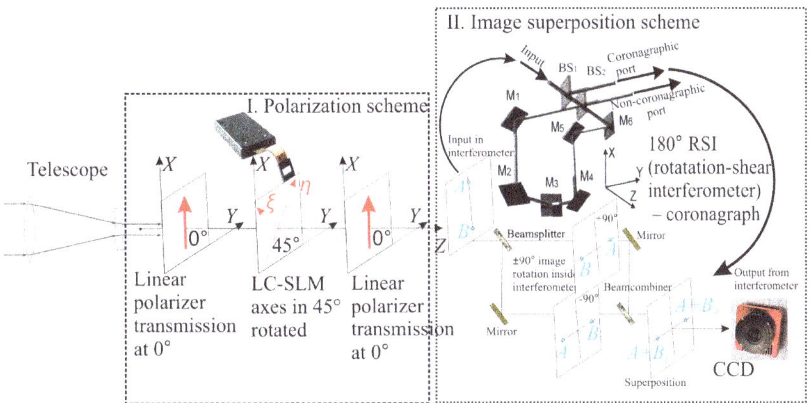

Figure 1. Optical schemes: I. Polarization scheme to orient LC-SLM (in dashed-line box) and II. Image superposition in rotation–shear interferometer (in dot-line box) simplified.

Math algebra of the polarization of Jones vectors and matrices was applied to describe the action of the *phase-only* LC-SLM being used as a phase and amplitude WFE corrector. We considered the output of the following polarization scheme: entrance linear polarization with the azimuth orientation at 45° to the X axis, LC-SLM with the phase modulation axis along X, and the final linear polarizer with transmission at 45° to X. See this set, in the dash-line box I in Figure 1.

Then, the phase–amplitude control of the WF is realized as:

$$\mathbf{E}_{out} = \mathbf{P}_{\beta=45°}\,\mathbf{SLM}\,\mathbf{E}_{in} \tag{1a}$$

$$= \frac{1}{2}\begin{pmatrix}1 & 1 \\ 1 & 1\end{pmatrix}\begin{pmatrix}e^{i\alpha} & 0 \\ 0 & 1\end{pmatrix}\frac{1}{\sqrt{2}}E\,e^{i\varphi}\begin{pmatrix}1 \\ 1\end{pmatrix} \tag{1b}$$

$$= \frac{1}{2\sqrt{2}}\left(e^{i\alpha}+1\right)E\,e^{i\varphi}\begin{pmatrix}1 \\ 1\end{pmatrix}, \tag{1c}$$

$$= \widetilde{E}\,e^{i\widetilde{\varphi}_A}\begin{pmatrix}1 \\ 1\end{pmatrix} \tag{1d}$$

where $\mathbf{E}_{in} = E\,e^{i\varphi}\frac{1}{\sqrt{2}}\begin{pmatrix}1 \\ 1\end{pmatrix}$ defines an entrance wave, with the polarization linear state with a 45-deg. azimuth respective X axis, with the complex amplitude $E\,e^{i\varphi}$ at entrance, where E denotes the modulus (or amplitude) and φ—the phase. At the output of box I in Figure 1, the complex amplitude is denoted by $\widetilde{E}\,e^{i\widetilde{\varphi}_A}$.

$\mathbf{SLM} = \begin{pmatrix}e^{i\alpha} & 0 \\ 0 & 1\end{pmatrix}$ defines the phase-only LC-SLM (liquid crystal spatial light modulator) with the phase angle modulation α along X, which is controlled by an externally applied electric voltage.

$\mathbf{P}_{\beta=45°} = \frac{1}{2}\begin{pmatrix}1 & 1 \\ 1 & 1\end{pmatrix}$ defines the linear polarizer oriented by its transmission axis at 45-deg. to the X axis.

From Equation (1c), it can be seen that the control of α leads to a change not only in the phase but also in the amplitude of the complex field. This useful property is next applied to a phase–amplitude correction of the wavefront.

Presented here, a phase–amplitude correction of the WFE is entirely developed for the optical schemes, which use the principle of a nulling interferometer with rotational shift functioning as a stellar coronagraph [11,13]. A shear interferometer has two mutually shifted waves being superposed. This allows two degrees of freedom: two controls (modulations) α_A and α_B, at two "coupled" points "A" and "B" of the initial wavefront, which are superimposed. We aim these controls to obtain the mostly accurate "dark field" destructive interference condition.

Let us consider that the shear interferometer superposes two electric fields E_{out_A} and E_{out_B} in points "A" and "B". This is illustrated in Figure 1; see the dot-line box II. Before the interferometer, at the initially separated WF symmetric points "A" and "B", by the SLM we control the complex amplitudes $E_{out_A}(\alpha_A) = \tilde{E}_A \, e^{i\tilde{\varphi}_A}$ and $E_{out_B}(\alpha_B) = \tilde{E}_B \, e^{i\tilde{\varphi}_B}$; see Equation (1d). The destructive interference regime requires both the following conditions (i) and (ii).

(i): A zero-phase difference before the interferometer while we count on an additional phase shift of π implemented inside the nulling interferometer [11,13]:

$$\Delta_\varphi = \tilde{\varphi}_A - \tilde{\varphi}_B \to 0. \tag{2a}$$

(ii): The equality of the moduli of the complex amplitudes:

$$\frac{\tilde{E}_A}{\tilde{E}_B} \to 1. \tag{2b}$$

Therefore, to satisfy both the phase (2a) and amplitude (2b) conditions, we have to consider the electric fields in two *coupled* (by following superposition) points A and B in a pupil before 180° RSI. Their waves interfere further in the conjugated pupil plane at the dark port (with implemented anti-phase) of the interferometer at the point denoted as $A + B$ (See Figure 1, dot-line box II).

Now, we recall the polarization scheme in Figure 1, in dashed-line box I. Then the complex amplitudes of the electric field at points A and B with respective α_A and α_B modulations via the LC-SLM control are defined according to (1a)–(1d):

$$E_{out_A}(\alpha_A) = \frac{1}{2\sqrt{2}}\left(e^{i\alpha_A} + 1\right) E_A \, e^{i\varphi_A} \begin{pmatrix} 1 \\ 1 \end{pmatrix} = \tilde{E}_A \, e^{i\tilde{\varphi}_A}, \tag{3a}$$

$$E_{out_B}(\alpha_B) = \frac{1}{2\sqrt{2}}\left(e^{i\alpha_B} + 1\right) E_B \, e^{i\varphi_B} \begin{pmatrix} 1 \\ 1 \end{pmatrix} = \tilde{E}_B \, e^{i\tilde{\varphi}_B}, \tag{3b}$$

We stress here, again, our notations for the complex amplitudes moduli: E_A, E_B and phases φ_A, φ_B (without *tilde*) which describe the initial wavefront (before the LC-SLM), while the complex amplitudes moduli $\tilde{E}_A$, $\tilde{E}_B$ and the phases $\tilde{\varphi}_A$, $\tilde{\varphi}_B$ (with tilde) denote the corresponding values after the LC-SLM and polarizer (after the polarization scheme $\mathbf{P}_{\beta=45°}\mathbf{SLM}\,\mathbf{E}_{in}$; see Equation (1a).

After passing the interferometer, on its dark port, by coherent wave subtraction, we have the complex field amplitude at point $A + B$ by the superposition of waves $\mathbf{E}_{out_A}$, $\mathbf{E}_{out_B}$:

$$E_{A+B} = E_{out_A} + E_{out_B} \tag{4a}$$

$$= \frac{E_A}{2\sqrt{2}}\left(e^{i\alpha_A} + 1\right)e^{i\varphi_A} + \frac{E_B}{2\sqrt{2}}\left(e^{i\alpha_B} + 1\right)e^{i(\varphi_B + \pi)}. \tag{4b}$$

The CCD detects the intensity I_{A+B} of the interference pattern, e.g., in the $A + B$ point after ensemble averaging (where $\langle \ldots \rangle$ defines the ensemble average over an exposure time):

$$I_{A+B} = \langle E_{A+B} \, \overline{E_{A+B}} \rangle$$
$$= \mathbf{E}_{out_A}{}^2 + \mathbf{E}_{out_B}{}^2 + \mathbf{E}_{out_A}\overline{\mathbf{E}_{out_B}} + \mathbf{E}_{out_B}\overline{\mathbf{E}_{out_A}}, \tag{5}$$

where the *upper underline* denotes the complex conjugate (in matrix form it denotes the Hermitian conjugate).

Substituting (4) into (5), we can express the detected intensity I_{A+B} through the complex amplitude moduli (E_A, E_B) and the phase difference $\Delta_\varphi = \varphi_A - \varphi_B$ of the electric fields in the entrance pupil WF and the corresponding SLM modulations (α_A, α_B):

$$I_{A+B} = I(E_A, E_B, \Delta_\varphi; \alpha_A, \alpha_B) \tag{6a}$$

$$= E_A{}^2 + E_B{}^2 + E_A{}^2 \cos(\alpha_A) + E_B{}^2 \cos(\alpha_B) - E_A E_B \left(\cos(\Delta_\varphi) + \cos(\alpha_A + \Delta_\varphi) + \cos(\alpha_B - \Delta_\varphi) + \cos(\alpha_A - \alpha_B + \Delta_\varphi)\right). \tag{6b}$$

2.1. Measurement of the Wavefront

For the sake of brevity, we show here the algorithm of the wavefront measurement to determine amplitude and phase difference $E_A, E_B; \Delta_\varphi$ by I_{A+B} intensity measurements, which is their function, (6a). We study Equations in (6): if one measures a series (j = (1..3) of three intensities $I^j_{A+B} = I_j(E_A, E_B, \Delta_\varphi; \alpha^j_A, \alpha^j_B)$ in the plane conjugated to the pupil (the SLM is set in the pupil) by providing certain SLM modulations as parameters α^j_A and α^j_B in the *coupled* points A and B, he determines the phase difference Δ_φ and the amplitudes E_A and E_B for the next step of WFE correction.

For example, a three-step j = (1..3) procedure to determine Δ_φ, a_A, a_B detects three intensities by SLM modulations set as $\{\alpha_A, \alpha_B\} = \{ \ (0,\,0), \ \ (\pi,\,0), \ \ (0,\,\pi) \ \}$:

$$\begin{cases} I_1(E_A, E_B, \Delta_\varphi; 0, 0) = I(0, 0), \\ I_2(E_A, E_B, \Delta_\varphi; \pi, 0) = I(\pi, 0), \\ I_3(E_A, E_B, \Delta_\varphi; 0, \pi) = I(0, \pi). \end{cases} \tag{7}$$

Entrance wavefront characteristics Δ_φ, E_A, E_B we determine from solving the system of Equation (7) by their expression as (6b):

$$E_A = \sqrt{\frac{I(0, \pi)}{2}}, \tag{8a}$$

$$E_B = \sqrt{\frac{I(\pi, 0)}{2}}, \tag{8b}$$

$$\Delta_\varphi = \arccos\left(\frac{I(0, 0) - 4E_A{}^2 - 4E_B{}^2}{8E_A E_B}\right). \tag{8c}$$

Certainly, the wavefront measurement method can be modified and improved in accuracy, e.g., by increasing the number j of measurements $I^j_{A+B} = I_j(E_A, E_B, \Delta_\varphi; \alpha^j_A, \alpha^j_B)$, similar to phase shifting interferometry [17].

We resume here that by means of Equations (8a)–(8c), it is possible to measure the wavefront error in terms E_A, E_B, Δ_φ at symmetric (or coupled by electric fields superposition) points A and B. Therefore, we can restore the spatial distribution of WFE pixel wise.

2.2. Wavefront Correction by Phase-Amplitude Modulation

We intend to show here a method to correct the wavefront error, taking into account three measured wavefront characteristics: E_A, E_B, Δ_φ. These measured WFE characteristics at the coupled pupil points A and B are associated with subsequent interference in the $A + B$ point. This correction method is easily extended to cover all the pupil points.

2.2.1. Phase Correction

To present the WFE correction problem, at first, we simplify considering the phase-only WFE: $\Delta_\varphi \neq 0$, $\Delta_E = (E_A - E_B) = 0$.

Figure 2 depicts schematically the principle of phase correction in the *coupled* points A and B. Here, the ψ_A and ψ_B denote the phases in the coupled points A and B (set diametrically opposite) in the pupil plane where the SLM is installed. After passing through the RSI interferometer, an optically conjugated pupil plane is formed. Here, in the points denoted $A + B$, L (in the left half-plane) and $A + B$, R (in the right half-plane), the interference signals intensities of interference pattern $I_{A+B,\,L}$, $I_{A+B,\,R}$ become proportional to the cosine of $\Delta_\varphi = \varphi_A - \varphi_B$:

$$I_{A+B,\,R} \sim cos(-\Delta_\varphi), \tag{9a}$$

$$I_{A+B,\,L} \sim cos(\Delta_\varphi). \tag{9b}$$

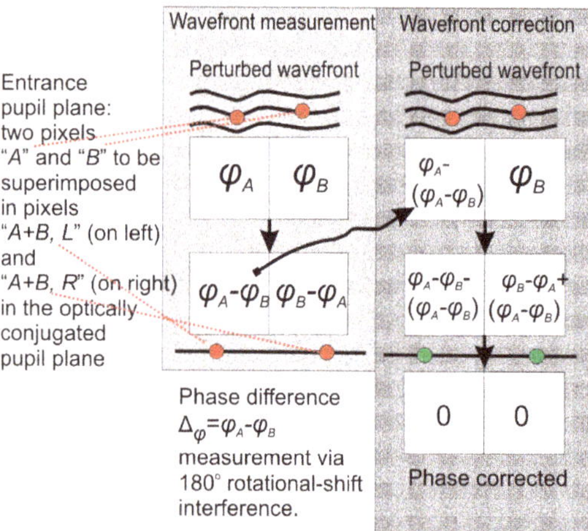

Figure 2. Algorithm to measure (see left column) and to correct (see left column) a phase-only wavefront error (WFE) in 180-deg. rotational–shift interferometer (interfero coronagraph).

Because the cosine function is even visually through the recorded intensity, there is no difference between the left and right half-planes: $I_{A+B,\,R} = I_{A+B,\,L}$, but physically, the relationship (9) turns out to be important to organize the proper wavefront correction.

The measured phase difference Δ_φ is added then as an additional phase shift to the phase of point A (on the left) or to the right point B (but with a negative sign of Δ_φ not shown in the figure). In Figure 2, next after passing through the interferometer RSI, at the two points $A + B$, L and $A + B$, R (shown in green), the corrected phase difference becomes $\widetilde{\Delta}_\varphi \equiv 0$, which means the phase-only aberrations (or wavefront errors) have been successfully corrected.

2.2.2. Phase- and Amplitude Wavefront Error Correction

We analyze here the algorithm to correct a more complex WFE by $\Delta_\varphi \neq 0$ (in phase), $\Delta_E = E_A - E_B \neq 0$ (and in amplitude), which in contrast to the phase-only correction (discussed above in Section 2.2.1), contains an additional amplitude imbalance correction option.

For simplicity, we denote $I_{A+B} = I_{A+B,\,R} = I_{A+B,\,L}$, see Equation (9a,9b). According to Equations (4) and (5), we search for such modulations α_A and α_B that cause a zero-intensity at this point: $I_{A+B} = \langle \mathbf{E}_{A+B}\, \overline{\mathbf{E}_{A+B}} \rangle \to 0$. We solve therefore the following equation:

$$\frac{E_A}{E_B}\left(e^{i\alpha_A} + 1\right)e^{i\varphi_A} = \left(e^{i\alpha_B} + 1\right)e^{i(\varphi_B)}, \tag{10}$$

aiming to find two unknowns, α_A and α_B, each of which depends on four variables: the moduli of complex amplitudes of fields E_A, E_B, and the phases φ_A, φ_B. We reduce four variables to two variables, which are physically relevant in the frame of our model, namely the amplitude ratio: $k = \frac{E_A}{E_B}$ and the phase difference: $\Delta_\varphi = \varphi_A - \varphi_B$. The k and Δ_φ are real numbers, and they were measured by Equations (8a)–(8c). Equation (10) has the following solution:

$$\alpha_A = \tan^{-1}\left(\frac{-1 + 2\frac{E_A}{E_B}\cos\Delta_\varphi - \left(\frac{E_A}{E_B}\cos\Delta_\varphi\right)^2 + \left(\frac{E_A}{E_B}\sin\Delta_\varphi\right)^2}{2\frac{E_A}{E_B}\left(-1 + \frac{E_A}{E_B}\cos\Delta_\varphi\right)\sin\Delta_\varphi}\right), \tag{11a}$$

$$\alpha_B = \tan^{-1}\left(\frac{\left(\frac{E_A}{E_B}\right)^2 - 2\frac{E_A}{E_B}\cos\Delta_\varphi + \cos 2\Delta_\varphi}{-2\left(\cos\Delta_\varphi - \frac{E_A}{E_B}\right)\sin\Delta_\varphi}\right). \tag{11b}$$

We did not simplify the arctangent arguments in $\alpha_A\left(\frac{E_A}{E_B}, \Delta_\varphi\right)$, Equation (11a), and in $\alpha_B\left(\frac{E_A}{E_B}, \Delta_\varphi\right)$, Equation (11b), and we left them in ratios to use the function $\tan^{-1}\left(\frac{Y}{X}\right) = \text{arctan2}(X, Y)$ [18].

3. Simulation

Figure 3 illustrates some numerical verifications to validate the analytical solution found by Equation (11). In the figure, an analytical solution via Equation (11) is shown by the *red cross* unnecessary mark and coincides visually with the minimum in intensity in the logarithmic scale $\log_{10}(I_{A+B}(\alpha_A, \alpha_B))$ of the interference pattern shown by the background colormap. The latter was evaluated by a trivial enumeration of α_A, α_B in the mesh region $\alpha_A, \alpha_B \in [-\pi, \pi]$ by the values $k = \frac{E_A}{E_B} = 1.11$; $\Delta_\varphi = -0.1$ radians, where the used k and Δ_φ have been chosen arbitrarily. The red cross mark position in $\{\alpha_A, \alpha_B\}$ coordinates mesh assigns the global minimum.

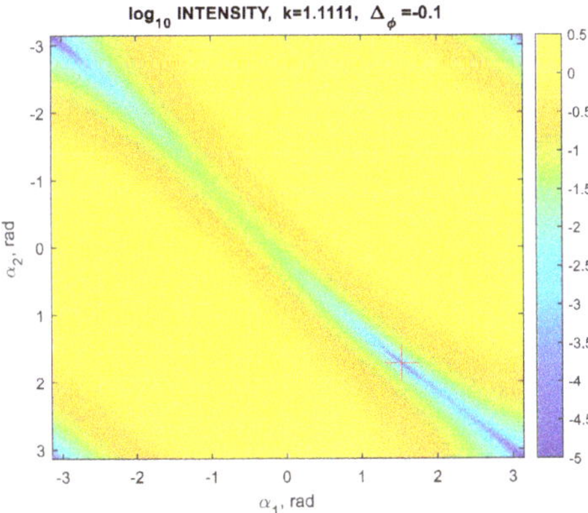

Figure 3. Check for an analytical solution Equation (11). In title are shown initial amplitude imbalance k and phase difference Δ_φ.

In Figure 4, we analyze an intensity transmission characteristic $T(\alpha_A, \alpha_B)$, which shows the attenuation of the signal (from a faint exoplanet) in a single pupil pixel because we corrected the amplitude imbalance of $k = 1.11$. In a pixel, the planetary intensity signal transmission $T(\alpha_A, \alpha_B) \approx 0.65$ becomes reduced relative to unitary because an amplitude correction attenuates the transmission of both signals from the star and planet. We have considered here the WFE in point $A + B$ by $k = \frac{E_A}{E_B} = 1.11$; $\Delta_\varphi = -0.1$. The Figure 4 background has the colormap that is the two-dimensional distribution of $T(\alpha_A, \alpha_B)$ being evaluated in the domain of $\alpha_A, \alpha_B \in [-\pi, \pi]$.

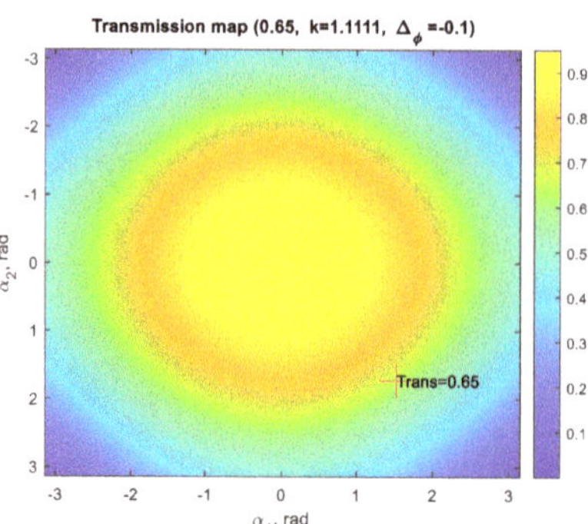

Figure 4. Transmission characteristic $T(\alpha_A, \alpha_B)$ can be associated with the planetary signal intensity i.3n a single pixel.

In Figure 5, in a $\pm15\,\lambda/D$ focal image domain we show several simulation graphs to compare the images (1) in non-coronagraphic mode, (2) in coronagraphic mode without any WFE correction, (3) in coronagraphic mode with WFE phase-only correction, and (4) in coronagraphic mode with WFE phase and amplitude correction. These images are resumed by azimuth-averaged stellocentric cross-sections (5).

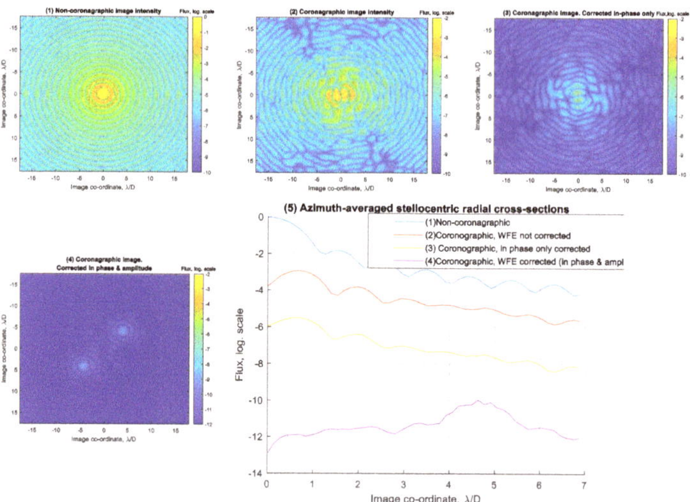

Figure 5. Non-coronagraphic (1) and coronagraphic images (2)–(4) simulated by phase error $\lambda/50$ rms and the amplitude error 5%; (3) phase-only corrected image; (4) phase and amplitude corrected image resolves a planet companion; (5) azimuth-averaged radial stellocentric profiles corresponding (1)–(4) shows gradual scattered stellar background suppression.

In simulations in Figure 5, one can see that, caused by diffraction and scattering on residual aberrations and surface micro-roughness, the stellar background shows the gradual suppression by means of the coronagraph without and with the wavefront control. The wavefront control was performed by correcting the phase-only wavefront errors at first, then correcting both the phase and amplitude WFE. In these simulations, the parameters were phase deviations of about $\lambda/50$ rms at $\lambda = 600$ nm and amplitude deviations of about 5%.

4. Laboratory Experiment

In the laboratory experiment, we studied the above proposed approach to correct the complex value wavefront error simultaneously in phase and amplitude at the wavelength 632.8 nm. In addition to our previously published efforts [9] where we performed the phase-only wavefront control, here, we succeeded in both measuring the complex number WFE, and then controlling the WFE in phase and amplitude operating by a spatial light modulator (SLM) being inserted in the optical and polarization scheme, as in Figure 1. We used a commercially available LC-SLM from HOLOEYE© (Berlin, Germany): model PLUTO-2.1-VIS-016. It has an 8 μm pixel size with a resolution up to 1920 × 1080 active pixels and a response time of ~66 ms [19]. The LC-SLM was capable of providing a maximum phase shift of about 5.2 π (on the working wavelength 630 nm), but it was calibrated to provide only reduced 2π of maximum phase shift, that was sufficient for the described correction method. The camera used in the setup was a CMOS monochrome camera from Edmund Optics© (Barrington, NJ, USA), model EO-5012M with a pixel size of 2.2 μm and a resolution of up to 2056 × 1920. It was synced with the LC-SLM by a dedicated frame sync signal. Several experimental results are shown in Figure 6, where we demonstrate the gradually reducing azimuth-averaged cross-sections of the non-coronagraphic PSF, the coronagraphic PSF without WFE correction, and the coronagraphic PSF with WFE correction in phase and amplitude; see panel (a). In Figure 6 in panels (b) and (c), we show noncorrected fragments of pupil intensities after the coronagraph, where the WFE was noncorrected and corrected. Corresponding histograms are shown in panels (d) and (e). If one analyzes pupil fragments (b) and (c), he finds at first the decreased intensity level; see the colorbar scale on the left. Integrally, this decrease corresponds in numbers to the ratio of the coronagraphic PSF without applied WFE correction (shown in the azimuth-averaged section by the red color graph in panel (a)) to the coronagraphic PSF with WFE correction (shown by the orange color graph). Additionally, the corresponding histogram (in panels (d) and (e)) demonstrate that the intensity deviations decrease after correction. Because of amplitude attenuation, we estimated the virtual planetary signal had its throughput at about 0.92 as averaged over all pixels.

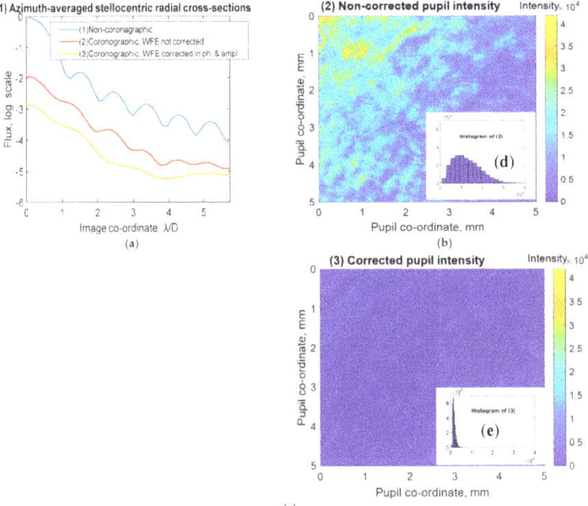

Figure 6. Experimental data of wavefront error correction in phase and amplitude regime. (**a**) Azimuth-averaged PSF cross-sections: blue (1) line—non-coronagraphic PSF; red (2) line—coronagraphic PSF without WFE correction; orange (3) line—coronagraphic PSF with WFE correction. (**b,c**) Coronagraphic pupil, the same areas before and after correction with corresponding histograms in panels (**d,e**).

5. Discussion

In the present article, we have continued the studies of our previous communication [9], where we just roughly sketched some general possibilities about how to correct wavefronts, not only in phase but simultaneously in amplitude. This is realized by the LC-SLM providing general polarization modulation at first, then it is converted in phase and in amplitude wavefront components by the linear polarizer mounted after the LC-SLM. Therefore, in the present communication, we have aimed to find and to test a practical analytical solution for wavefront error active correction both in phase and amplitude. This is shown analytically as well by numerical simulation and in a simple experiment in an optical lab.

Here, our goal was not to reach a high coronagraphic contrast in stellar light suppression, but we studied an appropriate optical architecture to control WFE. At first, we measured both the WFE phase and amplitude components. Then we operated by controlling voltages at an adaptive optics unit (LC-SLM) to correct wavefront error simultaneously in phase and amplitude. The advantages are the following: (i) both the wavefront measurement and the correction have been realized without any mechanical motion in the optical scheme and without its reconfiguration; (ii) such a WFE measurement was performed after the coronagraph, therefore we excluded possible non-common-path aberrations (NCPAs).

The simple approach presented here was found analytically and it was tested numerically by simulations using the Proper [20]. Finally, we designed an optical experiment in the lab, which has demonstrated the right tendency in deeper signal suppression in coronagraphic mode for an on-axis light source (that imitated starlight) by wavefront correction.

In future, we are optimistic about applying this technique onboard of an orbital telescope for direct exoplanet imaging. LC-SLM technology was *scientifically* tested under a thermo-vacuum environment [21], however not yet under the space requirement. A possible LC-SLM mounting can be in a chamber with an optical window, with stabilized normal temperature and pressure. Radiation tests are not known, but a telescope can be launched in a low orbit beneath the Van Allen radiation belt. Active adaptive optics competes with the static correction, but in space, an orbital telescope is stabilized statically up to certain level. Solar radiation plus heat generation by electronics causes a low speed mechanical instability of the primary mirror geometry and to optics having requirements greater than $\lambda/100$ in optical wavelength domain. Therefore, precise adaptive optics with the functions of measurement and the control of the wavefront in phase and simultaneously in amplitude remains advanced.

Author Contributions: Conceptualization, methodology, A.T. and A.Y.; part of math., A.V.; software, A.Y. and I.S.; experiment, A.T., A.Y. and A.K.; writing—original draft preparation, A.Y.; supervision, A.T. and O.K. All authors have read and agreed to the published version of the manuscript.

Funding: This research received no external funding.

Institutional Review Board Statement: Not applicable.

Informed Consent Statement: Not applicable.

Data Availability Statement: The data presented in this study are available on request from the corresponding author.

Conflicts of Interest: The authors declare no conflict of interest.

References

1. Traub, W.; Oppenheimer, B. Direct imaging of exoplanets. In *Exoplanets*; Seager, S., Ed.; University of Arizona Press: Tucson, AZ, USA, 2011; pp. 111–156. Available online: https://www.amnh.org/content/download/53052/796511/file/DirectImagingChapter.pdf (accessed on 16 January 2024).
2. Collisional Exozodi Simulation Catalog. Available online: https://asd.gsfc.nasa.gov/Christopher.Stark/catalog.php (accessed on 16 January 2024).
3. Goodman, J.W. *Introduction to Fourier Optics*, 3rd ed.; Roberts & Co. Publishers: Englewood, CO, USA, 2005; 491p.
4. Huang, Y.; Liao, E.; Chen, R.; Wu, S.-T. Liquid-Crystal-on-Silicon for Augmented Reality Displays. *Appl. Sci.* **2018**, *8*, 2366. [CrossRef]

5. Guyon, O.; Pluzhnik, E.A.; Kuchner, M.J.; Collins, B.; Ridgway, S.T. Theoretical Limits on Extrasolar Terrestrial Planet Detection with Coronagraphs. *Astrophys. J. Suppl. Ser.* **2006**, *167*, 81–99. [CrossRef]
6. Guyon, O. Extreme Adaptive Optics. *Annu. Rev. Astron. Astrophys.* **2018**, *56*, 315–355. [CrossRef]
7. Galicher, R.; Mazoyer, J. Imaging exoplanets with coronagraphic instruments. *Comptes Rendus Phys.* **2023**, *24*, 1–45. [CrossRef]
8. Lyon, R.; Clampin, M.; Woodruff, R.; Vasudevan, G.; Shao, M.; Levine, M.; Ge, J. Visible Nulling Coronagraphy for Exo-Planetary Detection and Characterization. *Proc. Int. Astron. Union* **2005**, *1*, 345–352. [CrossRef]
9. Yudaev, A.; Kiselev, A.; Shashkova, I.; Tavrov, A.; Lipatov, A.; Korablev, O. Wavefront Sensing by a Common-Path Interferometer for Wavefront Correction in Phase and Amplitude by a Liquid Crystal Spatial Light Modulator Aiming the Exoplanet Direct Imaging. *Photonics* **2023**, *10*, 320. [CrossRef]
10. Baudoz, P.; Rabbia, Y.; Gay, J. Achromatic interfero coronagraphy. *Astron. Astrophys. Suppl. Ser.* **2000**, *141*, 319–329. [CrossRef]
11. Frolov, P.; Shashkova, I.; Bezymyannikova, Y.; Kiselev, A.; Tavrov, A. Achromatic interfero-coronagraph with variable rotational shear: Reducing of star leakage effect, white light nulling with lab prototype. *J. Astron. Telesc. Instrum. Syst.* **2015**, *2*, 011002. [CrossRef]
12. Lyon, R.G.; Hicks, B.A.; Clampin, M.; Petrone, P.R., III. Phase-Occultation Nulling Coronagraphy. *arXiv* **2015**. [CrossRef]
13. Tavrov, A.V.; Kobayashi, Y.; Tanaka, Y.; Shioda, T.; Otani, Y.; Kurokawa, T.; Takeda, M. Common-path achromatic interferometer–coronagraph: Nulling of polychromatic light. *Opt. Lett.* **2005**, *30*, 2224–2226. [CrossRef] [PubMed]
14. Aime, C.; Rabbia, Y.; Carlotti, A.; Ricort, G. Reducing star leakage with a nuller coronagraph. An analytic approach for a Sonine apodized circular aperture. *A&A* **2011**, *530*, A52. [CrossRef]
15. Nishikawa, J.; Murakami, N. Unbalanced nulling interferometer and precise wavefront control. *Opt. Rev.* **2013**, *20*, 453–462. [CrossRef]
16. Chen, H.-M.P.; Yang, J.-P.; Yen, H.-T.; Hsu, Z.-N.; Huang, Y.; Wu, S.-T. Pursuing High Quality Phase-Only Liquid Crystal on Silicon (LCoS) Devices. *Appl. Sci.* **2018**, *8*, 2323. [CrossRef]
17. Malacara, D. *Optical Shop Testing*, 3rd ed.; John Wiley & Sons, Inc.: Hoboken, NJ, USA, 2006; ISBN 9780471484042.
18. Available online: https://en.wikipedia.org/wiki/Atan2 (accessed on 16 January 2024).
19. HOLOEYE. Spatial Light Modulators. Available online: https://holoeye.com/spatial-light-modulators (accessed on 16 January 2024).
20. PROPER. Optical Propagation Library. Available online: https://proper-library.sourceforge.net (accessed on 16 January 2024).
21. Manuel, S.-L.; Antonio, C.-J.; Alberto, Á.-H. Validation of a spatial light modulator for space applications. In Proceedings of the SPIE 11180, International Conference on Space Optics—ICSO 2018, 111806R, Chania, Greece, 12 July 2018. [CrossRef]

Article

Development of Cryogenic Systems for Astronomical Research

Yuri Balega [1], Oleg Bolshakov [2], Aleksandr Chernikov [2,3], Aleksandra Gunbina [2,4], Valerian Edelman [5], Mariya Efimova [2], Aleksandr Eliseev [2], Artem Krasilnikov [1,2], Igor Lapkin [2], Ilya Lesnov [2], Mariya Mansfeld [1,2], Mariya Markina [4,5], Evgenii Pevzner [2], Sergey Shitov [4,6], Andrey Smirnov [7], Mickhail Tarasov [4], Nickolay Tyatushkin [2], Anton Vdovin [1,2] and Vyacheslav Vdovin [1,2,7,*]

[1] Special Astrophysical Observatory RAS, Nizhnij Arkhyz 369167, Russia; balega@sao.ru (Y.B.); arkrasilnikov@ipfran.ru (A.K.); marym@ipfran.ru (M.M.); vdovinav@ipfran.ru (A.V.)
[2] A.V. Gaponov-Grekhov Institute of Applied Physics RAS, Nizhniy Novgorod 603600, Russia; bolshakov@ipfran.ru (O.B.); chern@nf.jinr.ru (A.C.); gunbina@ipfran.ru (A.G.); m.efimova@ipfran.ru (M.E.); aielis@ipfran.ru (A.E.); lapkin@ipfran.ru (I.L.); lesnov@ipfran.ru (I.L.); pevzner@ipfran.ru (E.P.); ttshkn@list.ru (N.T.)
[3] Joint Institute for Nuclear Research, Moscow 141980, Russia
[4] The Kotel'nikov Institute of Radio Engineering and Electronics of RAS, Moscow 101990, Russia; mamarkina@edu.hse.ru (M.M.); sergey3e@gmail.com (S.S.); tarasov@hitech.cplire.ru (M.T.)
[5] P.L. Kapitza Institute for Physical Problems RAS, Moscow 119334, Russia; edelman@kapitza.ras.ru
[6] Laboratory of Cryoelectronic Systems, The National University of Science and Technology MISIS, Moscow 119049, Russia
[7] P.N. Lebedev Physical Institute of RAS, Moscow 119333, Russia; smirnovavl@lebedev.ru
* Correspondence: vdovin@ipfran.ru

Abstract: The article presents a brief review of cooling systems that ensure various temperature levels (from 0.1 K to 230 K) for radio astronomical receivers of photonic and electronic (or optical and radio) devices. The features of various cooling levels and the requirements for the design of the cooling systems are considered in detail, as well as the approaches to designing interfaces for cooled receivers: vacuum, cryogenic, electrical, mechanical, optical, and other interfaces required for effective operation. The presented approaches to design are illustrated by a series of joint developments of the authors carried out over the past 45 years, including those produced over the past year.

Keywords: radio astronomy; cryogenic receivers; cryogenic systems; radio waves; IR and optics ranges; terahertz range

Citation: Balega, Y.; Bolshakov, O.; Chernikov, A.; Gunbina, A.; Edelman, V.; Efimova, M.; Eliseev, A.; Krasilnikov, A.; Lapkin, I.; Lesnov, I.; et al. Development of Cryogenic Systems for Astronomical Research. *Photonics* **2024**, *11*, 257. https://doi.org/10.3390/photonics11030257

Received: 24 November 2023
Revised: 4 March 2024
Accepted: 5 March 2024
Published: 13 March 2024

1. Introduction

Cooling systems, including those operated at deep cryogenic temperatures (LHe and lower levels), are frequently utilized to increase the sensitivity of the photonic and electronic devices used in the astronomic research [1–8]. Such cryogenic systems are represented by a wide range of hardware, from very big cryosystems [9,10] to extremely miniature refrigerators [11]. High sensitivity and an extremely small intrinsic noise are the main requirements for astronomical receivers. Deep cryogenic cooling meets these requirements like nothing else. "Why do low temperatures gain the attention of researchers? The fact is that the area near the absolute zero turned out to be not so "dead" as they thought a century ago: all movement stops, everything freezes, period! On the contrary, low-temperature physics can compete easily with any other field of natural science in terms of the abundance of physical phenomena, which are observed when substances are cooled to the temperature of liquid helium and lower". © [12].

Cooling systems are characterized by different temperature levels, which mainly determines their design. We should dwell on the very definition of the temperature and the scale on which we work. In this review, the Kelvin scale will be used in Dalton's infinite round-trip interpretation.

An equally important issue in the design of a cooling system for astronomical receivers is the operating frequency range. Partly, it defines the level of the required cooling. All these issues are discussed in the corresponding sections of the paper.

Astronomy has traditionally presented the most ambitious challenges in the development of highly sensitive receiving equipment for both optical astronomy and radio astronomy and their intersection in the terahertz range. Now, the cryogenics community publishes many papers about the development of cryogenic hardware for astronomy in its journals (for example, Cryogenics [13]), and astronomical journals regularly describe cryogenic astronomical equipment in their instrument-related sections (for example, the Astronomical Journal [14]). The pioneer of application of cryogenics in astronomy was G.C. Messenger [15]. Meanwhile, the active use of refrigerators in radio astronomy began with the work of W. Gifford and H. McMagon in 1959 [16] based on the technologies presented by Guy K. White [17]. Further, in 1966, the fundamentals for the development of a cryosystem, including its application in astronomy, were described by R.F. Barron [18].

In approximately the same years, Stirling machines were used actively in optical astronomy [19]. The basic principles of cooling the optical and IR detectors were outlined in [5,7,8,11,12].

In this paper, we analyze the development of the related equipment and methods and review the basic principles of designing cryogenic cooling systems for photonic and electronic receiving devices for astronomy, illustrating the discussion by both our own developments and the best examples of modern astronomical cryogenic equipment used in many observatories around the world. World leadership in astronomical apparatus developments is gained on the frontier associated with space technologies. Space technologies are actively represented in two series of regular conferences [20,21] held by the IEEE community and the Space Cryogenic Workshop [12,22] held by the Cryogenic Society of America. It is here where the cutting edge of the research and development of cryogenic equipment for astronomy is formed, where the authors of this paper work, and where the results are reported.

The presented summative review has not only the academic interest, but also a significant practical value. The development of terahertz astronomy, which fills the gap between the optical and radio ranges, involves the addition of several dozen existing telescopes for creation of an extensive network of instruments in new territories of the Earth, where there are currently no terahertz instruments at all. A series of projects for the development of THz astronomy in northeastern Eurasia provides for the development of an extensive line of cryogenic receivers.

2. Classification of Cooling Systems for Astronomy

Cooling systems are used in a wide range of human activities, from fundamental physics research and advanced technology to medicine and cosmetology [23–25]. Each area forms its own approaches to the design of cryosystems and chooses independently the most effective approaches and technologies from their vast variety. The cooling of optical and radio receivers has formed a certain development culture on its own. In this line, astronomers need devices with the most ambitious and advanced specifications. At the same time, the types and features of the cryosystems manufactured for various astronomical telescopes have fundamental differences depending on an extensive set of factors, ranging from the band of the electromagnetic spectrum, in which the receiving device operates, to the atmospheric and climatic conditions at the observatory location and the design of the telescope, where the cooled receiving device operates.

Cryogenics as a science emerged over 100 years ago. The foundations of approaches to engineering of cryogenic systems were also laid many decades ago. Development engineers devised certain approaches to classification and methodology in the related fields and have accumulated extensive experience and knowledge about the properties of various materials for cryogenics. All this knowledge is summarized in extensive bibliography. However, there are certain cryogenic Bibles which each developer trusts and holds in his or her bookcase.

In the English-language segment of the literature, such a Bible is the Brookhaven Reference Book [26], and in the Russian-language segment, it is the Malkov Reference Book [27]. After extensive discussions and disputes regarding which one is better, personal communication between the developers of cryosystems led eventually to the recognition that the books complement each other well and to the mutual exchange of available reference books. However, direct application of the available extensive knowledge and design approaches in specific fields and particular astronomy branches seems inconvenient. Not everything is equally suitable, and many things required, including new ones, simply do not exist. This publication is intended to systematize the current status of the developments in this area.

2.1. Frequency Range of Astronomical Instruments and Their Cooling Problems

Having started in the optical range of the electromagnetic wave spectrum due to the excellent photodetecting devices which each of us is endowed with, astronomy has acquired a multi-frequency character by now and is currently equipped with millions of eyes, i.e., detectors capable of operating at extreme sensitivity levels in the entire spectrum of electromagnetic radiation (EMR) (Figure 1), from ultra-low frequencies to the gamma ray part of the spectrum. In the areas at the spectrum edges, cryogenic cooling is not needed today in practice due to the perfection of semiconductor technologies in the low-frequency range and to the enormous energy of the X-ray and gamma quanta, which do not require cooling to be detected in the range of extremely high frequencies.

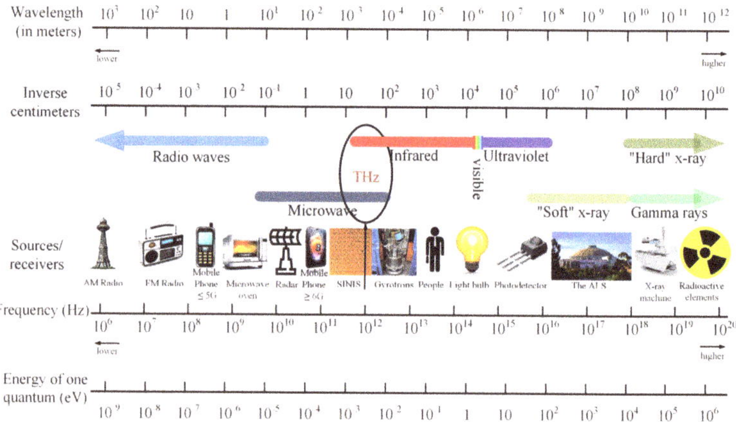

Figure 1. THz waves in the electromagnetic spectrum.

The EMR ranges, where the main scientific efforts of astronomers are focused today, cover two areas, specifically, the terahertz (THz) and the optical ranges, including the VUV (vacuum ultraviolet) band. These ranges require receivers with deep cryogenic cooling. This does not mean that there is no progress in the development of other ranges. It does exist and also requires cryogenic cooling sometimes. However, the scale of the need for cryogenics in that field is not comparable with the demand for photonic devices in optics or for electronic devices in the THz range. However, the THz range lies exactly between the traditional microwave radio electronics and the far-infrared optics combining the technologies and approaches inherent in both of them.

Therefore, the presentation below will focus on the problems of astronomy of these ranges.

2.2. Temperature Levels of Cooling for Astronomical Receivers

Astronomy uses a wide range of cooling from the boundary of cryotemperatures to extremely low sub-Kelvin levels. Table 1 shows the general classification of temperature levels of systems used.

Table 1. Classification of cryogenic systems by the temperature level.

Group Name	Temperature Range	Applications
Non-cryogenic temperatures	T > 100 (120) K	- climate chambers for testing astronomical receivers and their components - non-cryogenic cooling of photodetectors
High cryogenic temperatures	$\leq$100 K	- cooling of photodetectors to reduce the dark current
Nitrogen level	~77 K	- high-temperature superconductors - cooling of photodetectors - cooling of microwave radio receivers - radiation shields in deep cooling systems
Hydrogen level	~20 K	- cooling of photodetectors, radio receivers of the centimeter and millimeter (MM) range, semiconductor amplifiers up to the MM range
Helium level	~4 K (up to 1.6 K with pumping)	- detecting devices based on superconductors of the second kind, generally used for SIS junctions - superconducting readout electronics - antenna systems, cryogenic chambers for testing, and laboratory experiments
Sub-Kelvin cooling	<1 K	- superconductors of the first kind - superconducting electronics - quantum computers and quantum communications - highly sensitive receiving systems for modern astronomy, generally used for bolometers
Extremely low temperatures reached	~10^{-9} K	- problems of metrology - quantum standards of time and frequency, including those for VLBI astronomy - fundamental physics

The professional community divides the cryogenic temperatures into conventional levels. Cryogenic temperatures are the temperatures below 100 K (120 K in the USA). Non-cryogenic cooling is used quite widely in optical astronomy, as well as in climatic test chambers, in order to study the characteristics of devices and materials. The cryotemperature range is characterized by liquefaction and, for the most part, solidification of the majority of the gases present in the atmosphere. In particular, at T = 90 K, oxygen turns into liquid at atmospheric pressure. Nitrogen, which liquifies at 77 K, is extremely widely used in astronomy as both an independent and relatively cheap cryo-agent and an auxiliary agent for cooling radiation shields at deeper cooling. The main accepted temperature levels are given in Table 1.

The next characteristic level is the hydrogen level (~20 K). Here, not only the phase transition of the main hydrogen isotope occurs, but the Debye also temperature zone begins for the main structural materials, from which the components of cooled astronomical receivers are made. This leads to a dramatic change in their thermal characteristics (heat capacity, thermal conductivity, and thermal diffusivity). Despite certain fears caused by the explosive hazard of hydrogen mixed with atmospheric oxygen, liquid hydrogen is still used as a cryo-agent by the professional community. Also related to the hydrogen temperature level is a wide class of closed-cycle refrigerators, which have been and are used currently in astronomical applications, although more often as cryo-vacuum pumps of cryo-vacuum chambers for a wide variety of purposes, from vacuum deposition installations, where detectors for astronomy are manufactured, to testing chambers, where pre-launch studies of space equipment are carried out. However, hydrogen-level refrigerators are also used to cool astronomical receivers. It should be noted that they do not contain gaseous hydrogen, since the working gas is helium. Almost until the end of the 20th century, regenerators of cryorefrigerators, where heat exchange takes place due to gas expansion, were made of antimonial lead. The efficiency of such regenerators dropped dramatically in the vicinity of the hydrogen temperatures due to the transition of the material through the Debye temperature. When rare-earth regenerators began to be used instead of antimony–lead alloys, the problem was resolved, and the operating temperature of such refrigerators dropped to the helium level (4 K and lower, to the limit of 2 K). As is known, when the pressure decreases from the atmospheric level to vacuum, the temperature of the main helium isotope ^{4}He can reach 1.6 K. This is the helium temperature level.

Lower temperatures (at the sub-Kelvin levels, i.e., below 1 K) require different approaches to cooling, including the use of the ^{3}He isotope, thermomagnetic cycles, laser-based and other cryocooling technologies. Note that these temperatures are extremely actively used by astronomers for deep cooling of superconducting detectors in the terahertz range.

Even lower temperatures currently achieved by physicists are frequently used for other purposes, primarily the study of the extreme conditions not found in nature, such as Bose–Einstein condensate. Such a setup, operated at a temperature of about 10 nano-Kelvin [28], is installed in a laboratory adjacent to the laboratory of the authors of this paper. The application of the results of these studies in astronomy also has great prospects. On this basis, it is possible to construct highly stable time and frequency standards necessary for precision radio astronomical observations in the VLBI (very large baseline interferometer) mode. However, this is a distant prospect. Today, physicists have already passed the nano-Kelvin milestone. Among the many practical temperature scales in use [29–31] for astronomical instrumentation, the Kelvin–Dalton logarithmic scale, which is infinite towards absolute zero, is applied. The levels of the cryogenic temperatures listed in Section 2.1 will be further illustrated by examples of the development of the corresponding astronomical instruments. However, when we speaking about temperature of cryogenically cooled receivers, we should remember that the temperature of different components of the receiver is not the same and any thermometer measures only its own temperature.

The second fact upsets experimenters even more: although temperature is a measure of kinetic energy by definition, it must be measured in the equilibrium state. Since cryogenic receivers, as a rule, are not in the equilibrium state, their temperature in the strict sense is uncertain; this creates an extensive range of interesting scientific and engineering problems in the context of the radiophysical and optical properties of the receiving systems, which are discussed below, for the receiver elements that are not in the equilibrium state.

2.3. Main Types of Cryosystems for Astronomy

Cryosystems used to cool astronomical radiation receivers can be divided into the following groups: cryo-accumulators, cryorefrigerators, and their variations, as well as their combinations. Along with the cooling systems based on the use of gases and gas machines, various types of gas-free cooling systems are known, in particular, first of all, thermoelectric and thermomagnetic ones. In recent years, laser cooling systems have also appeared, though they have not yet found application in astronomical research. One of the obvious and very popular applications of the devices with deep (nano-Kelvin) laser gas cooling in the Bose–Einstein condensate state opens up enormous prospects in increasing the stability of time and frequency standards actively used to synchronize astronomical receivers operating in the regime of a VLBI.

2.3.1. Dewars or Cryo-Accumulators and Liquid Nitrogen Cooling Solutions

Cryo-accumulators use a pre-cooled cryo-agent in various phase states:

- gaseous agent (a cold flow of N_2, H_2 and He gases);
- solid agent (carbon dioxide); and
- liquefied gas (nitrogen < 77 K, hydrogen < 20 K, helium < 4 K).

For millimeter- and submillimeter-range (or subTHz) receivers, liquid nitrogen or helium is used most often (depending on the requirements). Receivers in cryo-accumulators are placed in either a boiling liquefied gas or a so-called "vacuum basement", i.e., the cavity between the outer and inner walls of the Dewar vessel. As a variant of the cryo-battery circuit, there is a flow circuit containing a separate Dewar transport vessel, as well as a closed or open line for supplying the cryo-agent to the cooling facility. The scheme of the cryo-accumulator with a "vacuum basement" is similar to the cryorefrigerator scheme, where a cryopanel with cooled receiver elements is attached to the cold stage of the refrigerator cooler. In essence, the energy required to maintain the temperature of the cooled object (elements of the receiving path of the astronomical receiver) below the state of thermodynamic equilibrium with the environment is stored in the form of a liquefied cooled gas and is easily estimated as the volume of the stored cryo-agent multiplied by the sum of its specific vapor formation temperatures and heat capacity at a temperature drop (see Formula (1)):

$$E = m \left(\lambda_\delta T + C \right), \tag{1}$$

where E is the energy or heat flow, which can prevent the supply of cold in a liquefied state, m is the cryo-agent mass, λ is the specific integral heat capacity of the cryo-agent, $_\delta T$ is the temperature difference used for cooling, and C is the specific heat of the phase transition.

The design of nitrogen Dewars (for example, Figure 2a shows a photo of the developed and manufactured nitrogen-filled cryostat) is quite simple, and design optimization is not difficult, whereas helium-filled Dewars require a more thorough thermal design. As a rule, they need nitrogen radiation shields to prevent radiation thermal inflows. To that end, in addition to the helium tank, the design provides a nitrogen tank with its own filling system. Estimates of heat flows are presented in Section 4.3.2. However, there is also the experience of creating nitrogen-free (Figure 2b) helium Dewars for astronomy, where radiation shields are cooled by evaporating helium of the appropriate temperatures.

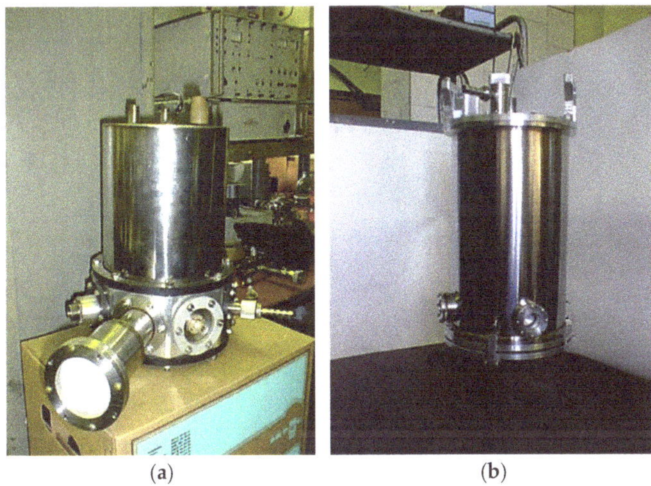
(a)   (b)

Figure 2. (**a**) Nitrogen-filled cryostat with a set of unified interfaces; (**b**) nitrogen-free helium optical filling cryostat.

The choice of the working gas and the volume of the Dewar is determined by the heat flows and the release of the cooled receiver itself. The analysis of the parameters of such systems, which are typical for astronomy, is given in Section 4.3.2.

Along with relatively simple liquid-nitrogen Dewars, many equally simple, efficient, and energy-saving closed-cycle refrigerators have been used in astronomy, mainly the Stirling machine. To reduce vibration loads, which are the main disadvantage of refrigerators as compared to Dewars, the split-Stirling version or other thermodynamic cycles were used more often, until the Cryotiger became a certain perfection of such machines. Comparable runs of cooling systems for various photodetectors using Cryotiger closed-cycle refrigerators were produced in various configurations for equipping optical telescopes [2,32,33] (Figure 3).

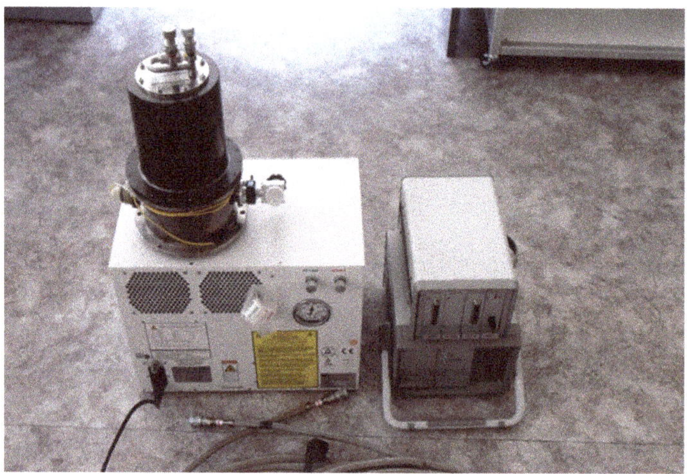

Figure 3. Optical cryostat based on the CryoTiger refrigerator.

2.3.2. Hydrogen Level Refrigerators for Astronomy and Telecommunications

The hydrogen level of cooling has been and is used now actively in astronomy and radio telecommunications. Various types of cm and mm radio frequency detectors and amplifiers can be cooled by such coolers, which are relatively cheap, compact, and economical compared to the helium coolers.

The success in using low-noise HEMT (high electron mobility transistor) amplifiers produced by Saturn Research Institute (A.M. Pilipenko) in astronomical satellites at SAO RAS (Special Astrophysical Observatory of the Russian Academy of Science) gave rise to their further extensive use with other antennas: RT-22 (Simeiz), RT-13 (Metsahovi), and RT-64 at Bear Lakes and Kalyazin. In particular, a unique set of equipment for astronomy and telecommunications was created on the basis of these HEMT amplifiers.

One should also mention the extremely efficient system of cooling at the hydrogen level, which was developed by the authors of this publication and is used for the X-band antennas at Bear Lakes and Kalyazin, both for astronomical observations and for deep space communications. It worked successfully and is working currently as a component of the Spektr-R and ExoMars missions [34]. A photo of the cryogenic system is shown in Figure 4.

Figure 4. Cryogenic system for cooling of the X-band receiver.

The key parameters that characterize the quality and, as a result, efficiency of such a receiver is the ratio of noise temperatures. If the receiver without cryogenics produces about 40 K of the noise at the input, and the antenna noise is 14 K, it is obvious that the atmosphere is not dominant, but makes a significant contribution. Cooling to a temperature of 6 K results in a decrease in the receiver noise to 5 K, and the system noise with the contribution of the atmosphere rises up to 19 K. This leads to a reduction in noise by almost three times and a similar increase in sensitivity. Strictly speaking, the system is closer to the helium level in terms of the temperature than to the level of liquid hydrogen, but it still does not belong to the class of helium refrigerators. The gain and the operating range of such a receiver are 30 dB and 8.4–8.5 GHz, respectively.

2.3.3. Cooling Systems of the Helium Level

Helium-level systems are actively used for cooling various coherent detectors (mixers), from Schottky barrier diodes (DBS) to mixers based on superconducting junctions. This approach has been presented both in classic papers, like [35–37], and in modern papers about receivers [38].

One of the most striking examples of an observatory, where detectors are cooled to the helium level, is SOFIA, a stratospheric aircraft-borne observatory [39]. The detecting devices are maintained at a temperature of 1.7–1.9 K.

However, it should be recognized that in recent years, with the transition to pulse tubes and rare-earth regenerators, helium-level refrigerators have been used more frequently.

Until recently, very little attention has been paid to the potential of deep cooling of the receivers to the helium level, which is one of the main types of the cryogenic systems used in optical telescopes. The reason was mentioned above. Specifically, it is that the BTA (Big Telescope Alt-Azimuthal) optical telescope operating in the optical and IR ranges does not need such systems, and on the RATAN-600 radio telescope, there is no need for cooling below the level of liquid hydrogen, since it operates at relatively long wavelengths in the conditions of a fairly significant atmospheric background. At the same time, the helium level was actively used in both the liquid and dry configurations of the refrigerator when developing the SAO receiver for RT-22 after the transition from the mixer to the DBS to the mixer at the superconducting junctions.

Currently, SAO RAS is preparing to switch to helium-level technologies for cooling the subTHz receivers on the BTA telescope. Moreover, the temperature of 4 K will be used only as a system of precooling for refrigerators of a deeper cooling level, which are presented in the next section. The cryogenic systems that are used in SAO RAS are described in more details in [40].

At the beginning of the cryogenic cooling era, cryo-accumulators and Dewars were mainly used in astronomy. However, the convenience of using refrigerators that do not require liquid gases has made them most popular in observatories. The ideology of unified cartridges with closed-loop 4 K refrigerators (Figure 5) should be recognized as a semi-industrial masterpiece of the approach to the design of cryosystems of astronomical receivers. It is implemented at the ALMA Observatory in an array of 64 telescopes, each of which is equipped with a set of cooled receivers of various frequencies of the THz range and replicated in hundreds of copies [41].

Figure 5. Cryogenic cartridges of ALMA receivers.

Cryogenic refrigerators, or gas fridge machines, operate by throttling or expanding the working gas in a cooler based on various thermodynamic cycles: Joule–Thomson, Stirling, Gifford–McMahon, etc., originating from the Carnot cycle and ultimately described by the equation of the state of the ideal gas (2):

$$PV = \nu RT, \tag{2}$$

where V is the volume, P is the pressure, $\nu = m/M$ is the amount of substance in moles, R is the gas constant, and T is the thermodynamic temperature.

Refrigerators of this kind are actively used to cool astronomical receivers down to the nitrogen, hydrogen, and helium temperature levels. Here, the energy of maintaining the temperature of the cooled object below the temperature of the thermodynamic equilibrium with the environment is received from an electrical power source. At the same time, energy has a very significant value. The fact is that following the basics of thermodynamics, the efficiency of the cycle drops sharply. Whereas nitrogen-level refrigerators having a cooling capacity of several watts consume only hundreds of watts, L-helium-level refrigerators with a cooling capacity of 1–2 watts require a compressor power of 4–7 kW. However, it should be recognized that the given values of nitrogen refrigerators relate to the Stirling cycle, which is essentially close to the ideal reverse Carnot cycle. Helium refrigerators employ other cycles and, in particular, the presented values relate to Gifford–McMahon machines (Figure 6) and the coolers based on pulse tubes.

Figure 6. 4 K cryostat based on a Gifford–McMahon cryomachine for a 2–3 mm superconducting receiver for the RT-13 radio telescope, Metsahovy Aalto High School.

There are two main methods of cooling detectors to a temperature below 0.3 K, specifically by using adiabatic demagnetization refrigerators or refrigerators based on dissolution of ^{3}He in ^{4}He. At the turn of and just below 0.3 K, sorption refrigerators also work on ^{3}He and ^{4}He.

2.3.4. Cryosorption Cryostats

Cryosorption devices are actively used in astronomy (for example, in [42]), when it is necessary to achieve sub-Kelvin (subK) temperatures. One of the first mentions of the development of a 0.3 K system for mounting a subTHz receiver on the BTA optical telescope is given in [43], and the system development was only at the design stage then. In that article, a cryogenic system was proposed that provided the operating temperature T~0.3 K of the receiving antenna arrays. This is a closed-cycle system; i.e., it does not require liquid refrigerants (nitrogen and helium). It is based on a refrigerator on pulse tubes and two sorption pumps on ^{3}He and ^{4}He. Despite the lack of funding for this project at that time, the development of this cryogenic system was continued, and the next stage of development was already presented in [44]. The photos of the elements of the 0.3 K cooling system are shown in Figure 7.

Figure 7. The 0.3 K sorption cryostat.

2.3.5. Dilution Cryostats

An even lower cooling level (0.1 K and below) required for radio astronomy can be provided by dilution cryostats. A mixture of two isotopes of helium ^{3}He and ^{4}He at a temperature below 0.8 K decays spontaneously into two phases, depleted and enriched ^{3}He, which ensures cooling in the dilution cryostat to transfer the ^{3}He atom from the enriched phase to the depleted one, it is necessary to spend energy (absorption of the heat of the dissolution process). If the ^{3}He atom crosses the phase interface continuously, the mixture will be effectively cooled. Thus, the principle of operation of the dilution cryostat is based on the circulation of ^{3}He and its transfer from one phase of the mixture to another.

Compared to the adiabatic demagnetization systems, the advantage of this system is that it is a continuously operating device and does not require the use of magnetic salts or an external source of a powerful magnetic field, which can negatively affect other components of the experiment (for example, detectors). In addition, its mass is very small at the coldest stages, and the cooling power can be distributed over a large focal plane, which avoids the narrow joints and heavy supports required by adiabatic systems. It is also possible to cool mechanical supports and electrical wires (by removing the heat that is transferred from higher temperatures through a reverse flow heat exchanger) without reducing the minimum temperature (for an example, see [45]).

The schematic diagram of operation of the classical dilution cryostat [46] is shown in Figure 8. The performance of the dissolution cycle (W/(mol/s)) is determined by the formula

$$\frac{dQ}{dt} = \left(94.5T^2 - 12.5T_c^2\right)\frac{dn}{dt},\tag{3}$$

where Q is the power supplied to the dissolution chamber (mixer); T and T_c are the temperatures of the mixer and concentrated ^{3}He at the entrance to the mixer, respectively; dn/dt is the circulation rate of ^{3}He, and the typical values of circulation rate are 10^{-4}–10^{-3} mol/s. At $T = 0.1$ K, these rates are $dQ/dt > 100$ µW, which is tens to hundreds of times more than what is required for the operation of low-temperature radiation and particle detectors or for most experiments in solid state physics and research in nanotechnology.

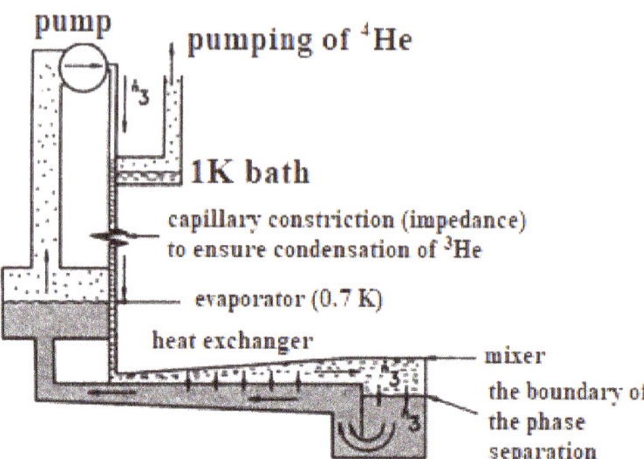

Figure 8. Schematic diagram of the operation of a classical dilution cryostat.

A sophisticated (and more commonly used) dilution refrigerator operates in a continuous cycle. In this case, a mixture of ^{3}He/^{4}He is liquefied in a condenser that is connected via a choke to the area of the mixing chamber rich of ^{3}He. The ^{3}He atoms, passing through the phase interface take energy from the system. Next, it is necessary to distinguish between dilution refrigerators with external and internal pumping. In the first case, the ^{3}He vapor is pumped out by a high-vacuum pump (turbomolecular or diffusion). In the second, it is carried out by a sorption pump.

Note that standard closed-cycle dilution refrigerators require gravity and cannot be used in space observatories. If there is an open loop, this is possible, as, for example, it was for the Planck mission [47–49] (see Figure 9). These issues are considered in some detail in dissertation [50]. As shown in [51], a modified version of a closed-loop dilution cryostat has already been proposed and will be discussed below.

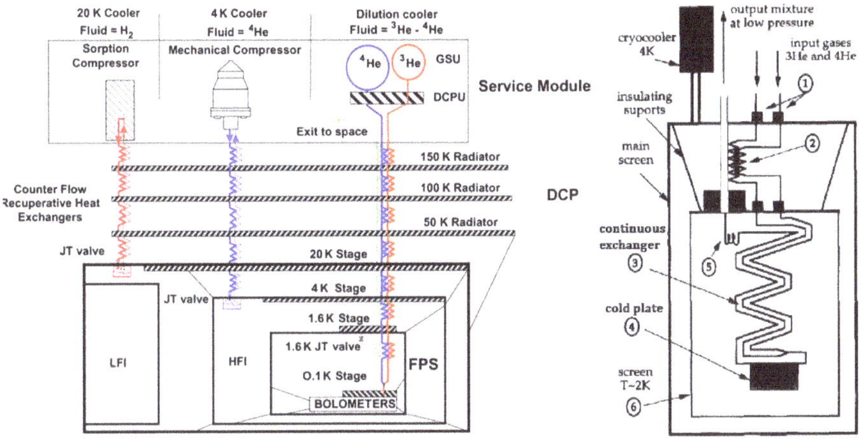

Figure 9. Schematic diagram of an open-cycle cryostat that was used onboard the Planck Space Observatory.

The design of a dilution refrigerator with an open cycle is completely different from the designs of "classic" dilution refrigerators: ³He and ⁴He circulate from different tanks through capillaries and are mixed in a mixing chamber, providing cooling, and then the mixture is not recycled, but thrown into space. The resulting mixture is used to pre-cool clean streams using a countercurrent heat exchanger before being released into space. The heat exchanger consists of three capillaries soldered in parallel and connected at one end by a junction forming a mixing chamber. This design works in zero gravity, because the separation of the liquid and vapor phases, for which gravity is required, is excluded, and in this mixing chamber design, surface tension replaces gravity to separate rich and lean phases. The isotopes are pre-cooled to 4.5 K by external cooling. Further cooling to a temperature below 1.6 K is achieved due to the internal expansion process using the Joule–Thomson (JT) effect in the return line of the mixing chamber.

As noted above, the author of [51] proposed a modified closed-loop cryostat that can be used for space missions, in which the mixture is not ejected into space, but is divided into components that are re-circulated in the system afterwards. A schematic representation of such a cryogenic system is described in detail [52]. The low-temperature part of the cooler (the countercurrent heat exchanger, the mixing chamber, and the loading heater) is much the same as that for the dilution cryostat with an open cycle, given the above. However, the main difference between this cooler and the open-cycle version is the addition of a ³He-⁴He separation/circulation system. The returned mixture enters the alembic, where the two components are separated. The double-layer spiral heater in the alembic evaporates almost pure ³He, which circulates with the help of a room temperature compressor. The ⁴He liquid in the alembic passes through a heavy-duty filter and circulates via a fountain pump operating at a temperature of about 2 K. The spent ³He and ⁴He are cooled first in a tank with a temperature of 1.7 K, and then in an alembic, before being sent down through a countercurrent heat exchanger. The mixing chamber is a Y-shaped connection of three capillaries forming a heat exchanger, so that capillary forces, rather than gravity, separate the enriched phase from the depleted one.

There exists now a modified closed-loop cryostat that can be used for space missions, in which the mixture is not ejected into space, but is divided into components that are then re-circulated into the system [51]. It should be recognized that commercial refrigerators of subK levels (Oxford [53], Bluforce [54] etc.) have been available on the market for more than

10 years. However, these machines are designed for laboratory experiments, and in radio astronomy, they are used only for testing concepts and samples, not for really working on telescopes' ultra-sensitive detectors of the subTHz frequency range. It is difficult to imagine the installation of such refrigerators into the focus of even a ground-based telescope. In this connection, their own cryogenic systems are manufactured for each of the tasks.

Special requirements for cryosystems are imposed by the subTHz observatories located not on Earth, but onboard various carriers, from airplanes and balloons to spacecraft. One of the most interesting examples of this kind follows.

2.3.6. SubK Systems for Balloon Missions

Balloon missions (balloon observatories) are less ambitious than space missions, but no less important. One should mention, for example, such well-known missions as the BOOMERANG experiment [9,55] and the "young" OLIMPO mission [56,57]. Strictly speaking, high-altitude balloons rising to altitudes above 40 km are located at the boundary between the atmosphere and space. It is physically almost *Space*. The pressure and density of atmospheric gases decrease exponentially to infinity. Formally too, in different countries, the boundary of space lies at different levels, from 40 to 100 km. In particular, the density of the atmosphere at the heights described in this subsection does not exceed 0.5% of the ground level. Such a low density of the atmospheric gases determines that the possibility of THz-wave observations from balloons is practically no worse than from spacecraft, but much less costly, including the possibility of returning, recharging, and reusing the cryosystem and receivers for subsequent flights.

One of the problems associated with a long balloon experiment is the requirement that the cryostat must operate autonomously during the entire measurement time. In this context, passive cryostats with cryogenic liquids are the best option. Publication [58] presents a description of such a cryostat, as well as a comparison of designs for the BOOMERANG and OLIMPO missions. For such projects, in addition to maintaining a stable temperature, important factors are the duration of work (more than two weeks) and the structural strength, which should be designed so as to withstand external thermal and mechanical loads. The low temperature of the outer stratosphere determines the design of vacuum seals in the outer shell of the cryostat, while the "shock" when the parachute opens determines the design of the support system for various internal cryostat tanks. The OLIMPO cryostat is an element of the telescope: the inner frame, which supports both the mirrors and the cryostat housing the detector system, can be tilted to set the viewing angle from $0°$ to $60°$. The photo is shown in Figure 10.

Figure 10. Cryostat for the OLYMPO balloon mission.

3. Specific Features of Cryogenic System Technology for Astronomical Instruments

The cooling systems for astronomical receivers are calculated and designed in the context of specific tasks and limitations. More than 45 years of experience in the development of such cooling systems by the team of the authors of this article has formed a set of approaches and technologies presented below. The main problem of such systems, especially deep-cooling ones, is the need for effective thermal insulation of the cold zone, including the cooled receiver and the cold part of the cooler. The best heat insulator of all materials is vacuum. That is why all cryostats for cooling of astronomical receivers are vacuum devices. The issues of vacuum maintenance are of crucial significance in the design of a cryosystem. On the one hand, along with effective thermal insulation, the structural elements of the cryosystem should ensure the reliable fixed position of the cooled receiver relative to the telescope located at the ambient temperature. On the other hand, there should be no significant heat flow along the elements of the supporting structure, which the cryosystem parries. Consideration of the full balance of heat flows is the key cryodesign task. Modern design packages have made this task much easier in recent years. However, there are quite a lot of uncertainties and sources of errors in the calculations.

Along with thermal insulation, a number of structural elements of the cryosystem requires extremely low thermal resistance of materials, such as, in particular, effective thermal conductors between the cooler and the cooling object, which cannot always be fully combined in space.

All these problems, as well as a number of less general and more specific, but sometimes more complex, issues, can be combined into the problem of interfaces: thermal, mechanical, electrical, optical, electrodynamic, vacuum, etc. In fact, if one has a cold source, an EMR receiver, and a telescope, it is only necessary to solve the problems of the interfaces between them.

3.1. Cryogenic Interface and the Basic Elements of the Cryo Design Calculation

The cooling system is selected basing on the balance of heat flows and the cooling capacity of the cooler. For cryo-accumulators, the minimum interval between cryo-agent refueling sessions is taken into account. When designing refrigerators, sufficient power should be provided for counteracting all heat inflows.

Let us estimate the parameters of the cryosystem's cooling capacity for local cooling of the elements of the receiving devices. The total heat inflow that the cryosystem should compensate is determined by the combination of the internal (Joule) heat release Q_J of the active elements of the device and the external heat inflows (convective Q_1, radiation Q_2 and conductive Q_3 (by elements)) of the load-bearing structures:

$$Q_\Sigma = Q_J + Q_1 + Q_2 + Q_3. \tag{4}$$

The Joule heat dissipation Q_J for superconducting detectors and mixers at typical DC modes (at the voltage and current being about 3 mV and 30 μA, respectively) is:

$$Q_J = VI \sim 0.1 \times 10^{-6} \text{ W}. \tag{5}$$

The convective component of the thermal conductivity of the residual gas in the cryostat can be reduced easily due to efficient gas pumping. At the pressures inside a vacuum cryostat with a cryosorption pump, which do not exceed 10^{-7} Torr, and the typical geometry of the receiver and cryosystem, the heat inflow through the residual gas will not exceed 10^{-6} W.

The radiation component of the heat inflow splits into two fractions: one flowing through the signal window (Q_{21}) and the other through the other surfaces of the device (Q_{22}):

$$Q_\Sigma = Q_{21} + Q_{22}. \tag{6}$$

In this case, the first term is expressed as:

$$Q_{21} = \varepsilon_1 \lambda_i \left(T_1^4 - T_2^4 \right) S_0, \tag{7}$$

where ε_1 is the reduced thermal radiation coefficient, λ_i is the blackbody radiation constant, T_1 and T_2 are the temperatures of the external environment and the cold surfaces of the receiver, respectively, and S_0 is the window surface area. This term is fundamentally unavoidable, since any filter that transmits the useful received radiation cannot completely avoid being heated by the radiation passing through it. The total heat input should not exceed the refrigerating capacity of the refrigerator or the capacity (volume) of the Dewar tank.

3.2. Vacuum Interface

Vacuum cryostats, which are the basis of any cooling system for astronomy, is characterized by the size, shape, depth, and method of providing vacuum. The density of connections between vacuum interfaces is ensured by their tightness. However, the vacuum level, and hence the vacuum pumping and control technologies, should be selected and maintained for reasons of reasonable sufficiency. In particular, the determining factor is the dominance of other components over the components Equations (5) and (6) in the overall balance.

At the same time, it is clear that different systems require different vacuum levels. For example, superconducting nanodetectors have negligible Joule heat dissipation and compact dimensions that minimize heat flows. The operating temperature is only one Kelvin, and the cooling capacity of such systems is less than one microwatt. Therefore, the vacuum levels required for such receivers reach values of 10^{-8} to 10^{-10}. At the same time, the productivity of the 4 K refrigerators based on pulse tubes is three orders of magnitude higher, and you can afford to limit yourself to $10^{-4} \div 10^{-5}$ in order to minimize the heat exchange between the cold and warm walls of the system by varying the values of Dewars characteristic of astronomical receivers. Moreover, in reality, such systems require preliminary pumping to $10^{-2} \div 10^{-3}$, after which the cryosystem is started and, as it enters the appropriate mode and passes the characteristic condensation temperatures of the atmospheric gases (90 K for oxygen, and 77 K for nitrogen) due to cryosorption, the refrigerator equipped with a small charcoal cryosorption panel begins to feed itself pumping the internal volume of the Dewar to values of 10^{-5} to 10^{-6}. Cryosystems operating at a temperature level higher than the nitrogen one, which are often used to cool the photodetector devices of optical telescopes, have big problems in terms of the vacuum interface. Cryosorption at such temperatures is not efficient enough and requires either good pre-pumping, with deep pre-cleaning and degassing of all internal components of the receiver, or the installation of a pumping device that is periodically switched on, such as a getter-based pump.

3.3. Mechanical Interface

The main purpose of the mechanical interface is the mechanical assembly of the cryosystem structure with a cooled receiver as a whole. Along with the standard set of technologies for cryosystems, the task associated with the vibrations generated by the mechanical system of the refrigerator is of key importance. Considering the noticeable power level of the compressor (up to 10 kW) and the vibration of the cooler engine (Gifford–McMahon cooler) or the valve switch (pulse tubes), this task is gaining significant importance. There are various forms of cryostats: cylindrical (Figure 2), rectangular (Figure 6), trapezoidal, as presented in Figure 11, and even more complex; see [40].

Figure 11. Trapezoidal mechanical interface for the optical cooling system.

3.4. Optical Interface

For photodetectors, the optical interface is of key importance in providing vacuum insulation. It transmits the received signal, the attenuation of which needs to be minimized, but also the adjacent section of the IR spectrum, thus bearing the main thermal load, which must be withstood by the cryosystem. The calculation and manufacture of high-quality strip illumination resolves this problem to a noticeable extent. This is also facilitated by a lower temperature and a greater cooling capacity at these temperatures, compared with the THz radio receivers and lower-frequency astronomical receivers. For radio receivers, quasi-optical or waveguide windows are used for the signal input. They are essentially optical and also have selective illumination. Sometimes, cooled radio receivers are also equipped with a control optical interface for visual control of the interior of the cryostat. Conventionally, the category of optical interfaces can also include sealed inputs of optical fibers, through which optical signals can be brought into the vacuum cryostats of cryogenic receivers.

3.5. DC, RF and Digital Interfaces

Various kinds of electrical interfaces of cryostatic systems are needed both for the supply and removal of DC (digital current) and RF (radiofrequency) signals of power supply systems, monitoring, etc. For this, there is an extensive set of the nomenclature of both vacuum inputs and thermally connected wires, sensors, electrically controlled keys, valves, regulators, alarms, etc., which are commercially available and periodically updated on the website and in the index issue of the journal *Cold Facts* [59]. However, in recent years, the appearance of cryogenic systems both in general and for astronomy specifically has changed radically. In the 1980s, the cryostat was a tank of an appropriate size, with a window and a wiring harness connecting the sensors and the devices of the outdoor monitoring system. A manometric condensation sensor was used to measure temperature as the most accurate device. Moreover, the engineers of the support group who helped the astronomers during the observations were on continuous duty monitoring the work of the cryogenics. Now, the cryosystem is primarily a computer with a proper hard drive and the special software, while the tank and the refrigerator have become less significant. In general, an astronomer, i.e., an operator, who is usually sitting thousands of kilometers from the telescope with a cryosystem, may not even know what it looks like and how it works, having a small and convenient interface that signals the state of the receiver as a whole, including its cryogenics, and provides the opportunity to conduct observations.

For the designers of cryosystems of astronomical receivers, all the lifehacks of electrical interfaces are still relevant, including the features of laying and organization of interceptions (anchoring) on wires and loops, through which signals are sent at intermediate temperature levels. If it is necessary to output microwave signals, specialized vacuum and thermal isolating coaxial lines and waveguides are used. The problem with quasi-optical supersized (multi-mode) waveguides operating under the temperature difference, rather than in the state of thermodynamic equilibrium, is rather difficult and requires a special approach to its analysis, which is described in Section 4.3. However, it has been solved. Among the novelties that are similar in the application of the technology, one should mention the fiber lines used for channeling optical radiation. However, all the approaches used for cables are applicable here, with an additional bonus that the thermal conductivity of the glass, from which the optical fibers are made, is several orders of magnitude lower than that of the metals used for RF and DC.

4. Development of Cryogenic Systems for Astronomy: Technical Solutions of Combined Optical and Radiophysical Problems

This section will briefly present the individual technical solutions for the problems of cryogenic cooling outlined above and related to the combination of optical and radio-physical problems. These solutions were taken as the basis for successful developments and subsequently replicated in various applications. A review of our own developments of cryosystems for astronomy is presented here. The unity of approaches is purely conditional, and it is not a matter of novelty. The experience of developing more than 100 different cryosystems of different temperature levels for astronomy, in my opinion, is worthy of presentation in terms of the most interesting examples. The main highlight of the presented developments is that after making cryosystems for radio astronomy receivers, and a little later, for cooling photodetectors in optical astronomy, for a long time, today we have moved on to the tasks of THz radio receivers in quasi-optical telescope systems, including those for THz radio receivers for the BTA optical telescope.

4.1. 4 K Cryostating Systems

An optical cryostat based on the Cryomech PT410 cryorefrigerator has been developed. It is designed for studying the thermophysical properties of materials (thermal coefficient of linear expansion, heat capacity, thermal conductivity), structures and mechanisms, and 4 K superconducting receivers, as well as for cooling samples and elements of the optical path with the possibility of using the Cryomech PT410 cryorefrigerator. A photo of the assembled cryogenic system and the system mounted on an antivibration rack is shown in Figure 12. This system has become practically replicable, and 5 similar versions have been created. As development progressed, it was possible to significantly improve the characteristics. An almost unprecedented result was achieved: one of the systems (among hundreds of cryostats made), which was created for the Institute of Microstructure Physics, achieved all the required performance characteristics immediately after assembly, without adjustments or modifications. A series of these developments has produced many interesting results for astronomy and space missions.

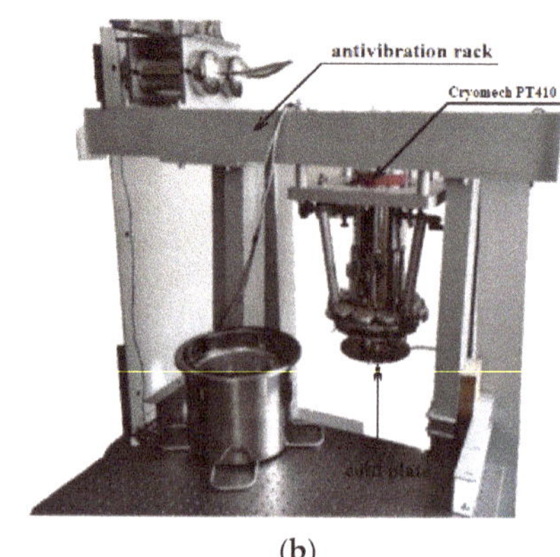

(a)   (b)

Figure 12. Photos of the 4 K cryostating system based on a pulse tube: (**a**) general view; (**b**) the cryosystem mounted on an antivibration rack in order to increase the accuracy of measurements.

The main characteristics of the closed-cycle microcryogenic unit system (MCS) RDK-408D2 (SHI) with optional Cryomech coolers are as follows:

- Operating temperature: 4 K $\pm$ 0.1 K; with a load: 0.8–1.5 W
- Vacuum level: 10^{-4} mbar; flange KF D25 for pumping
- 2 flanges for installation of optical windows with a diameter of 25 mm;
- Input of electrical and RF signals
- Size of the working cavity: diameter 185 mm, height 70 mm, overall dimensions not more than 1600 mm; the diameter is not less than 700 mm; there are several options
- A built-in interface for reducing the influence of temperature fluctuations with the possibility of observations when MCS is disabled
- Availability of optical windows
- The system should operate being installed on an antivibration rack

The main scientific results obtained using the equipment are as follows:

1. A prototype of the actuator that was designed for moving the main mirror of the Millimetron observatory at cryogenic temperatures has been tested already [60]. Photographs of the test sample are shown in Figure 13. Some results of the experimental studies are presented below

(a) (b)

Figure 13. The device under study: (**a**) the actuator; (**b**) the gearbox and the stem, the device under test.

The following characteristics of the actuator have been confirmed experimentally:

- Resolution in full-step mode: 0.64 µm;
- Resolution in microstep mode (practically achievable): 0.3 mm;
- Range of movement: $\pm$3 mm;
- Repeatability error: from 0.2 to 1.55 µm (0.8 µm, on average) in a range of 150 µm;
- Minimum consumption power (continuous operation): 600 µW (T = 8...10 K);
- Minimum consumption power when moving: 1 µm through 4 µW·s (mJ) (T = 8...10 K).

The research was carried out jointly by "Applied Mechanics" JSC and the Astro Space Center of Lebedev Physical Institute of the Russian Academy of Sciences (ASC LPI).

The result was that at ~10 K the temperature of the actuator-measured deviation was less than 0.6 µm and was true to type RMS at 18 µm.

2. Measurements of the thermal conductivity and heat capacity of beryllium samples intended for the manufacture of the switching mirror of the space cryogenic telescope "Millimetron" were carried out.

A photograph of the mounted sample and the results of experimental studies regarding its thermal conductivity and heat capacity are shown in Figure 14.

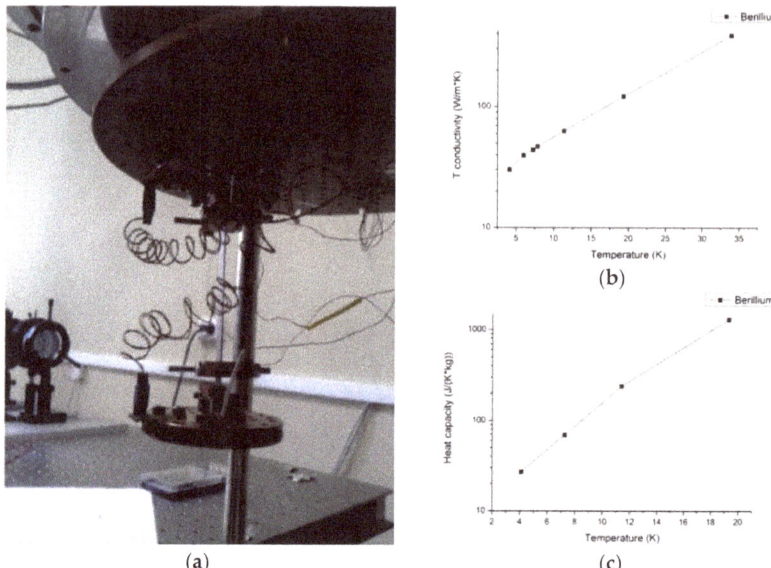

Figure 14. Study of thermal conductivity and heat capacity of beryllium at cryogenic temperatures: (**a**) photo of the sample mounted on the cold plate of a cryostat; (**b**) results of measuring the temperature dependence of thermal conductivity; (**c**) results of measuring the dependence of the heat capacity on the temperature.

4.1.1. Solving the Problem of Vibrations and Temperature Fluctuations in Cryogenic Refrigerators

Refrigerator cooling systems, along with an extensive list of advantages, have a very significant disadvantage that limits their use for astronomical applications. Due to the periodic operation of the refrigerator cooler, temperature pulsations and mechanical vibrations occur. This is unacceptable when operating sensitive astronomical equipment. This paper presents the results of vibration studies and suggests some measures for reducing these negative effects.

The vibroacoustic effect of mechanical MCS parts has an interfering effect on detectors, amplifiers, measurement systems, etc. A low-temperature closed-cycle cryorefrigerator refrigeration unit was selected as the test object for studying vibration levels. The purpose of the test was to measure the vibration levels created by the hardware at the points of the intended location of the tested and reference sensors. The tests were carried out using the following equipment:

- Accelerometers B&K 4371-2
- Charge amplifiers B&K 2651-2
- Power supply B&K 2805
- 2-channel ADC M-AUDIO Transit

The measurements were carried out at room temperature (thermal insulation shells were not installed on the installation). Two accelerometers with synchronous recording of signals on a laptop computer were used for measurements. One sensor was placed at the intended location of the electromagnetic radiation receivers on a low-temperature plate inside the working area of the installation. The second accelerometer was placed consecutively near the main sources of vibration activity included in the stand. The signals were recorded under three different modes of operation of the installation:

- All mechanisms were disabled (recording duration ~60 s).
- All main mechanisms (the pumps and compressors) were turned on; the cooler was turned off (recording duration ~60 s).
- All main mechanisms (the pumps and compressors) were "on"; the cooler was "on" (recording duration ~30 s).

Figure 15 shows an example of the time realizations of the vibration acceleration signal measured at the points at the center of the lower plate inside the working area of the installation and on the cooler during its operation. Figure 16 shows the amplitude spectra of vibration displacement at the same control points obtained by averaging over 30 samples with a length of 1 cm.

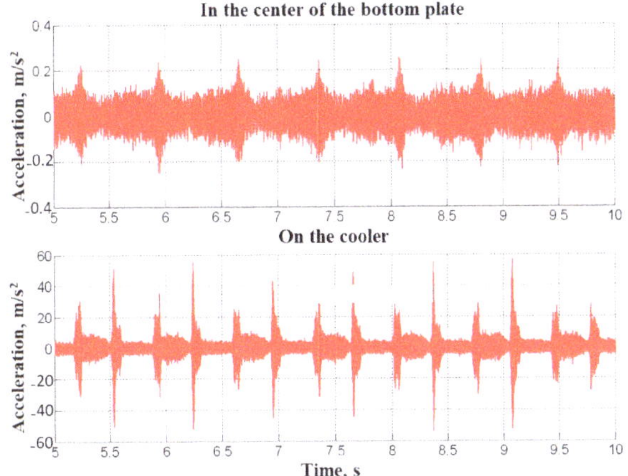

Figure 15. Temporary implementation of the vibration acceleration signal at the control points.

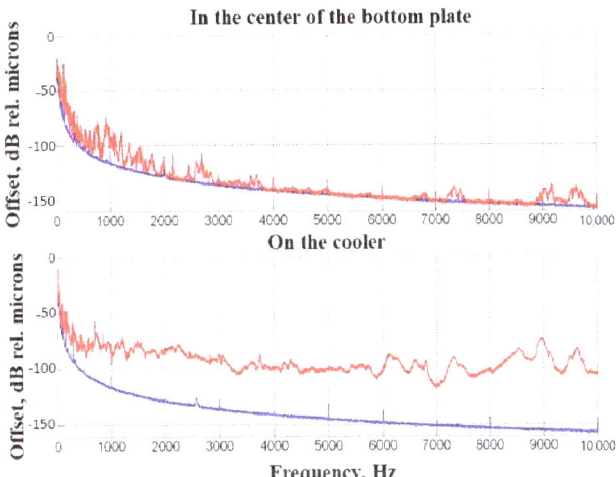

Figure 16. Averaged amplitude spectrum of vibration displacement. The blue curve is the background level (all mechanisms are off); the red curve shows the case where all mechanisms and the cooler are on. The analysis band is 1 Hz.

The excitation signal is a sequence of short pulses with a repetition period of about 0.6 s. The lower graph (Figure 15) corresponds to the spectrum of the source of the exciting force. The upper graph (Figure 15) illustrates the frequency dependence of the response to the impact from the cooler in the working area of the installation.

The graphs in Figure 16 demonstrate good vibration isolation of the load-bearing structures located inside the working area of the installation from the main vibro-active sources of the stand in the range above 1 kHz. The nature of the cooler operation determines the signal in the form of successive pairs of pulses. The pulses alternate along the carrier frequency, which is clearly seen in Figure 15, where only the low-frequency ones are visible (suppression of about 150 times), and the high-frequency ones are almost completely absorbed by the vibration isolation (suppression of at least 500 times).

The low-frequency range is the most energy-intensive and dangerous in terms of vibration displacement amplitudes. Figure 17 shows the averaged amplitude spectra of vibration displacement at the control point at the center of the lower plate in the working area of the installation in the 0–200 Hz range. The spectra were obtained by averaging 6 samples of a signal with a duration of 5 s. In the low-frequency range, the response at the control point is a set of discrete components that were created during the operation of the cooler (140 Hz, 176 Hz, etc.) and other installation mechanisms (24.8 Hz, 51.8 Hz, 78.8 Hz, etc.).

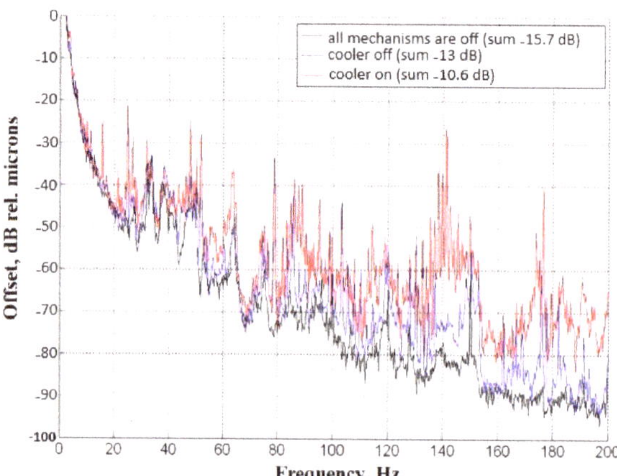

Figure 17. The averaged amplitude spectrum of vibration displacement (frequency resolution 0.2 Hz). The black curve is the background level (all mechanisms are off), the blue curve shows the case with all the main mechanisms except the cooler, and the red curve is the case where all the mechanisms and the cooler are on. The control point is at the center of the lower plate inside the working area of the installation. The legend gives the integral values of the amplitudes of vibration displacements in the band 8–200 Hz in dB relative to 1 micrometer.

As can be seen from Figure 17, at this point of the plate the level of each narrow-band component individually does not exceed 0.1 μm. The integral level of the vibration displacement amplitude in the 8–200 Hz band is given in the legend of the graph and is ~0.3 μm when all the mechanisms of the installation and the cooler were working.

The frequency resolution of the spectra shown in Figure 17 is 0.2 Hz. The width of the discrete components is significantly smaller, as can be seen from Figure 18, which shows the amplitude spectra at the control point on the lower plate obtained at different frequency resolutions.

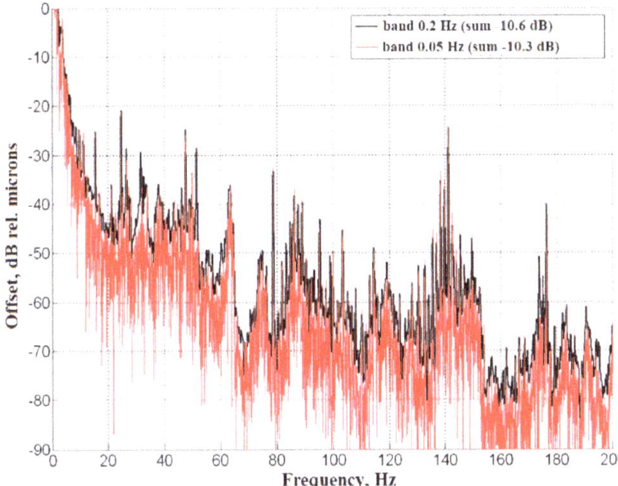

Figure 18. Averaged amplitude spectrum of vibration displacements. All mechanisms and the cooler were active. The black and red curves correspond to a frequency resolution of 0.2 Hz and 0.05 Hz, respectively. The control point is at the center of the lower plate inside the working area of the installation. The legend gives the integral values of the vibration displacement amplitudes in the 8–200 Hz range.

The sufficiently high level of the intrinsic noise of the measuring channel used at this stage and the limited registration time did not allow for reliable measurement of the vibration spectrum in the range near zero frequencies. It was possible to extend the frequency range into the low frequency region using accelerometers with a built-in amplifier. It was also recommended to consider the possibility of using an optical vibration measurement method (laser vibrometer), which is operable, in particular, at the operating temperature of the object.

Attention should be paid to the unevenness of the vibration level on the low-temperature plate where the receiver was located. Figure 19 shows the averaged amplitude spectra of the vibration displacement for the two control points at the center (red curve) and the edge (black curve) of the lower plate in the working area of the installation. The vibration levels in the narrow frequency bands could differ by ~20 dB at different points of the plate, and the integral levels, by 6 dB.

Since the equipment was installed on the 10th floor, where the cryogenic nanoelectronics laboratory is now located, the studied samples show the influence of people and buses passing nearby. During the measurements, a significant influence of external interference in the room on the measurement levels was noted, which is due to the lack of measures taken to vibro-isolate the installation from the floor where it was placed.

Based on the results of the vibration tests of the stand, the following conclusions can be made:

- The vibrations in the working area were determined by the periodic action of the cooler, and the width of the discrete components of the vibrations was obviously <0.1 Hz.
- The vibration displacement levels of the narrow-band components at the control points on the lower plate in the working area of the installation did not exceed 0.22 μm (<0.1 μm in the center of the plate). The integral level of vibration displacement in the 8–200 Hz band ranged from 0.3 μm to 0.6 μm at different points of the plate.

- To provide measurements in the frequency range of 0.5 Hz, it is recommended to use a laser vibrometer. It is also recommended to evaluate its measuring capability at the operating temperatures. The measurable levels achievable by using standard laser vibrometers will be no worse than 100 nm up to zero frequencies.
- To develop recommendations for reducing the vibration level of the structure in the working area of the installation, a more detailed measurement of the acoustic vibration characteristics of the stand (transmission coefficients from the source to the working area) should be carried out, and measures should be taken to ensure vibration isolation of the installation as a whole. For the following measurements, a measuring path with a lower noise level should be used.
- Of course, for the final verification of the data obtained, it is useful to conduct a test during cooling; for this, it is necessary to provide thermal and vacuum isolation of the measuring equipment, since the individual parameters, in particular, heat capacity, thermal conductivity, sound propagation velocity in the medium can vary depending on the temperature.

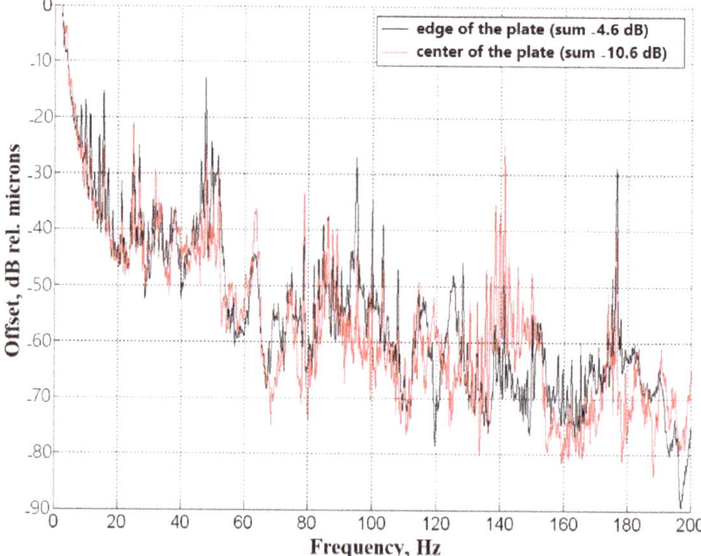

Figure 19. Averaged amplitude spectrum of vibration displacements (frequency resolution of 0.2 Hz). All mechanisms and the cooler are active. The red curve is at the control point in the center of the lower plate; the black curve is at the edge of the lower plate. The legend gives the integral values of vibration displacement amplitudes in the 8–200 Hz range.

4.1.2. Antivibration 4 K Cryostating System

The closed-cycle cryostating system with a reduced level of mechanical vibrations (vibration damping system) is designed to study samples of highly sensitive superconducting detectors, including not only THz but optical too. Photos of the cryosystem are shown in Figure 20. As a basic configuration (prototype), the 4 K system presented in the previous section was used. The vibration damping system includes the both design solutions and a specially designed structure of cooling pipes and screens, which soften the mechanical contact integrally, which leads to deterioration in the transmission of acoustic waves or vibration damping. Experiments have shown that the vibration damping system makes it possible to reduce the vibration level of the cryosystem panel, on which the elements of the receiver detector are supposed to be placed, by 7–8 dB in comparison with the vibrations of the basic configuration system. The vibration displacement level of the prototype ranged from 50 to 150 μm. In the experiments of this cycle, the vibration level did not exceed 1 μm, and was almost an order of magnitude lower than the requirements of the terms of reference and the measurement metric program, which is obviously sufficient for receiving systems of sub-Hz waves, where the criterion for the negligibility of the vibration effect is $A\_v = \lambda/20$ (here, A_v is the vibration amplitude and λ is the wavelength).

(a)

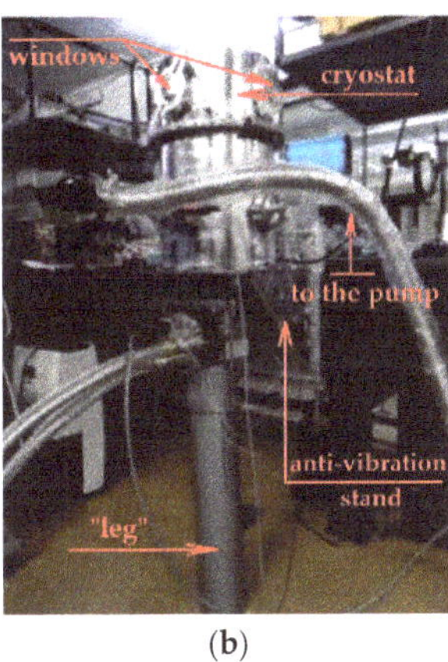

(b)

Figure 20. Photos of the antivibration 4 K cryostating system: (**a**) cryostat with windows (on both halves of the housing shell), an original easily removable system for connecting the body halves (in the middle) and a bellows vibration decoupler (at the top); (**b**) cryostat with the vibration damping system.

The main characteristics of the system are:

- Operating temperature: 4 K ± 0.1 K; with heat load: 1 W;
- Vacuum level: 10^{-4} mbar;
- Vibration displacement of the 4 K plate in the horizontal plane: no more than 10 μm;
- close-cycle MCS RDK-408D2 (SHI), mounting of MCS: cold head up
- 4 flanges KF D50 for mounting of optical windows with a diameter of 50 mm (optional), 3 flanges KF D50 or the mounting of electrical and RF connectors;
- Working cavity size: diameter 200 mm, height 90 mm;
- Overall dimensions of the product: diameter 400 mm, height 800 mm.

Additional options for the basic configuration are the presence of a vibration damping system, quick access to test samples and the possibility of quick installation of various equipment on 8 KF50 flanges.

Vibrations of cryogenic systems

Cryosystems with refrigerators based on pulse tubes are used for long-term experiments. The main disadvantage of such systems is the additional mechanical noise, which contributes to the general noise of receiving systems, which must be reduced in the case of high-precision/high-sensitivity experiments. There are various mechanisms for suppressing such noise vibrations, for example, compressor decoupling, damping pads, etc. The 4 K level cooling system with the vibration damping system, which was developed by the authors of this paper, has been presented above.

The problem of damping the vibrations of the refrigeration unit was resolved using both the special features in the design of the elements of the case and the layout of the cryostat, and the special design of the cold pipes and screens, which generally soften the mechanical contact, leading to a decrease in the transmission of acoustic waves or vibration damping. But this obviously cannot but affect the heat transfer mechanisms and even vacuum conditions, which can lead to undesirable consequences in the cooling characteristics of the receiver. In fact, a compromise is required between achieving the required thermal and acoustic performance. Vibration measurements of the sample simulator were carried out with a laser vibrometer in the horizontal direction. The laser beam was directed at the sample through an optical window in the housing. The cryosystem was mounted in a clean room on the optical table of the PEARL-10 multi-petawatt laser stand (see Figure 21a, right). The compressor with a capacity of 7 kW was installed in an adjacent room (with a separate foundation and no requirements for the cleanliness of the room) and connected to the cryostat and cooler by flexible metal hoses laid into the clean area of the silo, where the cryostat was located.

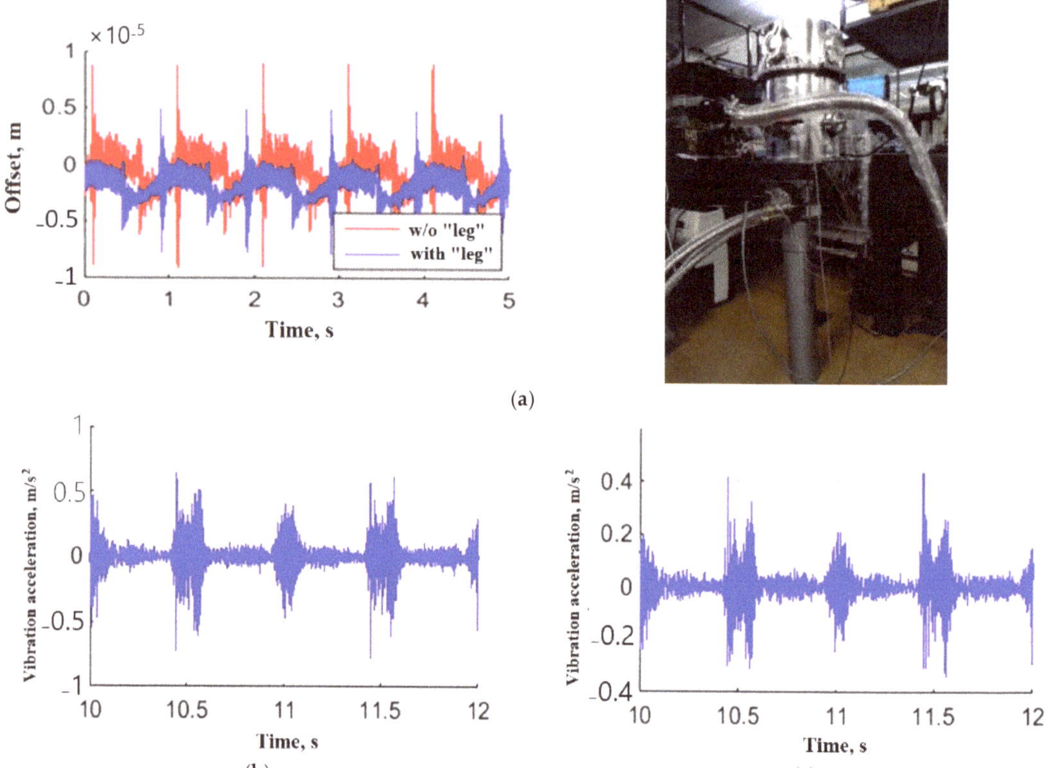

Figure 21. Some results of measuring the vibration level of various elements of the cryosystem: (**a**) Vibrations of the sample simulator in the horizontal direction and the presence/absence of an additional support; (**b**) vibrations of the bracket in the vertical direction; (**c**) vibrations of the table in the vertical direction.

After a successful vibroacoustic test, the cooling capacity and the maximum achievable temperature of the cryostat design upgraded from the point of view of acoustics were checked. As expected, the nominal characteristics of the cryostat were not achieved during the first test. The computational, theoretical, and experimental works were carried out to find the compromise mentioned above [2,3,25,40]. The main measures to achieve the nominal temperature in 4 K were as follows:

- To reduce the 4 K radiation losses, the support was covered with superinsulation (Figure 22a)—an additional cooling line made of a spring ring of soft annealed copper was installed under the 4 K panel and on the cold finger of the cryomachine to improve heat transfer between them ("cornflower") (Figure 22b). A great deal of this work regarding this element of the process has been carried out on microphonics and their treatment.

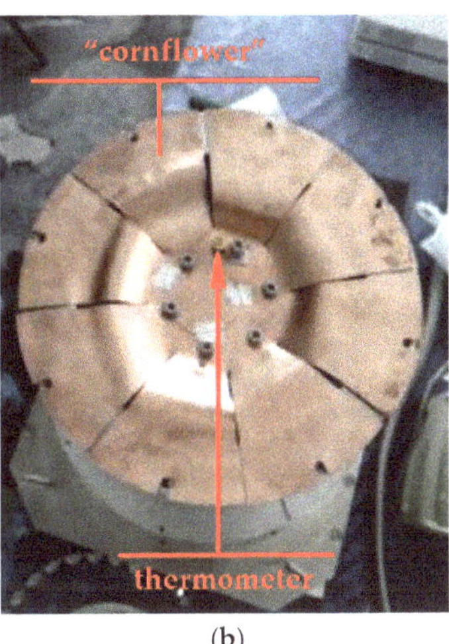

(a)   (b)

Figure 22. Additional measures to bring the cryogenic installation to the operating mode: (**a**) 4 K support in the super insulation; (**b**) soft copper "cornflower" cooling pipe having a thickness of 0.5 mm.

4.1.3. The Low-Vibration 4 K Cryostating System for Studying Thermal Deformations of the Panels of the Main Mirror of the Millimetron Space Mission at Cryogenic Temperatures

Together with the ASC LPI, a cryovacuum chamber was created for studying the thermal deformations of the panels of the main mirror of the "Millimetron" space telescope. A photo of the cryosystem is shown in Figure 23.

Figure 23. Photo of the 4 K cryosystem for studying thermal deformations of the mirror panels of the Millimetron space mission installed in P.N. Lebedev Physical Institute RAS, June 2023.

The main characteristics of the system are:

- Operating temperature: 4 K $\pm$ 0.5 K
- Residual pressure inside the cryostat: 10–5 mbar
- The level of vibration displacements on a cold plate: no more than $\pm$0.5 μm in the frequency range up to 300 Hz
- Vacuum inlets: 32 fiber optic, 2 KF25, KF16, and an optical window
- Overall dimensions: height 1450 mm, diameter 830 mm

4.1.4. Cryovacuum Resonator Complex

The cryovacuum resonator complex [61,62] (see Figures 24 and 25) is a unique new-generation laboratory hardware system for studying the dielectric parameters of gases and condensed matter (in the temperature range from $-50\ °C$ to $+ 200\ °C$), as well as the reflectivity of surfaces (reflection losses) in the frequency range from 100 to 520 GHz, and the pressure range from 10^{-3} Torr to atmospheric pressure at temperatures of 4–370 K. The radiation loss in the investigated substance determines the Q factor of two quasi-optical Fabry–Perot resonators located in the vacuum chamber. The length of the resonators differs by half: the long (symmetrical) resonator is formed by two identical spherical mirrors, and the short (asymmetrical) resonator is formed by one spherical and one flat mirrors. Deep cryogenic cooling (up to 4 K) is used to conduct reflectivity studies. A symmetrical resonator is used as a reference, and a test specimen of the reflector is mounted on the flat mirror of the asymmetric resonator. For thermal isolation from the resonator housing, the mirror with the specimen is fixed on fiberglass racks and is connected to the second stage of the cooler by a flexible cooling wire. The copper housing is connected to the first stage of the cooler and is also isolated from the chamber walls. The temperature deviation along the length of the housing does not exceed two degrees at 70 K. As the cryogenic basis of the equipment, a cryovacuum chamber with a closed-cycle RDK-415D cryorefrigerator manufactured by Sumitomo Heavy Industries, Ltd. with a Gifford–McMahon thermodynamical cycle of the helium temperature level is used. This refrigerator is able to provide cooling without a load to a temperature below 3 K in the second stage and up to 60 K in the first stage.

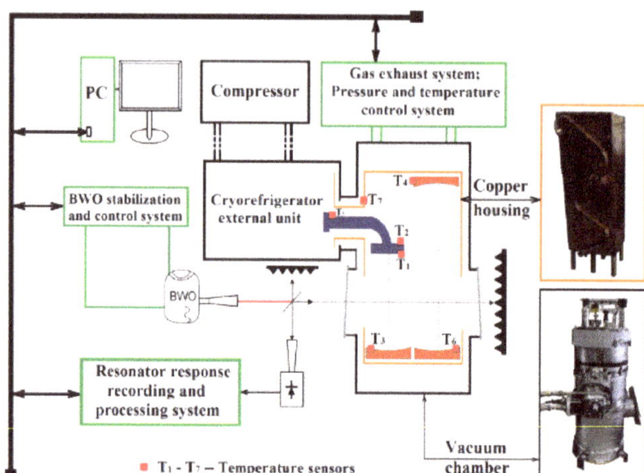

Figure 24. Schematic representation of a cryovacuum resonator complex (the figure from [62]).

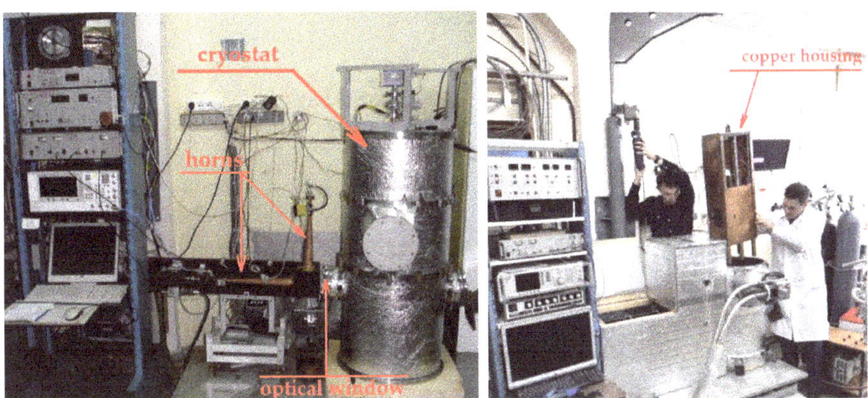

Figure 25. Photos of the cryovacuum resonator complex, IAP RAS, April 2022.

Also, the experimental setup is used in the temperature range from −50 °C to +200 °C to study the spectra of gases. However, in this article, we will not consider this research line because no deep cooling is provided in this mode, only down to −50 °C (more details can be found on the website of the Department of Microwave Spectroscopy of the IAP RAS—URL https://mwl.ipfran.ru/#/ (accessed on 3 March 2024). Recently, a new vacuum chamber has been developed and manufactured for such studies (see Figure 26) in order to increase the number of experiments.

Figure 26. A new vacuum stand for the study of gas spectra in the subTHz frequency range at temperatures from −50 °C to +200 °C, IAP RAS, May 2023; "the first launch" in November, 2023.

The chamber is equipped with a pressure and temperature monitoring system and is thermally insulated from the external environment. The vacuum system provides metered injection and pumping of gases. The temperature of the mirrors, the casing, and the sample is controlled by the automated LakeShore Temperature Monitor system (the sensors are marked T1–T7 in Figure 24).

Main characteristics of the equipment are as follows:

- Frequency range: $36 \div 520$ GHz
- Temperature range for gases: within 220–370 K with the possibility of long term stabilization at any temperature, within 10–220 K without temperature stabilization, and 4 K–900 K for dielectrics and metals
- Gas pressure: 0–1500 Torr
- Sensitivity to changes in absorption in gas: ~0.001 dB/km (4×10^{-9} cm^{-1})
- The range of measured values of the refractive index: 1–10 with a relative error up to 10^{-4}
- Measured thickness of the dielectric plane-parallel plates: 0.002–30 mm with an accuracy of up to 10^{-4}
- Minimum diameter of the solid sample under study: ~12 mm (on 140 GHz)
- Range of measured values (tgδ): $10^{-2} \div 10^{-7}$ with a relative error of up to 5%
- Range of measured values of reflection losses: $10^{-1} \div 10^{-4}$ with an average relative measurement error ~5% at the level of reflection losses ~10^{-3}

The main scientific results obtained with this equipment

A wide class of metal coatings for antennas of ground-based and space-based radio telescopes has been studied for the creation of highly sensitive cooled receivers. The reflectivity of mirrors with different internal structures made of ultrapure silver, copper, gold, aluminum [63], and beryllium alloy [63] has been studied. It has been shown that when specimens are cooled to the liquid helium temperatures, the reflection losses vary significantly depending on the structure of the surface of the reflecting layer and the presence of impurities, and a limit for reducing reflection losses is found. Reflection losses have been investigated for the film reflectors of the Millimetron Space Observatory [61].

The prospects of using materials with high-temperature superconductivity (HTS) [64] as well as classical superconductors of the second type based on Nb and NbTiN [62] are shown. Data on the losses of reflection of millimeter waves by these materials were obtained for the first time.

The dielectric properties of a wide class of dielectrics and semiconductors (crystals, ceramics, amorphous substances, plastics, etc.) have been studied. These studies have made it possible to create a reference database for the selection of materials used. In particular, a cycle of studies of dielectrics with ultra-low absorption in the millimeter and submillimeter ranges for the input and output windows of megawatt generators (within the framework of the ITER program) has been carried out. Based on the research performed, thermal calculation of the windows based on CVD diamonds for the output and input of heavy-duty radiation has become possible. A class of dielectric liquids which have a great potential for cooling energy output windows in high-power and heavy-duty generators of the millimeter and submillimeter ranges has been determined.

The main result of the research performed is to obtain data on the radio and thermal physical properties of new materials and composites used for cryogenic receivers, antennas, and other systems of astronomical radio telescopes. The data obtained were used in the design of the Millimetron space mission.

4.2. CCD Matrix Cooling Systems

CCDs (charge-coupled devices) are used in many areas, both in fundamental science and in military and astronomical applications. As mentioned in the introduction, one of the most effective ways to reduce noise characteristics and consequently increase the sensitivity of the receiving device is to cool the device itself. Nitrogen-level cooling (~80 K) is also quite popular for optical and infrared photodetectors [65,66]. In this subsection, we will present only a short description of the cooling device for CCD, while the lineage of CCD array cooling systems developed for BTA SAO RAS was presented in our previous paper [40]. However, for some CCD matrices, "soft" cooling up to -40 °C is usable [67].

Based on accumulated experience, specialized chambers were manufactured for cooling CCD matrix in the vacuum ultraviolet range with a temperature range from 80 K to

230 K. Since ground-based observations cannot be made in this range, it is assumed that it will be used for the Spectrum -UV space mission. Also, a cryogenic test system was created for cooling several objects with gaseous or liquid nitrogen to simulate the conditions of deep space, where the cameras manufactured for cooled CCD matrix of the VUV range were tested. This is a system consisting of a nitrogen Dewar vessel, insulated metal hoses, the interfaces for supplying cold to consumers, and a temperature control and stabilization unit. The characteristics and photos of the systems are given in Table 2.

Table 2. Characteristics of the cooling systems for the UV Spectrum project.

Photo	Characteristic	Advantages
	Chamber with remote cooling	
	- Chamber with remote cooling - Temperature range: 190 K–230 K - Type of cooling: liquid - Cryo-agent: liquid nitrogen - Vacuum level: 10^{-4} mbar - Availability of getter pumping - Dimensions: $240 \times 190 \times 154$ mm - Electrical connector: SNC - Optical window MgF Diameter: 90 mm	- Vibration-free system - Possibility of providing a high level of vacuum up to six months without pumping out with electromechanical pumps - The presence of a large-diameter window - No liquid cryo-agent inside the chamber
	Cryogenic transport system	
	- Temperature range: 163–168 K - The possibility of simultaneous temperature control of three cooled objects - The ability to control the temperature—step 1°, accuracy ± 0.5 degrees	- Ease of manufacture - High speed and precision temperature control The possibility of measuring the heat flow

4.3. Components of Cryogenic Astronomical Receivers That Are out of the State of Thermodynamic Equilibrium

A mandatory element of cryogenic systems for receiving devices is a sealed optical window (for signal input to the detecting cell), and often there are thermally coupled waveguide and quasi-optical lines inside the systems that supply the signal to a deeply cooled device (amplifier, mixer, detector). For the first time, these issues were addressed in publication in [68]. As noted above, these elements are characterized by the fact that they are not in a state of thermodynamic equilibrium, and strictly speaking, even the term "temperature" is not applicable to them without reservation. The related features in the calculation, analysis, and design of such systems are discussed below.

4.3.1. Physical Temperature and Noise of a Sealed Window

The task of analyzing and creating a cryoreceiver window is reduced to constructing a vacuum-dense and transparent entrance for the received radiation with a minimum of parasitic heat flows inside the cryostat. The heat flux and the temperature distribution over the window surface are calculated based on the thermal conductivity Equation (3):

$$\frac{\partial}{\partial x}\frac{k\partial T}{\partial x} + \frac{\partial}{\partial y}\frac{k\partial T}{\partial y} + \frac{\partial}{\partial z}\frac{k\partial T}{\partial z} + q(x,y,z) = C_p\rho\frac{\partial T}{\partial \tau}, \tag{8}$$

where x, y, z are the Cartesian coordinates, q is the heat source, C_p is the heat capacity, and ρ is the thermal conductivity of the material. As a result, it was found that under certain conditions, the temperature of the central zone of the window can differ quite significantly from the ambient temperature (by 5–15 K). At the same time, it is very likely that the temperature of the outer surface will be below the dew point for a certain range of ambient temperature and humidity, which will inevitably lead to the precipitation of water condensate, and that is extremely undesirable due to the significant absorption of water waves in the 0.1–1 THz range. The presence of an infrared filter [69] between the cold receiver input and the window improves the picture significantly. The infrared filter, which is a layered or loose structure fixed on the radiation screen of the cryostat, absorbs most of the radiation from the heated environment lying outside the band of received signals. Along with undesirable condensation, radiation cooling of the window has positive consequences: the colder central zone of the window, through which the received signal mainly passes, has slightly lower temperatures and losses. As a result, the overall noise temperature of the receiver decreases. Based on the obtained temperature distribution, in [68] the transfer equation is solved and a refined noise temperature is obtained, which is 5–7% less than the calculated one without cooling for the implemented configurations of windows and receivers.

Another important feature when designing a cryogenic system is that the larger the optical window is, the greater the thermal load becomes. For example, with a window diameter of about 100 mm and a background load of 300 K, the external power is about 9.5 watts. Various combinations of radiation filters, such as Zaytex, Fluorogold, HDPE, etc., are used for the solution.

The authors of the article have achieved success in the installation and operation of hermetic optical windows having various diameters (including "extra-large" ones) and made of various materials. Some examples are given in Figure 27.

Figure 27. Photos and characteristics (diameter and material) of optical windows.

4.3.2. Dissipative Transmission Line at the Temperature Drop

In addition to the window, a specific element of the cryogenic receiver is an incoming thermo-decoupling feeder, which, however, is sometimes absent if the horn of the antenna irradiator is placed inside the cryostat. The thermally coupled waveguide in the millimeter and submillimeter wavelength ranges is usually supersized and sometimes combines the function of transition to the main section of the waveguide. Here, we also solved a problem similar to the one discussed in the previous section. Specifically, the temperature distribution profile along the length of the line is determined on the basis of the thermal conductivity equation [68]. Knowing that the loss factor depends on the conductivity and the size of the waveguide, we can determine the attenuation constant as the product of the factor by the linear function $D(\lambda)$ of the working wavelength. The calculated distributions of the temperature and specific losses along the length of the line allow us to solve the transfer equation uniquely [70] for it and calculate its noise temperature (9):

$$T_n = \int_0^X T_a l(x) exp\left(\int_0^X l(\xi)d\xi dx \right), \tag{9}$$

where 0 and X are the coordinates of the waveguide ends, $T_a = T_a(x)$ is the law of distribution of the physical temperature aiong the length of the waveguide established from the solution of the equation of thermal conductivity, and $l(x)$ is the law of distribution of losses in a waveguide of a given section.

Another issue is the definition of the three-dimensional heat flux in a waveguide active device. The equation of thermal conductivity with allowance for the internal heat source was solved to estimate the temperature of an active device placed in a waveguide chamber. This configuration is typical for converters and amplifiers of millimeter and submillimeter waves. Consider the three-dimensional problem of thermal conductivity for a model of a cubic device cooled from one of the faces ($T|_{x=L} = T_0$) and located in vacuum without heat transfer along the remaining faces ($dT/dx|_{x=0} = 0$; $dT/dy|_{y=0,L} = 0$; $dT/dz|_{z=0,L} = 0$), with an internal heat source q having a heat dissipation characteristic of a diode (transistor) and characteristic dimensions l being approximately 2 orders of magnitude smaller than the external size L of the device (waveguide chamber). In this case, the distribution function of the source along the x coordinate has the form:

$$Q(x) = f(x) = \begin{cases} 0, & x - \frac{L}{2} > l/2 \\ q/l^3, & x - \frac{L}{2} \leq l/2 \end{cases} \tag{10}$$

For the rest of the coordinates, the distribution of the heat source is recorded similarly. Let us write the function

$$U_{mnk}(x,y,z) = \cos\left(\frac{m\pi}{L}y\right)\cos\left(\frac{n\pi}{L}z\right)\cos\left(\frac{2k+1}{L}\pi x\right) \tag{11}$$

satisfying the boundary conditions described above. Here, we introduce a variable t equal to the difference between the physical temperature and the temperature of the cooled wall, coupled with a cooler of conditionally infinite cooling capacity, while $t|_{x=L} = 0$ and $dt/dn = 0$. We define the Laplace operator of the Function (11)

$$\Delta U_{mnk} = -\frac{\pi^2}{L^2}\left[m^2 + n^2 + \left(\frac{2k+1}{2}\right)^2 \right] U_{mnk} \tag{12}$$

The solution of the thermal conductivity equation can be written as:

$$V(x,y,z) = \sum_{m,n,k=0}^{\infty} V_{mnk} U_{mnk}(x,y,z), \tag{13}$$

where

$$V_{mnk} = (2/L)^3 \int_0^L \int_0^L \int_0^L U_{mnk}(x,y,z)V(x,y,z)dxdydz. \qquad (14)$$

Solving the system of Equations (12)–(14) with respect to t, one can find the temperature of the source at the center of the cube analytically, without using numerical methods:

$$T_c = t + T_0 = t + N_f q/(kl) \qquad (15)$$

Here, N_f is the the form factor. For the "small cube in a large cube" model with a 1:100 aspect ratio, the form factor is $N_f = 0,19$.

Calculation (15) shows that the overheating of a standard planar millimeter-wave Shottky barrier diode in vacuum can reach 10–13 K, with heating up to 10 K due to the contribution of the planar structure itself.

5. Conclusions

This paper provides a brief overview of the principles of operation, basic designs, and features of the design and manufacture of cryogenic systems of various temperature levels used to cool detection devices of various frequency ranges for astronomy and other applications. The specific features of the cryodesign of such systems are considered: methods for thermal calculations of cryostats of various temperature levels for astronomical receivers were proposed, the fluctuation characteristics of refrigerators and their vibration were studied, measures were proposed to reduce the influence of these negative processes on the characteristics of the receivers, and original designs of optical inputs to cryostats with extremely low losses of the useful signal were proposed. Also, some transfer issues of radiation in a signal transmission channel that is nonstationary in temperature were considered. The main result of the article is the presented development of an extensive range of cryogenic astronomical equipment for different observatories. Based on these developments of both optical (photonic) and radio-astronomical (electronic) equipment, a new project by the authors of the article is being built to develop receiving equipment in the intermediate terahertz range for new telescopes that are expected to be built in northeastern Eurasia (the Caucasus and Siberia in Russia [71], Uzbekistan [72], and China [73]), or, for example, within the framework of the EHT project expansion program [74]. These observatories will be certainly equipped with cryogenic receivers based on the experience described in the article, and there is a close collaboration with international teams involved in the development of the above-listed observatories. The work not only summarizes the results of development of a series of cryogenic receivers for various observatories, but also presents a set of original technical solutions for both the cryogenic systems themselves and various interfaces that provide efficient cooling with a minimum of negative impacts of cryogenics on the operation of the receivers.

The work can also serve as a kind of a manual for designers of cryogenic systems for astronomical receivers of various wavelengths.

Author Contributions: Conceptualization, V.V.; methodology, Y.B., V.E., M.T., A.C. and S.S.; software, O.B. and I.L. (Ilya Lesnov); validation, M.E. and A.S.; formal analysis, M.M. (Mariya Markina); investigation, A.E., A.G., A.K., A.V., I.L. (Igor Lapkin)., S.S., M.M. (Mariya Mansfeld), M.M. (Mariya Markina); N.T. and E.P.; data curation, I.L. (Ilya Lesnov); writing—original draft preparation, A.G. and V.V.; writing—review and editing, V.V., A.G., A.V., M.E. and S.S.; supervision, V.V.; project administration, A.V.; funding acquisition, Y.B. and V.V. All authors have read and agreed to the published version of the manuscript.

Funding: This work is supported by the grant of the Russian Science Foundation # 23-79-00006.

Institutional Review Board Statement: Not applicable.

Informed Consent Statement: Not applicable.

Data Availability Statement: New scientific data obtained as part of the presented research are presented in the scientific report on the RSF project # 23-79-00006for 2023 and can be obtained on the website of the Russian Science Foundation https://rscf.ru/en/.

Acknowledgments: Cryogenic systems were manufactured on the basis of UNU No. 3589084 "CKP-7—Center for Microwave Research of Materials and Substances".

Conflicts of Interest: The authors declare no conflicts of interest.

References

1. Ladu, A.; Schirru, L.; Gaudiomonte, F.; Marongiu, P.; Angius, G.; Perini, F.; Vargiu, G.P. Upgrading of the L-P Band Cryogenic Receiver of the Sardinia Radio Telescope: A Feasibility Study. *Sensors* **2022**, *22*, 4261. [CrossRef]
2. Murzin, V.A.; Markelov, S.V.; Ardilanov, V.I.; Afanaseva, T.; Borisenko, A.; Ivashchenko, N.; Pritychenko, M.; Mitiani, G.; Vdovin, V. Astronomical CCD systems for the 6-meter telescop BTA (a review). *Appl. Phys.* **2016**, *4*, 500–506. (In Russian)
3. Vdovin, V.F. Problems of cryogenic cooling of superconductor and semiconductor receivers in the range 0.1–1 THz. *Radiophys. Quantum Electron.* **2005**, *48*, 779–791. [CrossRef]
4. Ulbricht, G.; De Lucia, M.; Baldwin, E. Applications for Microwave Kinetic Induction Detectors in Advanced Instrumentation. *Appl. Sci.* **2021**, *11*, 2671. [CrossRef]
5. Alzhanov, B.; Shafiee, M.; Kazykenov, Z.; Bekbolanova, M.; Smoot, G.; Grossan, B. The cryogenic detector for cosmology observation. *IOP Conf. Ser. Mater. Sci. Eng.* **2019**, *502*, 012060. [CrossRef]
6. 27th International Cryogenics Engineering Conference and International Cryogenic Materials Conference 2018 (ICEC-ICMC 2018) 3–7 September 2018, Oxford, United Kingdom. Available online: https://iopscience.iop.org/article/10.1088/1757-899X/502/1/011001/pdf (accessed on 24 November 2023).
7. Han, Y.; Zhang, A. Cryogenic technology for infrared detection in space. *Sci. Rep.* **2022**, *12*, 2349. [CrossRef] [PubMed]
8. Benford, D.J.; Moseley, S.H. *Cryogenic Detectors for Infrared Astronomy: The Single Aperture Far-InfraRed (SAFIR) Observatory*; NASA/GSFC, Infrared Astrophysics, Code 685: Greenbelt, MD, USA, 2004.
9. Archer, J.W. High-performance, 2.5-K cryostat incorporating a 100–120-GHz dual polarization receiver. *Sci. Instrum.* **1985**, *56*, 449–458. [CrossRef]
10. Lang, A.; De Bernardis, P.; De Petris, M.; Masi, S.; Melchiorri, F.; Aquilini, E.; Martinis, L.; Scaramuzzi, F.; Melchiorri, B.; Boscaleri, A.; et al. The BOOMERANG experiment. *Space Sci. Rev.* **1995**, *74*, 145–150. [CrossRef]
11. Little, W.A. Microminiature refrigeration. *Rev. Sci. Instrum.* **1984**, *55*, 661–680. [CrossRef]
12. Edelman, V.S. *Near the Absolute Zero*; Science: Moscow, Russia, 1983; 176p. (In Russian)
13. Space Cryogenics Workshop. 2021. Available online: https://www.sciencedirect.com/journal/cryogenics/special-issue/10KMQLLLNS1 (accessed on 24 November 2023).
14. Astronomical Journal. Available online: https://journals.aas.org/astronomical-journal/ (accessed on 24 November 2023).
15. Messenger, G.C. Cooling of Microwave Crystal Mixers and Antennas. *Trans. Microw. Theory Tech.* **1957**, *5*, 62–63. [CrossRef]
16. Timmerhaus, K.D. Advances in Cryogenic Engineering. In Proceedings of the 1959 Cryogenic Engineering Conference, Berkeley, CA, USA, 2–4 September 1959.
17. White, G.K. *Experimental Techniques in Low-Temperature Physics*, 3rd ed.; Clarendon Press: Oxford, UK, 1979; 331p.
18. Barron, R.F. *Cryogenic Systems*, 2nd ed.; Oxford University Press: Oxford, UK, 1985; 507p.
19. Köhler, J.W.L.; Jonkers, C.O. Fundamentals of the gas refrigeration machine. *Philips Tech. Rev.* **1954**, *16*, 69–78.
20. 33rd International Symposium on Space Terahertz Technology (Coming). Available online: https://web.cvent.com/event/ed54f48e-ce7d-4361-8d7f-7c2e997215a6/summary (accessed on 24 November 2023).
21. 32nd International Symposium on Space Terahertz Technology (The Last). Available online: https://www.showsbee.com/fairs/IEEE-ISSTT.html (accessed on 24 November 2023).
22. 30th Space Cryogenics Workshop. Available online: https://spacecryogenicsworkshop.org (accessed on 24 November 2023).
23. Wang, C. *Cryogenic Engineering and Technologies. Principles and Applications of Cryogen-Free Systems*; Zhao, Z., Ed.; Taylor & Francis Group, LLC.: Abingdon, UK, 2020; ISBN 978-1-4987-6576-3.
24. Jha, A.R. *Cryogenic Technology and Applications*; Elsevier: Amsterdam, The Netherlands, 2006; ISBN 13: 978-07506-7887-2. ISBN 10: 0-7506-7887-9.
25. Ventura, G.; Risegari, L. *The Art of Cryogenics: Low-Temperature Experimental Techniques*; Elsevier: Amsterdam, The Netherlands, 2008; ISBN 978-0-08-0444796.
26. Cryogenic materials. In *Handbook of Brookhaven National Laboratory*; Brookhaven National Laboratory: New York, NY, USA, 1966; p. 11973.
27. Malkov, M.P.; Danilov, I.B.; Zeldovich, A.G.; Fradkov, A.B. *Handbook on the Physical and Technical Foundations of Cryogenics*, 3rd ed.; Energoatomizdat: Moscow, Russia, 1985; p. 432.
28. Martiyanov, K.; Makhalov, V.; Turlapov, A. Observation of a Two-Dimensional Fermi Gas of Atoms. *Phys. Rev. Lett.* **2010**, *105*, 030404. [CrossRef] [PubMed]
29. Quinn, T.J. *Temperature*, 2nd ed.; Academic Press: Cambridge, MA, USA, 1990; ISBN 0-12-569681-7.
30. Hill, K.D.; Steele, A.G. The International Temperature Scale: Past, Present, and Future. *J. Meas. Sci.* **2014**, *9*, 60–67. [CrossRef]

31. Nicastro, A.J. Dalton's temperature scale. *Phys. Teach.* **1983**, *21*, 6. [CrossRef]
32. Masuda, T.; Hiramoto, A.; Ang, D.G.; Meisenhelder, C.; Panda, C.D.; Sasao, N.; Uetake, S.; DeMille, D.P.; Doyle, J.M.; Gabrielse, G.; et al. High-sensitivity low-noise photodetector using a large-area silicon photomultiplier. *Opt. Express.* **2023**, *31*, 1943–1957. [CrossRef]
33. Tan, G.H. Atacama Large Millimeter Array. In Proceedings of the 3rd ESA Workshop on Millimeter Wave Technology and Applications: Circuit, Systems, and Meas-urement Technique, Espoo, Finland, 21–23 May 2003; p. 107.
34. Vdovin, V.F.; Grachev, V.G.; Dryagin, S.Y.; Eliseev, A.I.; Kamaletdinov, R.K.; Korotaev, D.V.; Lesnov, I.V.; Mansfeld, M.A.; Pevzner, E.L.; Perminov, V.G.; et al. Cryogenically cooled low-noise amplifier for radio-astronomical observations and centimeter-wave deep-space communications systems. *Astrophys. Bull.* **2016**, *71*, 134–138. [CrossRef]
35. Raisanen, A.V. *Experimental Studies Cooled Millimeter Wave Mixers*; Acta Polytechnica Scandinavica, Electrical Engineering Series 46; Helsinki University of Technology: Espoo, Finland, 1980.
36. Weiner, S.; Pospieszalsky, M.R.; Norrod, R. Cryogenic HEMT low-noise receiver for 1.3–43 GHz. *IEEE Trans. Microw. Theory Tech.* **1987**, *MTT-35*, 1067–1069.
37. Zimmuidzinas, J.; Richards, P.I. Superconducting Detectors and Mixers for Millimeter and Submillimeter. *Proc. IEEE* **2004**, *92*, 1597–2006. [CrossRef]
38. Rudakov, K.I.; Khudchenko, A.V.; Richards, P.I.; Filippenko, L.V.; Paramonov, M.E.; Hesper, R.; Ronsó da Costa Lima, D.D.; Baryshev, A.M.; Koshelets, V.P. THz Range Low-Noise SIS Receivers for Space and Ground-Based Radio Astronomy. *Appl. Sci.* **2021**, *11*, 10087. [CrossRef]
39. Cryogenic Technologies for the Stratospheric Observatory for Infrared Astronomy (SOFIA) Science Instruments. Available online: https://ntrs.nasa.gov/api/citations/20205008777/downloads/20205008777.pdf (accessed on 24 November 2023).
40. Balega, Y.; Bolshakov, O.; Chernikov, A.; Edelman, V.; Eliseev, A.; Emelyanov, E.; Gunbina, A.; Krasilnikov, A.; Lesnov, I.; Mansfeld, M.; et al. Cryogenic Systems for Astronomical Research in the Special Astrophysical Observatory of the Russian Academy of Science. *Photonics* **2023**, *10*, 1263. [CrossRef]
41. About ALMA, at First Glance. Available online: https://www.almaobservatory.org/en/about-alma/ (accessed on 24 November 2023).
42. May, A.J. Sub-Kelvin Cryogenics for Experimental Cosmology. In Proceedings of the 28th International Cryogenic Engineering Conference and International Cryogenic Materials Conference 2022—ICEC28-ICMC 2022, Hangzhou, China, 25–29 April 2022; Volume 70, pp. 240–248.
43. Vystavkin, A.N.; Shitov, S.V.; Bankov, S.E.; Kovalenko, A.G.; Pestriakov, A.V.; Cohn, I.A.; Uvarov, A.V.; Vdovin, V.F.; Perminov, V.G.; Trofimov, V.N.; et al. High sensitive 0.13–0.38 THz TES array radiometer for the big telescope azimuthal of Special Astrophysical Observatory of Russian Academy of Sciences. In Proceedings of the 2007 Joint 32nd International Conference on Infrared and Millimeter Waves and the 15th International Conference on Terahertz Electronics, Cardiff, UK, 2–9 September 2007.
44. Chernikov, A.N.; Trofimov, V.N. Helium-3 adsorption refrigerator cooled with a closed-cycle cryocooler. *J. Surf. Investig. X-ray Synchrotron Neutron Tech.* **2014**, *8*, 956–960. [CrossRef]
45. Camus, P.; Vermeulen, G.; Volpe, A.; Triqueneaux, S.; Benoit, A.; Butterworth, J.; d'Escrivan, S.; Tirolien, T. Status of the Closed-Cycle Dilution Refrigerator Development for Space Astrophysics. *J. Low Temp. Phys.* **2013**, *176*, 5–6. [CrossRef]
46. Frossati, G. Experimental techniques: Methods for cooling below 300 mK. *J. Low Temp. Phys.* **1992**, *87*, 595–663. [CrossRef]
47. Benoit, A.; Pujol, S. A dilution refrigerator insensitive to gravity. *Phys. B Condens. Matter* **1991**, *169*, 457–458. [CrossRef]
48. Benoit, A.; Caussignac, M.; Pujol, S. New types of dilution refrigerator and space applications. *Phys. B* **1994**, *197*, 48–53. [CrossRef]
49. Sirbi, A.; Pouilloux, B.; Benoit, A.; Lamarre, J.-M. Influence of the astrophysical requirements on dilution refrigerator design. *Cryogenics* **1999**, *39*, 665–669. [CrossRef]
50. Volpe, A. Development of a Closed Cycle Dilution Refrigerator for Astrophysical Experiments in Space. Ph.D. Thesis, Universite de Grenoble, Saint-Martin-d'Hères, France, 2014.
51. Chaudhry, G.; Volpe, A.; Camus, P.; Triqueneaux, S.; Vermeulen, G.-M. A closed-cycle dilution refrigerator for space applications. *Cryogenics* **2012**, *52*, 471–477. [CrossRef]
52. Haziot, A.; Vermeulen, G. *A New Type of Mixing Chamber for a Zero Gravity Dilution Refrigerator*; CEC-ICMC и: Hartford, CT, USA, 2019.
53. Oxford Instruments Group/Cryogenics. Available online: https://nanoscience.oxinst.com/cryogenics (accessed on 24 November 2023).
54. Blueforce. Available online: https://bluefors.com (accessed on 24 November 2023).
55. Masi, S.; de Bernardis, P.; De Troia, G.; Giacometti, M.; Iacoangeli, A.; Piacentini, F.; Polenta, G.; Ade, P.A.R.; Mauskopf, P.D.; Bock, J.J.; et al. The BOOMERanG experiment and the curvature of the universe. *Prog. Part. Nucl. Phys.* **2002**, *48*, 243–261. [CrossRef]
56. Nati, F.; Ade, P.; Boscaleri, A.; Brienza, D.; Calvo, M.; Colafrancesco, S.; Conversi, L.; de Bernardis, P.; De Petris, M.; Delbart, A.; et al. The OLIMPO experiment. *New Astron. Rev.* **2007**, *51*, 385–389. [CrossRef]
57. Masi, S.; de Bernardis, P.; Paiella, A.; Piacentini, F.; Lamagna, L.; Coppolecchia, A.; Ade, P.A.R.; Battistelli, E.S.; Castellano, M.G.; Colantoni, I.; et al. Kinetic Inductance Detectors for the OLIMPO experiment: In-flight operation and performance. *J. Cosmol. Astropart. Phys.* **2019**, *2019*. [CrossRef]
58. Nati, L.; Calvo de Bernardis, P.; Fiadino, P.; Lamagna, L.; Masi, S.; Piacentini, R.; Rispoli, R. Cryogenic systems for Long Duration Balloon experiments. *Mem. Della Soc. Astron. Ital.* **2008**, *79*, 878–882.

59. Cold Facts Issued by the Cryogenic Society of America, Inc. Available online: https://www.cryogenicsociety.org/cold-facts#:~:text=Cold%20Facts,%20an%20internationally%20recognized,%20every%20member%20of%20the%20CSA (accessed on 24 November 2023).

60. Yusov, A.V.; Kozlov, S.A.; Ustinova, E.A.; Arkhipov, M.Y.; Kostrov, E.A.; Smirnov, A.V.; Kuznetsov, I.V.; Mansfeld, M.A.; Tyatushkin, N.V.; Kovalev, F.N.; et al. Testing high-precision electromechanical actuators used for adjustment of deployable antennas of astronomy space missions. *Cryogenics* **2021**, *118*, 103346. [CrossRef]

61. Parshin, V.V.; Serov, E.A.; Bubnov, G.M.; Vdovin, V.F.; Koshelev, M.A.; Tretyakov, M.Y. Cryogenic resonator complex. *Radiophys. Quantum Electron.* **2014**, *56*, 554–560. [CrossRef]

62. Gunbina, A.A.; Serov, E.A.; Mineev, K.V.; Parshin, V.V.; Vdovin, V.F.; Chekushkin, A.M.; Khan, F.V.; Koshelets, V.P. Experimental Study of the Reflectivity of Superconducting Nb-Based Films in the Subterahertz Frequency Band. *Radiophys. Quantum Electron.* **2022**, *65*, 471–480. [CrossRef]

63. Serov, E.A.; Parshin, V.V.; Bubnov, G.M. Reflectivity of metals in the millimeter wavelength range at cryogenic temperatures. *IEEE Trans. Microw. Theory Tech.* **2016**, *64*, 3828–3838. [CrossRef]

64. Parshin, V.V.; Serov, E.A.; Bubnov, G.M.; Vdovin, V.F.; Nikolenko, A.S.; Lesnov, I.V.; Gunbina, A.A.; Smirnov, A.V.; Malginov, V.A.; Dolzenko, D.E.; et al. Terahertz Reflectivity of $YBa_2Cu_3O_{7-\delta}$ at Cryogenic Temperatures. *IEEE Trans. Appl. Supercond.* **2020**, *30*, 9001705. [CrossRef]

65. Gardner, D.V. Does Your CCD Camera Need Cooling? Available online: https://www.photonics.com/Articles/Does_Your_CCD_Camera_Need_Cooling/a12810 (accessed on 24 November 2023).

66. Cease, H.; DePoy, D.; Derylo, G.; Diehl, H.T.; Estrada, J.; Flaugher, B.; Kuk, K.; Kuhlmann, S.; Lathrop, A.; Schultz, K.; et al. Cooling the dark energy camera CCD array using a closed-loop two-phase liquid nitrogen system. In *Modern Technologies in Space and Ground-Base Telescopes and Instrumentation*; SPIE: Bellingham, DC, USA, 2010; p. 77393N.

67. Afanasieva, I.; Murzin, V.; Ardilanov, V.; Ivaschenko, N.; Pritchenko, M.; Moiseev, A.; Shablovinskaya, E.; Malginov, E. Astronomical Camera Based on a CCD261-84 Detector with Increased Sensitivity in the Near-Infrared. *Photonics* **2023**, *10*, 774. [CrossRef]

68. Vdovin, V.F.; Korotaev, D.V.; Lapkin, I.V. *Collection of Abstracts of the 11th International School of Microwave Radiophysics and RF Electrodynamics*; IAP PAS: Nizhniy Novgorod, Russia, 1993; p. 21.

69. Lamb, W. Miscellaneous data on materials for millimetre and submillimetre optics. *Int. J. Infrared Millim. Wave* **1993**, *14*, 951. [CrossRef]

70. Siegel, R.; Howell, J.R. *Thermal Radiation. Heat Transfer*; McGraw-Hill Book Company: New-York, NY, USA, 1972.

71. Marchiori, G.; Rampini, F.; Tordi, M.; Spinola, M.; Bressan, R. Towards the Eurasian Submillimeter Telescope (ESMT): Telescope concept outline and first results. In *Proceedings of the Ground-Based Astronomy in Russia. 21st Century*; Romanyuk, I.I., Yakunin, I.A., Valeev, A.F., Kudryavtsev, D.O., Eds.; Special Astrophysical Observatory: Nizhnii Arkhyz, Russia, 2020; pp. 378–383.

72. Hojaev, A.; Shanin, G.I.; Artyomenko, Y.N. Suffa Radio Observatory in Uzbekistan: Progress and radio-seeing research plans. *Proc. Int. Astron. Union* **2007**, *5*, 177–182. [CrossRef]

73. Wang, N. Xinjiang Qitai 110 m radio telescope. *Sci. Sin. Phys. Mech. Astron.* **2014**, *44*, 783–794. [CrossRef]

74. Doeleman, S.S.; Barrett, J.; Blackburn, L.; Bouman, K.; Broderick, A.E.; Chaves, R.; Fish, V.L.; Fitzpatrick, G.; Fuentes, A.; Freeman, M.; et al. Reference Array and Design Consideration for the next-generation Event Horizon Telescope. *arXiv* **2023**, arXiv:2306.08787. [CrossRef]

Article

SAINT (Small Aperture Imaging Network Telescope)—A Wide-Field Telescope Complex for Detecting and Studying Optical Transients at Times from Milliseconds to Years

Grigory Beskin [1,2,*], Anton Biryukov [2,3,4], Alexey Gutaev [1,2], Sergey Karpov [5], Gor Oganesyan [6,7], Gennady Valyavin [1], Azamat Valeev [1], Valery Vlasyuk [1], Nadezhda Lyapsina [1] and Vyacheslav Sasyuk [2]

[1] Special Astrophysical Observatory, Russian Academy of Sciences, Nizhniy Arkhyz 369167, Russia; algutaev@sao.ru (A.G.); gvalyavin@sao.ru (G.V.)
[2] Engelhardt Observatory, Kazan Federal University (KFU), Kazan 420008, Russia
[3] Sternberg Astronomical Institute, Moscow State University, Moscow 119992, Russia
[4] Faculty of Physics, HSE University, Moscow 101000, Russia
[5] Central European Institute for Cosmology and Fundamental Physics (CEICO), Institute of Physics, Czech Academy of Sciences, 182 21 Prague, Czech Republic; karpov@fzu.cz
[6] Gran Sasso Science Institute, 67100 L'Aquila, Italy
[7] Istituto Nazionale di Fisica Nucleare (INFN)—Laboratori Nazionali del Gran Sasso, 67100 L'Aquila, Italy
*   Correspondence: beskin@sao.ru

Citation: Beskin, G.; Biryukov, A.; Gutaev, A.; Karpov, S.; Oganesyan, G.; Valyavin, G.; Valeev, A.; Vlasyuk, V.; Lyapsina, N.; Sasyuk, V. SAINT (Small Aperture Imaging Network Telescope)—A Wide-Field Telescope Complex for Detecting and Studying Optical Transients at Times from Milliseconds to Years. *Photonics* **2023**, *10*, 1352. https://doi.org/10.3390/photonics10121352

Received: 28 October 2023
Revised: 30 November 2023
Accepted: 1 December 2023
Published: 7 December 2023

Abstract: In this paper, we present a project of a multi-channel wide-field optical sky monitoring system with high temporal resolution—Small Aperture Imaging Network Telescope (SAINT)—mostly built from off-the-shelf components and aimed towards searching and studying optical transient phenomena on the shortest time scales. The instrument consists of twelve channels each containing 30 cm (F/1.5) GENON Max objectives mounted on separate ASA DDM100 mounts with pointing speeds up to 50 deg/s. Each channel is equipped with a 4128×4104 pixel Andor Balor sCMOS detector and a set of photometric *griz* filters and linear polarizers. At the heart of every channel is a custom-built reducer-collimator module allowing rapid switching of an effective focal length of the telescope—due to it the system is capable of operating in either wide-field survey or narrow-field follow-up modes. In the first case, the field of view of the instrument is 470 square degrees (39 sq.deg. for a single channel) and the detection limits (5σ level at 5500 Å) are 12.5, 16.5, 19, 21 with exposure times of 20 ms, 1 s, 30 s and 20 min, correspondingly. In the second, follow-up (e.g., upon detection of a transient of interest by either a real-time detection pipeline, or upon receiving an external trigger) regime, all telescopes are oriented towards the single target, and SAINT becomes an equivalent to a monolithic 1-meter telescope, with the field of view reduced to $11' \times 11'$, and the exposure times decreased down to 0.6 ms (1684 frames per second). Different channels may then have different filters installed, thus allowing a detailed study—acquiring both color and polarization information—of a target object with the highest possible temporal resolution. The telescopes are located in two pavilions with sliding roofs and are controlled by a cluster of 25 computers that both govern their operation and acquire and store up to 800 terabytes of data every night, also performing its real-time processing using a dedicated fast image subtraction pipeline. Long-term storage of the data will require a petabyte class storage. The operation of SAINT will allow acquiring an unprecedented amount of data on various classes of astrophysical phenomena, from near-Earth to extragalactic ones, while its multi-channel design and the use of commercially available components allows easy expansion of its scale, and thus performance and detection capabilities.

Keywords: optical observations; wide-field instruments; photometry; transients; gamma-ray bursts; fast radio bursts; meteors; red dwarfs; white dwarfs

1. Introduction

A special place among the studied astronomical phenomena belongs to non-stationary ones. In a sense, all astronomy is the science of change (evolution) of the Universe with time, both as a whole and its parts of various scales, from meteors to clusters of galaxies. It is no coincidence that in recent years a new direction has been formed and is rapidly developing—astronomy in the time domain ("Time domain astronomy"). It combines methods, tools and ideas focused on the study of non-stationary phenomena in the Universe at different time and space scales. The website of the IAU working group (URL: http://timedomainastronomy.net/resources.html#master (accessed on 1 December 2023), Master list of Time Domain Surveys section) provides information on 62 instruments that investigate (or have investigated) variable objects. As long as the list of telescopes is, so is the range of specific tasks for which these instruments are intended. This includes observations of meteors, comets and asteroids, artificial satellites of the Earth, space debris, searches for the effects of microlensing and transits of exoplanets, the study of variable stars, studies of optical afterglows of gamma-ray bursts and searches for optical flares synchronous with these bursts, searches for supernovae, and the study of brightness variations of galactic nuclei, search and characterization of optical counterparts of fast radio bursts and gravitational wave signals. The latter direction is all the more important and exceptionally interesting in connection with the discovery of the gravitational wave event GW170817 [1], caused by the merger of two neutron stars and the associated short gamma-ray burst GRB 170817A [2–4], the kilonova emission [5,6] and subsequently with a long multi-wavelength afterglow [7–12]. At the same time, there are considerations that precursors of short gamma-ray bursts can precede gravitational-wave pulses [13], thereby marking the region of their localization, which gives hope for searching for electromagnetic radiation synchronously with the pulse itself. When searching for variability, as a rule, two modes of observation are implemented (we note a certain conventionality of such a division): survey and monitoring. In the first case, the radiation of relatively large areas of the sky is periodically recorded, while the ratio between the time of their exposure (determining the depth of the survey) and the size of the fraction of the celestial sphere observed during the night is determined by the nature of the required variability. Some of these projects and tools are SDSS [14]; Sky Mapper [15]; QUEST-La Silla survey [16]; ASAS-SN [17]; Catalina Surveys [18]; ZTF [19], LSST under construction [20]. In the second observation mode, we are talking about the longest possible registration of the radiation of a certain object in anticipation of changes in its characteristics (intensity, spectrum, polarization, etc.). An outstanding example of this kind of TAOS II program is the search for trans-Neptunian planets by their occultations of 10,000 stars, whose long-term simultaneous monitoring of high temporal resolution (exposure—50 ms) in individual subapertures is carried out using three telescopes with a diameter of 1.3 m [21]. Another classic example is the monitoring of UV Cet-type stars to study their flare activity using the largest telescopes. In particular, subsecond polarized spikes of synchrotron origin were first detected during a powerful UV Cet flare with the 6-meter telescope of SAO RAS [22]. In essence, both modes consist of monitoring, spatiotemporal in the first case, and temporal in the second.

A special place among the non-stationary phenomena studied within the framework of Time Domain Astronomy is occupied by unexpectedly appearing (and also suddenly disappearing) objects, whose localization in space and/or time is not known in advance, and whose duration is rather short—the so-called transients (transient sources). They manifest themselves as stochastic fluctuations of electromagnetic radiation of different frequencies (from radio to gamma-rays), low (tens of MeV) and high (TeV-PeV) energy neutrino events, cosmic rays, and gravitational waves. It should be emphasized that despite the variety of forms of energy release in these events, the possibilities of understanding their nature and constructing their models are determined to a certain extent by the detection and study of their manifestations in the optical range—optical transients. These include non-stationary phenomena in the near-Earth space and the Earth's (exoplanet's?) atmosphere, non-moving phenomena such as auroras, transient luminous events (TLE)—elves, sprites, jets [23], and moving phenomena—comets, asteroids, meteors, satellites, space debris. Finally, undoubtedly, such transients can be optical flashes (including periodic) of artificial origin—signals (transmissions) of extraterrestrial civilizations [24,25]. The characteristic duration of these phenomena ranges from minutes-hours to weeks-months (transits of exoplanets, outbursts of novae, supernovae, variable stars, microlensing effects) and from milliseconds to tens of seconds (gamma and radio bursts, gravitational waves, TLE, flyby meteors and satellites). In essence, we are talking about two different types of transients—long and short—the detection and study of which require the use of different methods and tools. At the same time, within the framework of the prevailing ideas and conceptual apparatus, events of the first type are often referred to as fast transients. For example, one of the most extensive modern programs "Wider Faster Deeper" [26], which unites about 40 ground and space instruments and is dedicated to the study of all of the fastest transients, uses the follow-up mode in the optical range (repointing of the instruments to the area of the already detected fast transients). With the use of the 4-meter Blanco telescope and DECam, at an exposure of 20 s, information on the subsecond fast transients is completely lost, but their environment and slow transients are successfully studied—gamma-ray burst afterglows, localization regions of fast radio bursts and gravitational wave events, bursts of novae, supernovae, red dwarfs [27].

There is a real contradiction between the need to use instruments with an extremely high temporal resolution in the search and study of fast optical transients and the real characteristics of telescopes, both existing and under construction. Figure 1 and Table A1 from Appendix A demonstrate this perfectly. We present the characteristics of 30 survey instruments focused on the search and study of non-stationary objects of various types, whose fields of view exceed 4 square degrees.

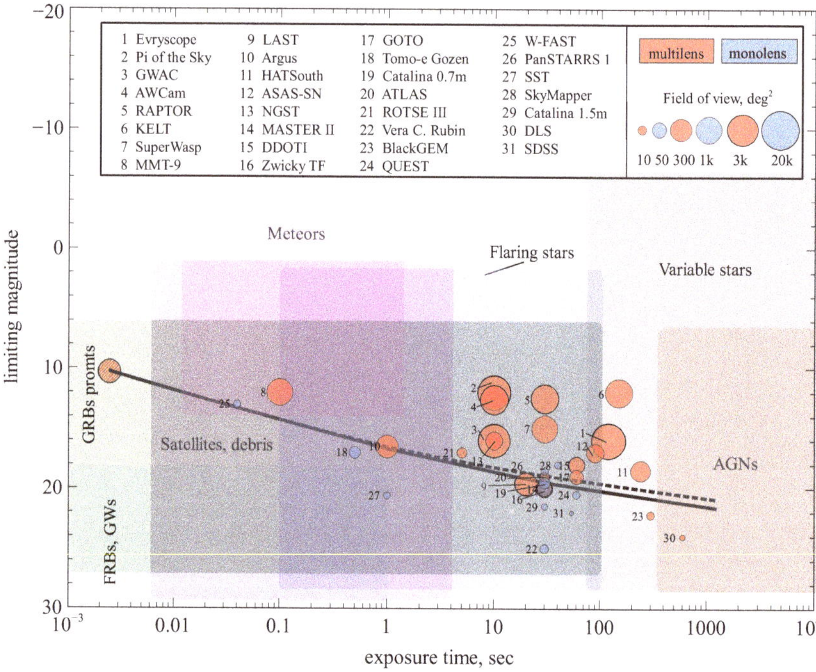

Figure 1. Exposure times and detection limits of various sky surveys mentioned in the text and Table A1. Black solid and dashed lines represent the theoretical detection limits for the SAINT complex in follow-up and monitoring regimes, respectively. Shaded circle represents the full field of view of SAINT.

The minimum exposure duration exceeds 10 s for the vast majority of these telescopes, which, given the possibility of summing up a sequence of individual frames, fully provides a depth sufficient for detecting and studying variability up to magnitudes 19–25. At the same time, these limits for systems with close fields of view may differ significantly due to the difference in the diameters of the mirrors. Note, that most of the telescopes in our list are multi-element systems. Indeed, to implement wide-angle spatial-temporal monitoring with a single aperture, a combination of initially mutually contradictory conditions is necessary—a sufficiently high detection limit (large diameter of the supply optics) and a wide field of view (short focus). This circumstance significantly limits the set of possible optical schemes of instruments that determine the optimum of such a combination. So, with an aperture diameter of D > 1–2 m, the size of the field of view is 2–3 degrees, and with D < 1 it can reach 5–10 degrees [28,29]. This implies the need to move to multi-telescope systems with fields of view of hundreds and thousands of square degrees, which also have a number of other advantages—relative cheapness, the ability to change the configuration, and (most importantly), since the dimensions of the detectors can be quite small, the use of high-temporal resolution instruments. [29,30]. However, as can be seen from Figure 1 and Table A1 of Appendix A, only one multi-aperture system is currently being developed—Argus [31], which has a time resolution of 0.05 s. It consists of 38 (as a starting point, with at least 900 planned in the future) 20 cm telescopes with a total field of view of 344 (8000 in the final version) sq. degrees. Unfortunately, Argus is not able to change the configuration of the field of view, orienting all telescopes to one region of the sky, which will not allow it to be used as a single telescope with an effective diameter of about 5 m. The remaining three instruments have the limiting dimensions of the fields of view (3–5 degrees) for a monolithic configuration in the Schmidt scheme and are equipped with sufficiently fast receivers. Nevertheless, it is clear that due to the small size of the fields of view when searching for fast transients, they can function only in the alert (follow-up) mode, like many large telescopes with standard fields of view of <1 degree (see, for example, [32] and references therein). At the same time, this mode does not allow one to detect optical radiation synchronously with the fastest transients—gamma ray bursts, fast radio bursts, and gravitational waves—for this, it is necessary to continuously monitor the celestial sphere using instruments that have the widest possible fields of view of hundreds and thousands of square degrees. In other words, it is necessary to independently detect and study fast optical transients and only afterward look for their connection with events in other spectral ranges and of a different nature, comparing the time of occurrence and the position of both. We emphasize that the understanding of their origin, and the choice of models that describe them, will largely be determined by the solution of this most complex problem of modern practical astrophysics. The areas of monitoring of gravitational waves, gamma and radio telescopes when searching for transients cover almost half of the area of the celestial sphere (the exception is the BAT detector of the Swift instrument with a field of view of about 5000 square degrees). This means that even with a relatively high accuracy (arc minutes–degrees) of determining the coordinates of an event for gamma detectors [32], positioning optical telescopes according to an alert will take tens of seconds–minutes. At the same time, the duration of 90% of gamma-ray bursts lies in the 20 ms–100 s interval, and about 30% of the events last less than 2 s [33]. The latter is the result of the merger of two neutron stars or a black hole and a neutron star and form a class of short gamma-ray bursts (SGRB), in contrast to long-duration GRBs (LGRB), caused by the collapse of some massive stars to form black holes [34,35]. Thus, in the alert mode, it is impossible to detect the optical companion of the short gamma-ray burst itself—gamma radiation is no longer detected by the time the telescope is pointed at the event localization region. When a long burst is detected, it is sometimes possible to begin optical observations of its location zone until the gamma-ray emission disappears. As a rule, in both cases, in the alert mode, it is possible to detect and study only the so-called afterglow—optical radiation resulting from the interaction of the plasma ejected during the generation of a gamma-ray burst with interstellar gas [36].

It should be emphasized that so far, despite numerous works on observations and modeling of the phenomenon of gamma-ray bursts, no self-consistent theory is yet available. It is generally accepted that the study of joint optical radiation with gamma emission (prompt emission), the comparison of their characteristics and temporal structure can provide the key to solving this problem [34,37–40]. At the same time, out of 900 optically identified gamma-ray bursts, only 21 have scant optical information (only 1–3 brightness measurements) simultaneous with gamma-ray emission, and only six events have truly informative optical light curves obtained during the same period, usually in its final phase [38]. Finally, for the single GRB 080319B (Naked Eye Burst) event, we detected a joint optical burst with a temporal resolution of 0.13 s during the entire GRB by the wide-field monitoring independent of gamma-ray telescope data [41,42]. These unique observations with the wide-angle (600 sq. degrees) high-temporal resolution camera TORTORA [43] made it possible to establish the essential features of the mechanisms of gamma-ray burst generation [44]. The prototype of this camera is our similar instrument FAVOR (FAst Variability Optical Registrator), which was used for observations in 2003–2009 [45,46].

Further, in the course of developing the program for wide-angle high temporal resolution monitoring, SAO RAS (together with KFU and OOO Parallax) created a 9-channel system with a field of view of 900 square degrees and a temporal resolution of 0.1 s, Mini-MegaTORTORA (MMT) [47]. It was used to detect optical flares synchronous with the bursts GRB 160625B [48] (however, with a time resolution of 30 s) and GRB 210619B. In the latter case, radiation was recorded in 5 channels simultaneously with exposures of 1, 5, 10, and 30 s in white light (3 channels) and in filters B and V. The implementation of such a research mode is a fundamental feature of our approach, it allowed us to compare the structures of light curves in the gamma and optical ranges, build spectra in these ranges, and finally show that in this case, the burst emission is due to a backward shock wave propagating in a relativistic jet [49].

Note, that the MMT system is also equipped with polaroids with different orientations for measuring the linear polarization of optical flares, which should provide unique information about the physical properties of the bursts, namely, the structure of the emitting regions, the characteristics of the magnetic field, and the details of the mechanisms for generating radiation of different energies [34,50,51]. Space-telescope measurements of the polarization of hard burst radiation have low accuracy (30–50%) and temporal resolution [52], which does not allow one to choose between models, providing information only about the most general features of the phenomenon [37]. Because of this circumstance, polarization studies of optical flashes accompanying gamma-ray emission are of particular importance. Such observations are practically absent—a small number of works on this topic contain the results of measurements of the polarization of optical afterglows, at the level of 10–20%, but minutes–hours after the burst itself ([53] and references therein). In [54] an upper limit for linear polarization at a level of 12% has been obtained for the second episode of the GRB 140439A burst 2.5 min after the alert. Thus, the polarization of optical radiation simultaneous with the bursts, as well as its spectral characteristics, have yet to be investigated in high-resolution wide-angle surveys.

At present, one of the unsolved (and most important) problems of observational astrophysics is the detection of optical companions of fast radio bursts—the choice of adequate models of these phenomena from a large number of the ones considered requires the study of their possible optical counterparts, which, unlike gamma-ray bursts, are the subject of only theoretical constructions. Some of the several dozen models predict the generation of sufficiently bright optical flares [55–57]. On the other hand, the frequency of occurrence of fast radio bursts throughout the sky can reach several thousand per day [58], and, consequently, the optical flares accompanying them can be quite frequent. At the same time, as we know, unlike radio observations, optical surveys to search for these events with a high (1 ms) time resolution are not carried out at the moment. As a rule, studies are focused on searching for stationary optical burst hosts [59], or searching for millisecond optical flashes in repetitive bursts. About 20 such objects, which erupted about 200 times, were found among the almost 800 bursts proper. Their coordinates were determined, which made it possible to observe the regions of localization of these sources using large telescopes (see, for example, [60,61]), in which optical flares were not detected—the same result was obtained by us with a 6-meter telescope [62]. Gravitational wave events can also be accompanied by optical flares ranging from a few milliseconds to a few seconds, especially when they come from merging neutron stars, like the recently discovered event GW 170817 [63,64]. At the same time, since the accuracy of localization of gravitational waves is hundreds of degrees, such flares can only be detected by monitoring the celestial sphere independent of the data of gravitational detectors using instruments with comparable fields of view and millisecond time resolution. Moreover, in the constantly updated catalog of fast radio burst models [65], probable deep connections between them, the gravitational waves and gamma-ray bursts are clearly visible, based on the universal cause of these phenomena—the interaction and evolution of relativistic objects. It is all the more important to look for signatures of similar phenomena in the optical range using wide-angle monitoring of high temporal resolution. Based on the above considerations, and on the experience of developing and using the FAVOR, TORTORA, Mini- MegaTORTORA tools, we proposed the concept of a multi-telescope complex with high temporal resolution for wide-angle monitoring of the celestial sphere—SAINT (Small Aperture Imaging Network Telescope) [62,66–68] using telescopes with a diameter of 30–50 cm, with the limiting magnitude at the level of single- component instruments (see Figure 1 and Table A1 in the Appendix A) and capable of detecting and investigating non-stationary objects at times from milliseconds to months-years. This paper is devoted to the development of this project using modern devices, the choice of the design of the complex, its components, mode of operation, and the assessment of its parameters.

2. Small Aperture Imaging Network Telescope (Saint) Project

The performed analysis demonstrates the obvious need to create a multi-channel optical monitoring system that would be capable of conducting complex searches and studies of non-stationary objects and phenomena within the framework of universal instrumental and methodological approaches and would not have the disadvantages of existing systems.

The features of this kind of instrument should be as follows:

- a large field of view of several hundred degrees, which requires a multi-channel design;
- high temporal resolution in the range of tenths to hundredths of a second;
- detection limit of 18–20 mag at time scales of 20–30 s to be at the level of the limits of existing survey telescopes (Figure 1);
- a combination of monitoring (wide-field) and follow-up (narrow-field) modes with their rapid change;
- the possibility of obtaining maximum information (spectral, polarimetric, photometric) about non-stationary objects in follow-up mode;
- processing of accumulated information in real time to detect, characterize and classify transient phenomena, and make decisions on the transition to the follow-up mode;
- preservation of all raw and reduced data, as well as maintenance of databases obtained as a result of its a posteriori analysis;
- complete robotization of the complex operation using information from external sources (meteo station, network, other instruments) and a system to control its condition.

As a practical implementation of these principles, we propose the project of SAINT (Small Aperture Imaging Network Telescope)—12-telescope complex with high temporal resolution, built mostly from off-the-shelf commercially available components that may be implemented in about $2 M cost (plus additionally about $500 K for labor costs). Its overall parameters are listed in Table 1, schematic view of a single channel is shown in Figure 2, and details on individual components and details of its operation modes are given below.

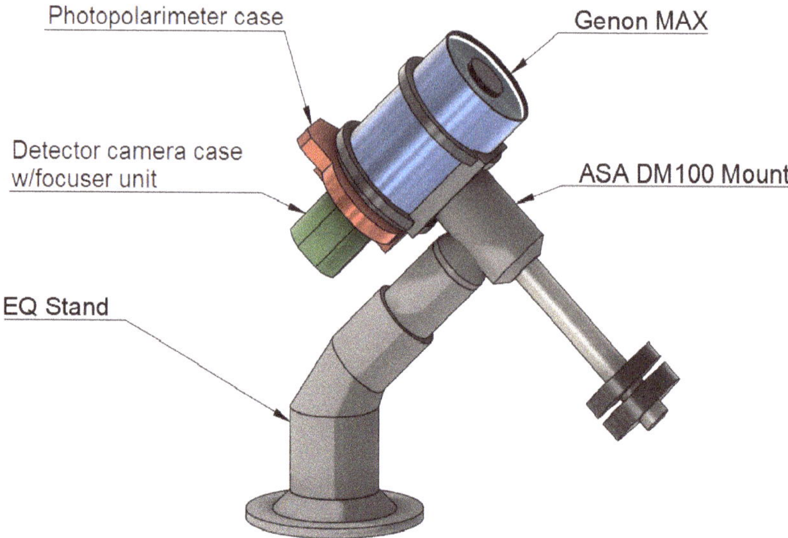

Figure 2. Schematic view of a single assembled channel of SAINT complex.

Table 1. Main characteristics of the SAINT complex.

Number of telescopes/mounts	12/12
Telescope aperture	300 mm
Telescope focal length	450 mm
Optical scheme	Schenker-Terebizh
sCMOS array size	49.5×49.2 mm
Fullframe readout speed	54 fps
Pixel scale	5.5 ″/pix
Telescope FOV	$6.3° \times 6.3° = 39$ sq.deg.
System FOV	468 sq.deg.

2.1. Telescope

The basic component of the complex is GENON Max—a catadioptric two-mirror Schenker–Terebizh telescope with five corrective lenses [29] with a diameter of 30 cm (F/1.5) and a field of view of about 40 square degrees (Table 2). In its Cassegrain focus a multimode photopolarimeter of a rather complex design is installed, capable of operating both in wide-field monitoring and in the study of a single object (Section 2.4).

Table 2. Main characteristics of the GENON Max telescope (design by G. Borisov).

Scheme with Cassegrain focus	Schenker-Terebizh
Diameter, mm	300
Focal length, mm	450
Wavelength range, Å	4500–8000
The coefficient of central screening by diameter	0.50
Linear diameter of the field of view, mm	70
Angular diameter of the field of view, deg.	8.9
Relative illumination across the field, center—edge of the field	1.0–0.7
Telescope weight (tube with optics assembled), kg	29
Optical system length, mm	430

2.2. Mount

The telescopes are mounted on ASA DDM100 mounts based on direct drive technology (see Table 3). This design has a number of important advantages listed below, which are especially significant for the proposed complex.

- High acceleration and, as a result, high pointing speed up to $50°/s$, which will allow changing the field of view of the complex and the observation mode in several seconds.
- Due to the absence of drive belts and gears, the rotation of the axes is very uniform and silent, which ensures the accuracy of the guidance and the absence of vibrations, which determine the high stability of the position of sources in the focal plane of telescopes.
- Due to the absence of gearboxes, the absence of mechanical backlash and hysteresis is ensured. For the same reason, moving parts wear out little, which increases the period of maintaining the stability of the mount's characteristics.
- All movable structural elements are equipped with 28-bit encoders and feedback systems, which ensures the accuracy of rotation angles and their determination up to $0''.004$.
- Maintaining the stability of the functioning of all elements of the mount and its control is provided along with feedback systems by a high-speed controller online with a computer.

Table 3. ASA DDM100 mount specifications.

Type	Equatorial
Motor	Direct drive
Max load	100 kg
Weight (without counterweight)	56 kg
Power supply	24 Volts
MAX slew speed	50 deg/s
Pointing Accuracy ($20°$ to $85°$)	$<8''$ rms
Tracking Accuracy ($20°$ to $85°$)	$<0.25''$ rms

2.3. Detector

Every channel of SAINT is equipped with a photopolarimeter (Section 2.4), which is the primary instrument in the SAINT complex. The detector used in it is Andor sCMOS Balor 17F-12 camera, whose characteristics are listed in Table 4.

Table 4. Characteristics * of the Andor Balor sCMOS camera.

Sensor size W × H, mm (px)	49.5 × 49.2 (4128 × 4104)
Cell size, μm	12 × 12
Readout noise, e^-	2.9
Full well depth, e^-	80,000
Frame rate of full frame reading, Hz	54
Maximum quantum efficiency (at 6000 Å), %	61
ADC, bit	16
The highest frame rate for the reading area 512 × 512, Hz	431

* All characteristics are given according to official technical specifications from Andor website, available at https://andor.oxinst.jp/assets/uploads/products/andor/documents/andor-balor-17F-12-specifications.pdf (accessed on 1 December 2023).

This detector currently has the best combination of a large size of 16.9 Mpix (70 mm diagonal), high temporal resolution when reading a full-frame (about 20 ms) and a fairly low readout noise ($2.9\ e^-$), which is unattainable in CCD matrices for several data transmission channels [69]. A sufficiently large pixel size (12 μm) allows the use of the maximum temporal resolution in observations even with relatively low image quality ($>1''$), nevertheless concentrated in one pixel, which keeps the readout noise to a minimum. At the same time, a high degree of signal response linearity (>99.7%) is combined with a large electron well depth (80,000) and a special way of amplifying and digitizing the signal, which ensures its continuous recording in the maximum dynamic range. These features make it possible to detect sources with an intensity from the read-out noise level to the saturation limit without distortion in a single camera frame, which is fundamental in wide-field observations. Moreover, reducing the read-out area (Region of Interest, ROI) (Table 5) allows for improving the temporal resolution down to 2.5 or 0.6 ms which may be used in the regime of targeted observations of individual objects.

Table 5. Maximum frame rate depending on the size of a read-out region of a sensor (region of interest, ROI).

ROI Size, Pixels	Max Frame Rate, fps
4128 × 4104	54
2048 × 2048	108
1920 × 1080	205
1024 × 1024	216
512 × 512	431
128 × 128	1684

The microlens array on top of the sensor in the Balor camera allows lossless operation for the relative apertures of up to F/0.3 at cone angles up to 110°, which makes it possible to avoid light losses even at the outer margins of the telescope field of view at aperture ratio of F/1.5. An important point is what the camera allows to timestamp and synchronize the acquired frames using the signal from the GPS receiver which is crucial for the combination of data from different channels pointed toward a single target. On the other hand, in contrast to traditional CCD sensors, the question of long-term spatio-temporal stability of individual pixels of CMOSes poses a potentially significant problem for the co-addition of images from such detectors, as it introduces additional per-pixel noise that cannot be mitigated by classical calibration methods (bias and dark frame subtraction, non-linearity correction and flat fielding). The limits imposed by these low-frequency effects on frame co-addition are still poorly studied and have to be thoroughly investigated prior to the decision on an exact strategy of frame co-addition in the operation of the complex.

2.4. Photopolarimeter and Different Modes of Operation

The primary instrument of every SAINT channel, and the only custom-built part of it, is a photopolarimeter which provides observations in both monitoring and research modes. It is attached to the Cassegain focus of the telescope and consists of two parts—an optical-mechanical unit and a camera unit, its general view with installation on the telescope is shown in Figure 2.

The camera unit houses the detector, which is mounted on sliding rails that allow it to move along the optical axis, thereby focusing images on the sensor. The hoses of the water cooling system of the chamber, not shown in the figure, are inserted into the rear end of the casing of the unit.

The structure of the photopolarimeter is shown in the diagrams of Figure 3, its internal view is in Figure 4, its longitudinal view is in Figure 5. Control electronics units, power supply, motors that ensure the movement of moving elements of the device, and connecting cables are not shown here.

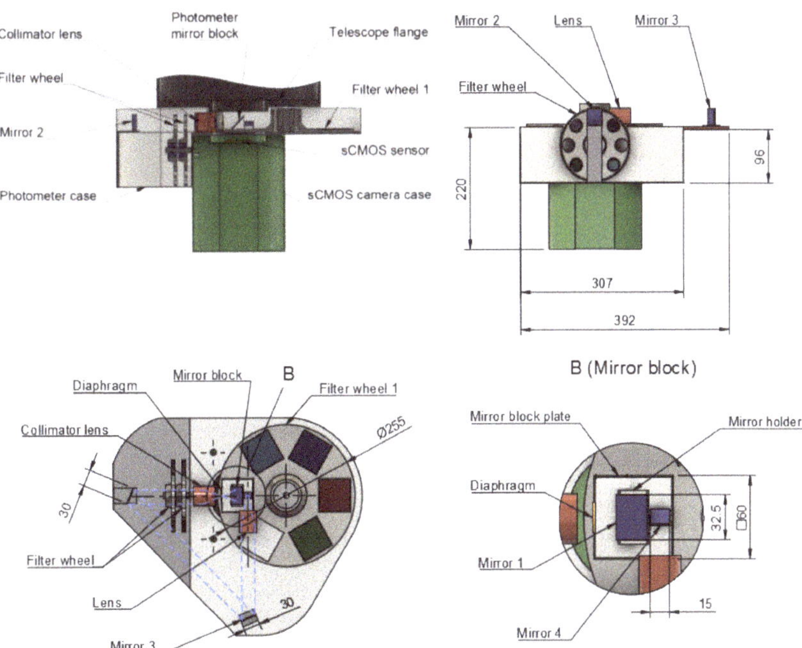

Figure 3. Schematic view of a photopolarimeter installed in the Cassegrain focus of every channel of SAINT complex.

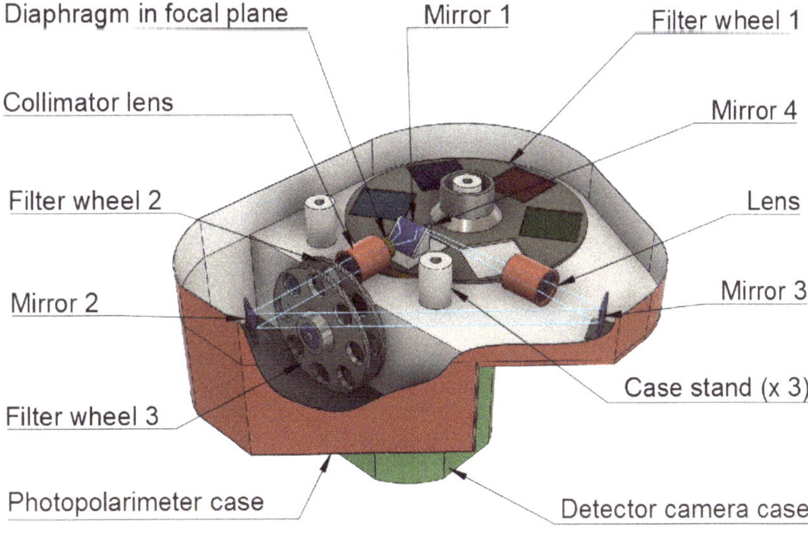

Figure 4. Internal view of the photopolarimeter and the course of rays in the follow-up mode.

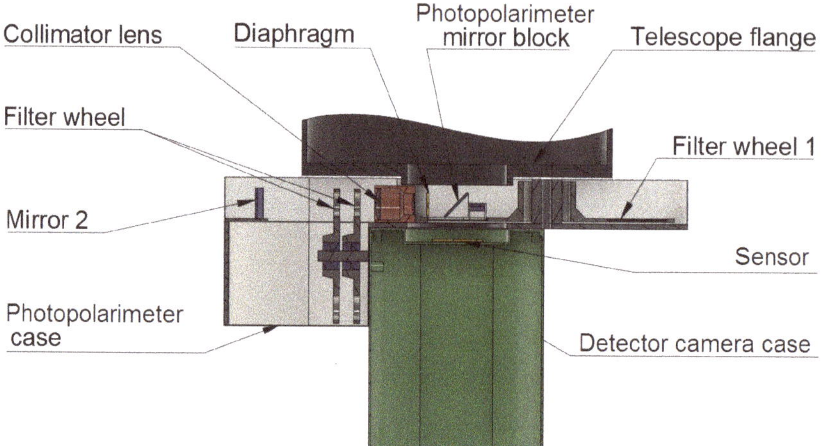

Figure 5. Longitudinal view of the SAINT photopolarimeter.

The front part of the photopolarimeter, located in the gap between the back focus of the telescope and the entrance of the camera, contains the following components: a turret with *griz* filters, an empty window, and a narrow-field block containing mirrors 1 and 4, as well as a field-of-view diaphragm (the optical axis of the telescope passes through the centers of the turret holes); input (COMPUTAR M2514-MP2 with a focal length of 25 mm and aperture ratio of 1:1.4) and output (COMPUTAR V5014-MP with the focal length of 50 mm) lenses of the collimator; mirror 3.

The side part of the photopolarimeter is located outside the camera casing. It contains a turret with narrow-field *griz* filters and an empty window, as well as a turret coaxial with it with three polaroids of different orientations (with polarization plane rotated by $120°$ in respect to each other) and an empty window; mirror 2.

During observations in the wide-field (monitoring, or primary survey) mode, an empty turret window 1 or one of its filters is installed on the optical axis of all 12 telescopes (depending on the specific tasks of the survey, observations of different areas with different filters, etc. are possible). The cameras acquire a sequence of full-frame images (4128×4104 pixels, or $6.3° \times 6.3°$ each) with a rate of 54 fps, and the entire complex monitors the overall field of view at 470 sq. degrees.

In the narrow-field (research) mode, a block with mirrors 1, 4 and a field diaphragm measuring $11' \times 11'$ (1.45×1.45 mm) of turret 1 (see Figure 3) is installed on the optical axis, while mirror 1 focuses the redirected beam on the diaphragm placed at the focus of the input lens of the collimator. The parallel beam constructed by it passes through one of the *griz* filters (or empty window) of turret 2 and one of the polaroids (empty window) of turret 3. After being reflected by mirrors 2 and 3, the parallel beam is focused by the output lens of the collimator and redirected by mirror 4 to the sensor of the camera. As a result, the light from the $11' \times 11'$ sky region with a linear size of 3×3 mm (240×240 pixels) is imaged with a scale of $2.75''/\text{pix}$, twice the original ($5.5''/\text{pix}$)— the scale is determined by the focal length of the collimator output lens.

This image can be recorded with a time resolution of 20 to 2.5 ms, and its central $6' \times 6'$ part (128×128 pixels)—with a resolution of 0.6 ms (see Table 5). The transition from monitoring to research mode takes a few seconds and is determined by both the rotation speed of the photopolarimeter turrets and the repointing speed of the mount. In particular, when an optical transient is detected during real-time monitoring, the following operations are performed:

- its characteristics are determined (initial brightness, characteristic duration, structure of the light curve);

- rough initial classification is performed (as one of noise, meteor, satellite, new or already known astrophysical object classes);
- the follow-up mode for every channel is selected (photometry and/or polarimetry, filters to use, exposure time);
- in parallel, all telescopes are repointed to the source, placing it in a small central field (with the exception of the instrument that detected the transient, it retains the original mode used for detection, in order to keep an uninterrupted sequence of data);
- observations start.

Figure 6 illustrates this process.

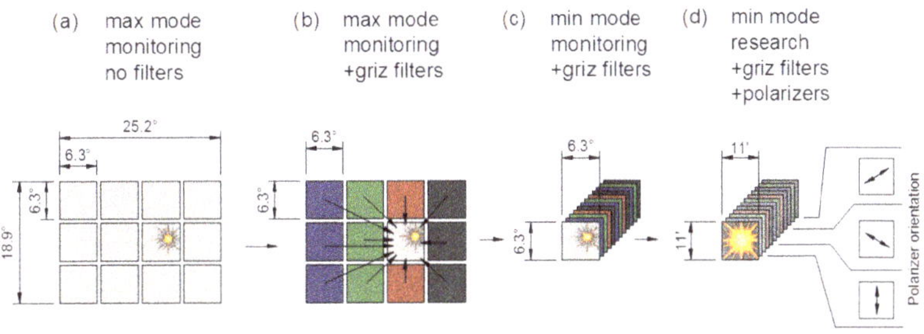

Figure 6. Change of observation modes of the complex after detecting an optical transient.

Table 6 shows the estimated detection limits of the complex in different modes and with different time resolutions. Estimates were made for a single exposure in white light at a characteristic wavelength of 5500 Å, a sky background of 21.5 mag/sq.arcsec, an optical and atmospheric transmission of 0.25, the characteristics of the telescope and detector from Tables 2 and 4, and assuming that the object flux is fully contained inside a single pixel (i.e., significantly undersampled PSF). It should be noted that the real limits when using filters and polaroids will be 0.5–1.0 magnitude worse.

Table 6. Detection limits in different modes.

Mode	Pixel Scale	Size	Mode *	System FOV	Detection Limit at Exposure					
	Channel FOV	Pixels		sq.deg.	0.6 ms	2.5 ms	20 ms	1 s	30 s	20 min
Monitoring	5.5″/pix	4128 × 4104	Max	468			12.6	16.5	18.9	20.9
	6.3° × 6.3°		Min	39			14.0	17.9	20.2	22.2
Follow-up	2.75″/pix	240 × 240	Max	432	8.8	10.3	12.6	16.7	19.5	21.6
	11′ × 11′		Min	36	10.2	11.6	14.0	18.0	20.8	23.0

* Possible modes of operation: - Max—telescopes are pointed at different areas (12 × 39 sq. deg.); - Min—telescopes are pointed at one area (39 sq. deg.).

2.5. Shelter and Infrastructure of the Complex

Twelve telescopes of the SAINT complex are installed in two cylindrical shelters housing six channels each (see Figure 7). They consist of a reinforced concrete supporting slab mount and a steel truss frame sheathed with sandwich panels on which it is mounted. Its free part serves as a flyover to accommodate the cylindrical roof of the shelter in the working state of the complex, which, moving along rail guides, closes the plate with the telescopes in the non-working state. The frame is fixed on a reinforced concrete base covered with metal-plastic cladding, and installed on a concrete pad dug into the ground.

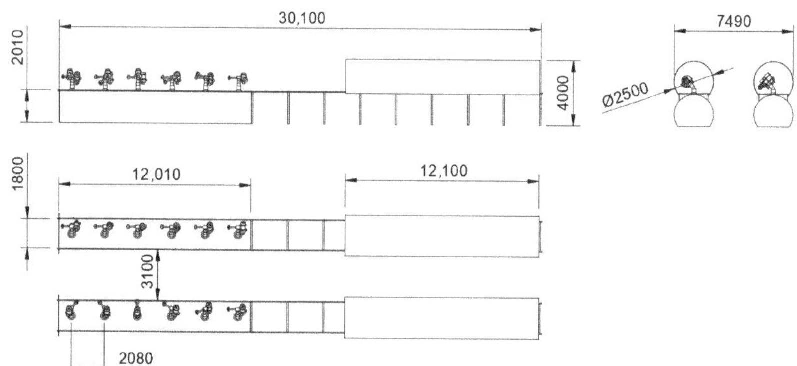

Figure 7. The scheme and general view of the shelter of the SAINT complex.

The internal volumes of the frame are the technical compartments of the SAINT complex. They house the following equipment and infrastructure components:

- power supply input unit of the complex network, including power drives of all its mechanical elements, electronic components of the telescopes and instruments, computer cluster, and uninterruptible power supplies;
- system for providing water cooling of detectors;
- climate control system based on dehumidifiers COTES CR300.

Both parts of the shelter are equipped with video cameras for monitoring the internal and external space. The complex is also equipped with a meteorological system, a fish-eye all-sky camera and an IR cloud sensor, which allows to automatically monitor weather conditions and make a decision on the start and end of observations.

The movable roofs of the shelters are covered with a sun-protection coating, and flat screens with uniform illumination are placed on their inner surface to calibrate the sensitivity of the detectors.

3. Data Acquisition, Processing and Storage

During the primary monitoring mode of operation, the complex will acquire the data for every sky field (470 square degrees) for 20 min in a row, thus covering up to 12,700 square degrees in a typical 9-hour observational night. The data will be processed in real time, and a number of data products (transients on various time scales, co-added images with different temporal resolutions, etc.) will be derived and stored.

As both the computational hardware and data analysis algorithms evolve very quickly, it is not possible (and not necessary) to specify the IT infrastructure of the SAINT complex here in full detail as we did for the telescope itself. Therefore, below we will outline its generic structure, requirements and possible approaches for handling an enormous amount of data the complex will produce.

In general, the operation of the complex will be carried out using the software that is based on that we used in observations with the FAVOR, TORTORA and Mini-MegaTORTORA instruments (see Beskin et al. [47] and references therein). On the shortest time scales, where timing constraints are the most demanding, transient detection will be performed in real time using fast image subtraction methods that are based on sufficient local stability of image PSF, as described e.g., in Karpov et al. [70]. On longer time scales, where PSF stability cannot be ensured, the methods developed for image subtraction with PSF matching developed over the last 20 years will be used, such as Alard and Lupton [71] family of methods, ZOGY [72] algorithm, or SFFT [73]. These methods generally allow parallelization using modern GPU hardware and thus may be expected to run fast enough to be used for analyzing the data in real time on minutes to hours time scale. An alternative to image subtraction methods may also be implemented using convolutional neural networks (CNN, see e.g., Sedaghat and Mahabal [74] or Acero-Cuellar et al. [75]). CNNs will also be employed in order to reliably separate the image subtraction artifacts, cosmic rays, as well as, e.g., the effects of stellar scintillations on short time scales from bona fide transients using the methods developed in e.g., Cabrera-Vives et al. [76], Gieseke et al. [77], Carrasco-Davis et al. [78], Makhlouf et al. [79].

Image co-addition will also allow simultaneously achieving the highest possible temporal resolution (on individual frames) and going for deeper objects (in running coadds of various lengths, corresponding to, e.g., every 100 or 1000 original images). Unfortunately, without a thorough laboratory study of the stability of the detector, it is impossible to define the limits for such co-addition, as at some point the effects of pixel parameter drift specific for CMOSes will be more significant than the further sensitivity gain. However, we may foresee that implementing some tailored observational strategy in the data sequence— e.g., dithering using some pre-defined pattern—will allow us to mitigate such effects to some degree.

As a minimal architecture for the data processing cluster for SAINT, we propose a hierarchical scheme with two identical computers per telescope channel—one for data acquisition from the CMOS and real-time image processing, second for the channel hardware and high-level controls, intermediate data storage and image processing—plus a central computer that handles the overall operation of the complex, including survey scheduling and reaction to the detected transients, communication with external networks, as well as various databases. Such architecture also allows for the use of all cluster machines for more computationally-intensive data processing tasks during the day time, when no observations are performed and no real-time data processing is necessary.

Let's outline the basic requirements for data processing and storage infrastructure. They are defined by the extremely large amount of data being acquired from several large format detectors (16.9 million pixels each for the Balor camera) routinely operating at frame rates of up to 54 frames per second (or even up to 1684 fps in follow-up regime), thus amounting to 1.8 gigabytes of data per second for every channel, or up to 800 Terabytes per night for the whole complex. As a minimal realistic configuration using present-day commercially available hardware, we may use a camera connected to the PC through a four-channel CoaXPress interface, and a RAID-0 array consisting of five SATA-III hard drives 20 terabytes each. Due to parallel writing, a speed of up to 4 gigabytes per second may be achieved for such RAID configuration, which is sufficient for handling the real-time data flow. Longer-term data storage will require datacenter-class hardware which is also commercially available.

4. Expected Results

The proposed complex is able to operate in still poorly studied region of the parameter space shown in Figure 1, and thus may provide a number of results related to previously unknown classes of rapid optical transients, as well as a vast amount of data on known ones. Below we will briefly outline some of the fields where such instruments may provide important developments.

Flaring stars are now being routinely discovered in modern time-domain sky surveys like Kepler [80], TESS [81,82], EvryScope [83], NGTS [84], ZTF [85] or Tomo-e Gozen [86] what perform high cadence continuous observations of the fixed sky regions. However, most of these experiments do not provide temporal resolution better than half a minute, or in rare cases—better than a second, and are being performed in a single photometric filter that somehow reduces the scientific content of collected statistical results. Thus, using a specifically tailored monitoring mode with SAINT which is possible due to its multi-channel architecture—i.e., simultaneous multicolor observations—would allow us to augment the statistics of white-light flares with at least some estimations of their temperature. Moreover, employing the polarimetric mode for targeted (or follow-up, for the flares detected by real-time processing pipeline) observations would also allow placing statistically reliable upper limits on (or even obtain a detection of) the polarized components of stellar flares of various durations, including the shorter ones not typically resolved by existing surveys (see also Figure 8 for details).

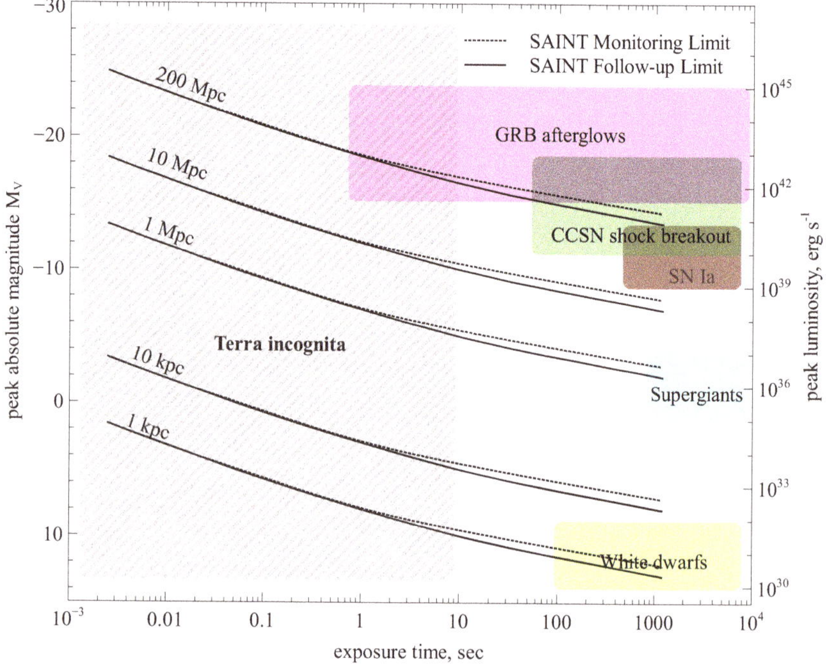

Figure 8. Detection limits of the SAINT according to the Table 6 for various distances, timescales and peak luminosities.

Aizawa et al. [86] reported 22 flares detected in 40 h of 1 s cadence observations with Tomo-e Gozen camera (21 square degrees simultaneous field of view). While the depth of the SAINT survey is not as good, due to the superior field of view we may expect up to 12 flares per hour in a wide-field mode, thus obtaining a significantly larger amount of flashes, covered with better temporal resolution.

The observations of faint meteors with the Mini-MegaTORTORA system [47] typically resulted in the detection of several hundreds of events per observational night, with maximum brightness down to 8–10 magnitudes, which is 2–4 magnitudes deeper than the limits of most modern meteor detection experiments (see, for example, Weryk et al. [87], Jenniskens [88]) and comparable to the capabilities of the Tomo-e Gozen camera [89]. Such faint, and even fainter meteor events are caused by micrometeoroids—particles of interplanetary (interstellar?) dust with masses of 0.1 µg–0.1 g [90]—that, according to direct transatmospheric studies, make up the majority of the cosmic matter falling to the Earth, 100–200 tons/day [91]. On the other hand, observations of meteors give an order of magnitude smaller volume [92]—this discrepancy is most probably due to an underestimation of the role of ablation and fragmentation processes of micrometeoroids, as well as inaccuracy in determining their velocities [92]. These problems are largely associated with the difficulty of interpreting radar observation data, which is the main means of studying such faint meteors with optical magnitudes of 10–14 [93]. Thus, carrying out a long-term survey of a population of faint meteors, covering the entire range of velocities from 12.3 to 71.9 km/s, will allow us to study the features of their optical emission, taking into account possible fragmentation, and compare it with dynamic models of micrometeoroids. Another direction of meteor study might be the analysis of their mass distribution both in the showers and in the sporadic component, as well as massive colorimetry of events during the peaks of different meteor showers in order to compare their spectral properties.

UV Ceti type flare stars belong to the main sequence red dwarfs, making up 70–75% of the population of the Galaxy [94]. The duration of their stay on the main sequence exceeds the age of the Universe, and during this time their main physical characteristics remain unchanged [95]. These features determine great interest in red dwarfs as possible regions where life arises [96]. The probabilities of the occurrence of Earth-like planets orbiting these stars in habitable zones with a characteristic size of about 0.2 AU [97], estimated from various observations, are in the range of 30–50% [98,99]. This gives about 100 habitable Earth-like planets in the vicinity of 10 pc, and several billion in the entire Galaxy [96]. The small masses and sizes, as well as the low luminosity of red dwarfs, as a result of which the habitable zones are located close to the star, determine the relatively high probability of detecting Earth like planets during transits. Their depth for earth-like planets ranges from 0.002 to 0.0084 magnitudes with the radius of the host star from 0.6 to 0.1 solar, and the probability of a successful "edge-on" orientation of the orbit of the star-planet system for an observer is 0.5–1% [100]. The photometric accuracy required for transit detection is close to the limit for ground-based instruments. Nevertheless, it is achieved in various programs, thanks to the optimal strategy of observations and selection of objects, the use of effective methods of data analysis and methods of modeling the processes of obtaining them. Thus, when studying several dozen red dwarfs of 12–15 magnitudes with systems of distributed telescopes with diameters of 0.2 to 2 m at exposures of 30–120 s, measurement accuracies in the range of 0.001–0.004 magnitudes were achieved, leading to the detection of effects with a depth of 0.005–0.007 magnitudes for planets of 1–3 Earth masses located in habitable zones, with a probability of 2.5–8% [101–103]. These results were obtained from large data sets using BLS [104,105] and TLS [106] folding algorithms in the space of periods and filling factors.

In this context, it seems that such studies using the SAINT complex should be very effective. Instead of simultaneously monitoring up to 10 red dwarfs (according to the number of instruments used) within the framework of the mentioned programs, SAINT will be able to observe more than a thousand stars. With a minimum brightness of objects with potentially habitable Earth-like planets at a level of 16 magnitude, they will be located at a distance of up to 100 pc, their number in the northern sky will be about 100,000 [96], and transits of these planets may be detected at several hundred red dwarfs. Note, that when observing each star about 50 times over a visibility period of 100 days and a transit duty cycle of 0.001–0.005 [100] the accuracy of intensity determination in phased light curves increases by tens of times relative to its initial estimates in individual exposures. Thus, as a result of monitoring the sky with the SAINT complex, it is possible to increase the sample of Earth-like planets in the habitable zones by 4–6 times relative to its current volume (See Habitable Exoplanet Catalogue at http://phl.upr.edu/projects/habitable-exoplanets-catalog (accessed on 1 December 2023)).

Similar results within the framework of the strategy for detecting periodic signals in wide-angle surveys can be obtained when searching for transits of Earth-like (and not only) planets around white dwarfs. These objects may also host planetary systems and Earth-like planets located in habitable zones. Indeed, with a luminosity of 10^{-3}–10^{-4} and a mass of 0.5–0.6 solar, their radii are close to those of the Earth, and the temperature lies in the range of 4000–9000 K, which leads to the existence of a habitable zone in the 0.005–0.02 AU range of distances from stars over several billion years [97,107,108]. On the other hand, there is evidence of the existence of various proto-planetary and post-planetary structures around white dwarfs—disks, planetesimals, asteroids [109,110]. Finally, four giant planets were discovered in systems with white dwarfs [111], one of them by the transit method [112]. Thus, one can hope also for the presence of Earth-like (rocky) planets with a mass of 1–2 of Earth one in the habitable zones of white dwarfs, as established for 63 G-M dwarfs (according to Habitable Exoplanet Catalogue). The depth of their transits, repeated with periods in the interval of 4–30 h (corresponding to the size of the habitable zone), is 10–100%, the duration is 1–2 min, and the probability of the best "edge-on" orientation of the orbital plane relative to the observer is close to 1–2% for planet radii 1–2 Earth's [107,108,113]. In ground-based observations, even with small-diameter telescopes and standard photometric accuracy, it is possible to search for such effects. The primary difficulty in such studies is the needed relatively high temporal resolution at the level of several seconds. In particular, Faedi et al. [113] searched for transits in long-term monitoring of 194 white dwarfs of 9–15 mag with a time resolution of 30 s (an 8-channel SuperWASP telescope was used—see Figure 1 and Table A1) and obtained only upper limits at 10% for the frequency of occurrence of planets and brown dwarfs. The authors come to the conclusion that it is necessary to increase the studied sample, and the duration of observations, and to improve the temporal resolution. Similar programs have been planned, and are being partially implemented using space-borne instruments [114,115], ground-based telescopes [116,117] and their combinations [118].

In the northern sky, according to various estimates, there are 10–15 thousand white dwarfs brighter than 18 magnitude (see, for example, Kleinman et al. [119] or Gentile Fusillo et al. [120]), and for approximately 100–150 ones the transits with periods between 3 h and 50 days, with duty cycles of 0.03–0.0006 and depths 0.1–1, may be detected by SAINT. Note, however, that this estimate is solely for the objects where it is possible to detect planets if they are there—and the probability of it is unknown to us. On the other hand, in the absence of an effect, it will be possible to obtain an estimate of the upper limit for this probability.

Finally, there are several other (high risk) directions placed in the area of "Terra incognita" in Figure 8 where SAINT observations may be potentially crucial:

1. detection of optical flashes accompanying gamma-ray bursts (about five events per year?),
2. registration of optical companions of fast radio bursts, in particular, the ones with repeating activity, where the same approach as for exoplanets may be used to improve the performance,
3. registration of optical companions of gravitational wave events or setting upper limits on their electromagnetic counterparts' energy,
4. search for signals of terrestrial civilizations.

5. Conclusions

In this paper, we present a project of a SAINT (Small Aperture Imaging Network Telescope)—robotic 12-telescope wide-field complex with high temporal resolution, aimed towards detection and studying optical transient phenomena on the shortest possible time scales. Its multi-channel nature allows different modes of operation depending on the task—e.g., a wide-field survey or a narrow-field follow-up. Using modern large-format

CMOS detectors allows operation with the exposures as short as 20 milliseconds, and to observe the whole sky once every two nights. A novel reducer-collimator that may be rapidly installed into the light beam allows for increasing the spatial resolution and effectively converting the complex to a photopolarimeter able to simultaneously acquire both color and polarimetric information.

The flexible nature of the complex, and its operation in the poorly studied region of parameter space for optical transients faster than a second, will provide an overwhelming amount of data on various classes of both astrophysical and near-Earth objects. On the other hand, as the complex is built from mostly off-the-shelf components, it may be easily scaled horizontally, to either increase the performance of an individual installation or to expand it to several sites over the globe able to provide a continuous view of the entire sky, thus realizing the "all sky all time" concept which is extremely important when searching for any transient events, especially short ones.

Author Contributions: Conceptualization, G.B., G.O., S.K. and A.G.; methodology, G.B. and A.G.; software, S.K.; validation, G.V., A.V., V.V. and N.L.; formal analysis, A.G. and A.B.; investigation, N.L. and V.S.; data curation, S.K. and N.L.; writing—original draft preparation, G.B., G.O., A.G., S.K. and A.B.; visualization, A.G. and A.B.; supervision, G.B. and G.O.; project administration, G.B., G.O. and G.V.; funding acquisition, G.B., S.K., G.O. and G.V. All authors have read and agreed to the published version of the manuscript.

Funding: The work was partially supported by the European Structural and Investment Fund and the Czech Ministry of Education, Youth and Sports (project CoGraDS CZ.02.1.01/0.0/0.0/15 003/0000437) and the Federal Program for Improving Competitiveness of the Kazan (Volga Region) Federal University.

Institutional Review Board Statement: Not applicable.

Informed Consent Statement: Not applicable.

Data Availability Statement: The data presented in this study are available on request from the corresponding author.

Acknowledgments: The work was carried out within the framework of the state assignment of the SAO RAS, approved by the Ministry of Science and Higher Education of the Russian Federation. The authors are grateful to Lisa Chmyreva and Tatiana Sokolova for their help with the preparation of the manuscript.

Conflicts of Interest: The authors declare no conflict of interest.

Appendix A. Current Survey Wide-Field Telescopes and Systems

In the table below main parameters of the current wide-field observational systems are presented. Remarks to the table content are as follows: (a) For multilens systems etendue $A\Omega$ have been calculated as sum of etendues of all individual lenses in the system. (b) The limit is given for the 8.5 cm camera. (c) There is also another regime with $t_{exp} = 30$ s and $m_{lim} = 19.5$. (d) The limit is given in m_{AB} and depend on the band chosen. (e) Exposure time varies from 30 to 45 s depend on the band chosen. Typically sum of 10 successive images is then used to achieve deeper limit. (f) The exposure time is given, while the total image cycle time is equal to 6 s. (g) Effective exposure time is given, since TDI is adopted in SDSS. (h) The project "Pi of the Sky" stopped gathering data in 2017 [121] (i) Currently only one (ROTSE III-d) telescope is in operating.

Table A1. Summary of wide field survey telescopes currently at operation. The parameters of the SAINT are also included to the bottom of the table.

N	Survey	t_{exp}, s	FOV, deg^2	m_{lim}	D, m	$A\Omega$, m^{2}deg^2	References
1	Evryscope	120	8150	16	22×0.06	23.0	[122]
2	Pi of the Sky	10	6400	12	16×0.085	36.3	[123]
3	Ground-based Wide Angle Cameras (GWAC)	10	5000	16	40×0.18	127	[124]
4	Arctic Wide-field Cameras (AWCam)	10	1799	12.6	$0.085 + 0.050$	5.40	[125]
5	RAPTOR	30	1500	12.5	8×0.085	8.51	[126]
6	Kilodegree Extremely Little Telescope (KELT)	150	1352	12	2×0.042	1.87	[127]
7	SuperWasp	30	1000	15	16×0.2	31.4	[128]
8	Mini-MegaTORTORA (MMT-9)	0.1	950	12	9×0.071	3.76	[47]
9	Large Array Survey Telescope (LAST)	20	355	19.6	48×0.28	2.96	[129,130]
10	Argus	1	344	16.5	38×0.203	11.1	[131]
11	HATSouth	240	201	18.5	12×0.18	5.11	[132]
12	ASAS-SN	90	108	17	24×0.14	1.66	[17]
13	The Next Generation Transit Survey (NGST)	10	96	16	12×0.2	3.02	[133]
14	MASTER-II	30	80	20	20×0.18	2.04	[134]
15	Deca-Degree Optical Transient Imager (DDOTI)	60	72	18	6×0.28	4.43	[135]
16	Zwicky Transient Facility (ZTF)	30	47	20	1.2	53.2	[19]
17	GW Optical Transient Observatory (GOTO)	60	40	19	8×0.40	5.03	[136]
18	Tomo-e Gozen	0.5	21	17	1.0	16.5	[89]
19	Catalina Sky Survey 0.7m	30	19.4	19.5	0.7	7.47	[18,137]
20	ATLAS	30	15	19	2×0.5	2.95	[138]
21	ROTSE III	5	13.7	17	4×0.45	2.18	[139]
22	Vera C. Rubin observatory (LSST)	30	9.6	25	6.5	318	[20]
23	BlackGEM	300	8.1	22.2	3×0.6	2.29	[140]
24	La Silla-QUEST Variability survey	60	7.5	20.5	1.0	5.89	[16]
25	Weizmann Fast Astronomical Survey Telescope (W-FAST)	0.04	7	13	0.55	1.66	[141]
26	PanSTARRS-1	30	7	18.5	1.8	17.8	[142]
27	Space Surveillance Telescope (SST)	1	6	20.6	3.5	57.7	[143]
28	SkyMapper	40	5.5	18	1.35	7.87	[15]
29	Catalina Sky Survey 1.5m	30	5	21.5	1.5	8.84	[18,137]
30	Deep Lens Survey	600	4	24	2×4	50.3	[144]
31	Sloan Digital Sky Survey (SDSS)	54	3	22	2.5	14.7	[14]
32	SAINT	0.020	468	12.6	12×0.3	33	This work

References

1. Abbott, B.P.; Abbott, R.; Abbott, T.D.; Acernese, F.; Ackley, K.; Adams, C.; Adams, T.; Addesso, P.; Adhikari, R.X.; Adya, V.B.; et al. GW170817: Observation of Gravitational Waves from a Binary Neutron Star Inspiral. *Phys. Rev. Lett.* **2017**, *119*, 161101. [CrossRef] [PubMed]
2. Abbott, B.P.; Abbott, R.; Abbott, T.D.; Acernese, F.; Ackley, K.; Adams, C.; Adams, T.; Addesso, P.; Adhikari, R.X.; Adya, V.B.; et al. Gravitational Waves and Gamma-Rays from a Binary Neutron Star Merger: GW170817 and GRB 170817A. *Astrophys. J. Lett.* **2017**, *848*, L13. [CrossRef]
3. Goldstein, A.; Veres, P.; Burns, E.; Briggs, M.S.; Hamburg, R.; Kocevski, D.; Wilson-Hodge, C.A.; Preece, R.D.; Poolakkil, S.; Roberts, O.J.; et al. An Ordinary Short Gamma-Ray Burst with Extraordinary Implications: Fermi-GBM Detection of GRB 170817A. *Astrophys. J. Lett.* **2017**, *848*, L14. [CrossRef]
4. Savchenko, V.; Ferrigno, C.; Kuulkers, E.; Bazzano, A.; Bozzo, E.; Brandt, S.; Chenevez, J.; Courvoisier, T.J.L.; Diehl, R.; Domingo, A.; et al. INTEGRAL Detection of the First Prompt Gamma-Ray Signal Coincident with the Gravitational-wave Event GW170817. *Astrophys. J. Lett.* **2017**, *848*, L15. [CrossRef]
5. Pian, E.; D'Avanzo, P.; Benetti, S.; Branchesi, M.; Brocato, E.; Campana, S.; Cappellaro, E.; Covino, S.; D'Elia, V.; Fynbo, J.P.U.; et al. Spectroscopic identification of r-process nucleosynthesis in a double neutron-star merger. *Nature* **2017**, *551*, 67–70. [CrossRef] [PubMed]
6. Tanvir, N.R.; Levan, A.J.; González-Fernández, C.; Korobkin, O.; Mandel, I.; Rosswog, S.; Hjorth, J.; D'Avanzo, P.; Fruchter, A.S.; Fryer, C.L.; et al. The Emergence of a Lanthanide-rich Kilonova Following the Merger of Two Neutron Stars. *Astrophys. J. Lett.* **2017**, *848*, L27. [CrossRef]
7. Abbott, B.P.; Abbott, R.; Abbott, T.D.; Acernese, F.; Ackley, K.; Adams, C.; Adams, T.; Addesso, P.; Adhikari, R.X.; Adya, V.B.; et al. Multi-messenger Observations of a Binary Neutron Star Merger. *Astrophys. J. Lett.* **2017**, *848*, L12. [CrossRef]
8. Margutti, R.; Berger, E.; Fong, W.; Guidorzi, C.; Alexander, K.D.; Metzger, B.D.; Blanchard, P.K.; Cowperthwaite, P.S.; Chornock, R.; Eftekhari, T.; et al. The Electromagnetic Counterpart of the Binary Neutron Star Merger LIGO/Virgo GW170817. V. Rising X-Ray Emission from an Off-axis Jet. *Astrophys. J. Lett.* **2017**, *848*, L20. [CrossRef]
9. Troja, E.; Piro, L.; van Eerten, H.; Wollaeger, R.T.; Im, M.; Fox, O.D.; Butler, N.R.; Cenko, S.B.; Sakamoto, T.; Fryer, C.L.; et al. The X-ray counterpart to the gravitational-wave event GW170817. *Nature* **2017**, *551*, 71–74. [CrossRef]
10. D'Avanzo, P.; Campana, S.; Salafia, O.S.; Ghirlanda, G.; Ghisellini, G.; Melandri, A.; Bernar ini, M.G.; Branchesi, M.; Chassande-Mottin, E.; Covino, S.; et al. The evolution of the X-ray afterglow emission of GW 170817/ GRB 170817A in XMM-Newton observations. *Astron. Astrophys.* **2018**, *613*, L1. [CrossRef]
11. Cowperthwaite, P.S.; Berger, E.; Villar, V.A.; Metzger, B.D.; Nicholl, M.; Chornock, R.; Blanchard, P.K.; Fong, W.; Margutti, R.; Soares-Santos, M.; et al. The Electromagnetic Counterpart of the Binary Neutron Star Merger LIGO/Virgo GW170817. II. UV, Optical, and Near-infrared Light Curves and Comparison to Kilonova Models. *Astrophys. J. Lett.* **2017**, *848*, L17. [CrossRef]
12. Alexander, K.D.; Berger, E.; Fong, W.; Williams, P.K.G.; Guidorzi, C.; Margutti, R.; Metzger, B.D.; Annis, J.; Blanchard, P.K.; Brout, D.; et al. The Electromagnetic Counterpart of the Binary Neutron Star Merger LIGO/Virgo GW170817. VI. Radio Constraints on a Relativistic Jet and Predictions for Late-time Emission from the Kilonova Ejecta. *Astrophys. J. Lett.* **2017**, *848*, L21. [CrossRef]
13. Wang, J.; Liu, L. Electromagnetic Precursors of Short Gamma-Ray Bursts as Counterparts of Gravitational Waves. *Galaxies* **2021**, *9*, 104. [CrossRef]
14. York, D.G.; Adelman, J.; Anderson, J.E.J.; Anderson, S.F.; Annis, J.; Bahcall, N.A.; Bakken, J.A.; Barkhouser, R.; Bastian, S.; Berman, E.; et al. The Sloan Digital Sky Survey: Technical Summary. *Astron. J.* **2000**, *120*, 1579–1587. [CrossRef]
15. Keller, S.C.; Schmidt, B.P.; Bessell, M.S.; Conroy, P.G.; Francis, P.; Granlund, A.; Kowald, E.; Oates, A.P.; Martin-Jones, T.; Preston, T.; et al. The SkyMapper Telescope and The Southern Sky Survey. *Publ. Astron. Soc. Aust.* **2007**, *24*, 1–12. [CrossRef]
16. Coppi, P.; Baltay, C.; Girard, T.; Rabinowitz, D.; Zinn, R. The La Silla QUEST Variability Survey. In Proceedings of the 237th American Astronomical Society Meeting, Online, 11–15 January 2021; Bulletin of the American Astronomical Society; Volume 53, p. 324.03.
17. Kochanek, C.S.; Shappee, B.J.; Stanek, K.Z.; Holoien, T.W.S.; Thompson, T.A.; Prieto, J.L.; Dong, S.; Shields, J.V.; Will, D.; Britt, C.; et al. The All-Sky Automated Survey for Supernovae (ASAS-SN) Light Curve Server v1.0. *Publ. ASP* **2017**, *129*, 104502. [CrossRef]
18. Drake, A.J.; Djorgovski, S.G.; Mahabal, A.; Beshore, E.; Larson, S.; Graham, M.J.; Williams, R.; Christensen, E.; Catelan, M.; Boattini, A.; et al. First Results from the Catalina Real-Time Transient Survey. *Astrophys. J.* **2009**, *696*, 870–884. [CrossRef]
19. Bellm, E.C.; Kulkarni, S.R.; Graham, M.J.; Dekany, R.; Smith, R.M.; Riddle, R.; Masci, F.J.; Helou, G.; Prince, T.A.; Adams, S.M.; et al. The Zwicky Transient Facility: System Overview, Performance, and First Results. *Publ. ASP* **2019**, *131*, 018002. [CrossRef]
20. Ivezić, Ž.; Kahn, S.M.; Tyson, J.A.; Abel, B.; Acosta, E.; Allsman, R.; Alonso, D.; AlSayyad, Y.; Anderson, S.F.; Andrew, J.; et al. LSST: From Science Drivers to Reference Design and Anticipated Data Products. *Astrophys. J.* **2019**, *873*, 111. [CrossRef]
21. Castro-Chacón, J.H.; Lehner, M.; Reyes-Ruiz, M.; Hernández, B.; Alcock, C.; Huang, C.K.; Cook, K.H.; Geary, J.C.; Figueroa, L.; Guerrero, C.; et al. TAOS II: The Robotic Operations. *RMxAC* **2021**, *53*, 137–139. . [CrossRef]
22. Beskin, G.; Karpov, S.; Plokhotnichenko, V.; Stepanov, A.; Tsap, Y. Discovery of the Sub-second Linearly Polarized Spikes of Synchrotron Origin in the UV Ceti Giant Optical Flare. *Publ. Astron. Soc. Aust.* **2017**, *34*, e010. . [CrossRef]
23. Adams, J.H.; Ahmad, S.; Albert, J.N.; Allard, D.; Anchordoqui, L.; Andreev, V.; Anzalone, A.; Arai, Y.; Asano, K.; Ave Pernas, M.; et al. Science of atmospheric phenomena with JEM-EUSO. *Exp. Astron.* **2015**, *40*, 239–251. [CrossRef]

24. Schwartz, R.N.; Townes, C.H. Interstellar and Interplanetary Communication by Optical Masers. *Nature* **1961**, *190*, 205–208. [CrossRef]
25. Howard, A.W.; Horowitz, P.; Wilkinson, D.T.; Coldwell, C.M.; Groth, E.J.; Jarosik, N.; Latham, D.W.; Stefanik, R.P.; Willman, A.J., Jr.; Wolff, J.; et al. Search for Nanosecond Optical Pulses from Nearby Solar-Type Stars. *Astrophys. J.* **2004**, *613*, 1270–1284. [CrossRef]
26. Andreoni, I.; Cooke, J. The Deeper Wider Faster Programme: Chasing the Fastest Bursts in the Universe. In Proceedings of the Southern Horizons in Time-Domain Astronomy, Cape Town, South Africa, 17 November 2017; Griffin, R.E., Ed.; International Astronomical Union: Paris, France, 2019; Volume 339, pp. 135–138. [CrossRef]
27. Strausbaugh, R.; Cucchiara, A.; Dow, M., Jr.; Webb, S.; Zhang, J.; Goode, S.; Cooke, J. Finding Fast Transients in Real Time Using a Novel Light-curve Analysis Algorithm. *Astron. J.* **2022**, *163*, 95. [CrossRef]
28. Ackermann, M.; McGraw, J.; Zimmer, P. An Overview of Wide-Field-Of-View Optical Designs for Survey Telescopes. In Proceedings of the Advanced Maui Optical and Space Surveillance Technologies Conference, Wailea, HI, USA, 14–17 September 2010; p. E41.
29. Terebizh, V.Y. New designs of survey telescopes. *Astron. Nachrichten* **2011**, *332*, 714–742. [CrossRef]
30. Beskin, G.; Bondar, S.; Karpov, S.; Plokhotnichenko, V.; Guarnieri, A.; Bartolini, C.; Greco, G.; Piccioni, A.; Shearer, A. From TORTORA to MegaTORTORA—Results and Prospects of Search for Fast Optical Transients. *Adv. Astron.* **2010**, *2010*, 171569. [CrossRef]
31. Law, N.; Vasquez Soto, A.; Corbett, H.; Galliher, N.; Gonzalez, R.; Machia, L.; Walters, G. The inside-out, upside-down telescope: The Argus Array's new pseudofocal design. In Proceedings of the Ground-Based and Airborne Telescopes IX, Montreal, QC, Canada, 17–23 July 2022; Society of Photo-Optical Instrumentation Engineers (SPIE) Conference Series; Marshall, H.K., Spyromilio, J., Usuda, T., Eds.; SPIE: Bellingham, WA, USA, 2022; Volume 12182, p. 121824H. [CrossRef]
32. Tsvetkova, A.; Svinkin, D.; Karpov, S.; Frederiks, D. Key Space and Ground Facilities in GRB Science. *Universe* **2022**, *8*, 373. [CrossRef]
33. Kouveliotou, C.; Meegan, C.A.; Fishman, G.J.; Bhat, N.P.; Briggs, M.S.; Koshut, T.M.; Paciesas, W.S.; Pendleton, G.N. Identification of Two Classes of Gamma-Ray Bursts. *Astrophys. J. Lett.* **1993**, *413*, L101. [CrossRef]
34. Kumar, P.; Zhang, B. The physics of gamma-ray bursts & relativistic jets. *Phys. Rep.* **2015**, *561*, 1–109. [CrossRef]
35. Bošnjak, Ž.; Barniol Duran, R.; Pe'er, A. The GRB Prompt Emission: An Unsolved Puzzle. *Galaxies* **2022**, *10*, 38. [CrossRef]
36. Mészáros, P.; Rees, M.J. Optical and Long-Wavelength Afterglow from Gamma-Ray Bursts. *Astrophys. J.* **1997**, *476*, 232–237. [CrossRef]
37. Covino, S.; Gotz, D. Polarization of prompt and afterglow emission of Gamma-Ray Bursts. *Astron. Astrophys. Trans.* **2016**, *29*, 205–244. [CrossRef]
38. Oganesyan, G.; Nava, L.; Ghirlanda, G.; Melandri, A.; Celotti, A. Prompt optical emission as a signature of synchrotron radiation in gamma-ray bursts. *Astron. Astrophys.* **2019**, *628*, A59. [CrossRef]
39. Yu, Y.W.; Gao, H.; Wang, F.Y.; Zhang, B.B. Gamma-Ray Bursts. In *Handbook of X-ray and Gamma-ray Astrophysics*; Springer: Singapore, 2022; p. 31. [CrossRef]
40. Panaitescu, A.D.; Vestrand, W.T. The Synchrotron Low-energy Spectrum Arising from the Cooling of Electrons in Gamma-Ray Bursts. *Astrophys. J.* **2022**, *938*, 132. [CrossRef]
41. Karpov, S.; Beskin, G.; Bondar, S.; Bartolini, C.; Greco, G.; Guarnieri, A.; Nanni, D.; Piccioni, A.; Terra, F.; Molinari, E.; et al. GRB 080319B: TORTORA synchronous observation. *GRB Coord. Netw.* **2008**, *7452*, 1.
42. Racusin, J.L.; Karpov, S.V.; Sokolowski, M.; Granot, J.; Wu, X.F.; Pal'Shin, V.; Covino, S.; van der Horst, A.J.; Oates, S.R.; Schady, P.; et al. Broadband observations of the naked-eye γ-ray burst GRB080319B. *Nature* **2008**, *455*, 183–188. [CrossRef]
43. Molinari, E.; Bondar, S.; Karpov, S.; Beskin, G.; Biryukov, A.; Ivanov, E.; Bartolini, C.; Greco, G.; Guarnieri, A.; Piccioni, A.; et al. TORTOREM: Two-telescope complex for detection and investigation of optical transients. *Nuovo C. B Ser.* **2006**, *121*, 1525–1526. [CrossRef]
44. Beskin, G.; Karpov, S.; Bondar, S.; Greco, G.; Guarnieri, A.; Bartolini, C.; Piccioni, A. Fast Optical Variability of a Naked-eye Burst—Manifestation of the Periodic Activity of an Internal Engine. *Astrophys. J. Lett.* **2010**, *719*, L10–L14. [CrossRef]
45. Karpov, S.; Beskin, G.; Biryukov, A.; Bondar, S.; Hurley, K.; Ivanov, E.; Katkova, E.; Pozanenko, A.; Zolotukhin, I. Optical camera with high temporal resolution to search for transients in the wide field. *Nuovo C. C Geophys. Space Phys. C* **2005**, *28*, 747. [CrossRef]
46. Beskin, G.; Karpov, S.; Bondar, S.; Guarnieri, A.; Bartolini, C.; Greco, D.; Piccioni, A. Rapid optical variability of the gamma-ray burst grb 080319b and its central engine. *Astrophys. Bull.* **2010**, *65*, 223–229. [CrossRef]
47. Beskin, G.M.; Karpov, S.V.; Biryukov, A.V.; Bondar, S.F.; Ivanov, E.A.; Katkova, E.V.; Orekhova, N.V.; Perkov, A.V.; Sasyuk, V.V. Wide-field optical monitoring with Mini-MegaTORTORA (MMT-9) multichannel high temporal resolution telescope. *Astrophys. Bull.* **2017**, *72*, 81–92. [CrossRef]
48. Zhang, B.B.; Zhang, B.; Castro-Tirado, A.J.; Dai, Z.G.; Tam, P.H.T.; Wang, X.Y.; Hu, Y.D.; Karpov, S.; Pozanenko, A.; Zhang, F.W.; et al. Transition from fireball to Poynting-flux-dominated outflow in the three-episode GRB 160625B. *Nat. Astron.* **2018**, *2*, 69–75. [CrossRef]
49. Oganesyan, G.; Karpov, S.; Salafia, O.S.; Jelínek, M.; Beskin, G.; Ronchini, S.; Banerjee, B.; Branchesi, M.; Štrobl, J.; Polášek, C.; et al. Exceptionally bright optical emission from a rare and distant gamma-ray burst. *Nat. Astron.* **2023**, *7*, 843–855. [CrossRef]

50. Gill, R.; Granot, J.; Kumar, P. Linear polarization in gamma-ray burst prompt emission. *Mon. Not. R. Astron. Soc.* **2020**, *491*, 3343–3373. [CrossRef]
51. Gill, R.; Kole, M.; Granot, J. GRB Polarization: A Unique Probe of GRB Physics. *Galaxies* **2021**, *9*, 82. [CrossRef]
52. McConnell, M.L. High energy polarimetry of prompt GRB emission. *New Astron. Rev.* **2017**, *76*, 1–21. [CrossRef]
53. Brivio, R.; Covino, S.; D'Avanzo, P.; Wiersema, K.; Maund, J.R.; Bernardini, M.G.; Campana, S.; Melandri, A. GRB 080928 afterglow imaging and spectro-polarimetry. *Astron. Astrophys.* **2022**, *666*, A179. [CrossRef]
54. Kopač, D.; Mundell, C.G.; Japelj, J.; Arnold, D.M.; Steele, I.A.; Guidorzi, C.; Dichiara, S.; Kobayashi, S.; Gomboc, A.; Harrison, R.M.; et al. Limits on Optical Polarization during the Prompt Phase of GRB 140430A. *Astrophys. J.* **2015**, *813*, 1. [CrossRef]
55. Kumar, P.; Lu, W.; Bhattacharya, M. Fast radio burst source properties and curvature radiation model. *Mon. Not. R. Astron. Soc.* **2017**, *468*, 2726–2739. [CrossRef]
56. Burke-Spolaor, S. Multiple messengers of fast radio bursts. *Nat. Astron.* **2018**, *2*, 845–848. [CrossRef]
57. Petroff, E.; Hessels, J.W.T.; Lorimer, D.R. Fast radio bursts. *Astron. Astrophys. Rev.* **2019**, *27*, 4. [CrossRef]
58. Thornton, D.; Stappers, B.; Bailes, M.; Barsdell, B.; Bates, S.; Bhat, N.D.R.; Burgay, M.; Burke-Spolaor, S.; Champion, D.J.; Coster, P.; et al. A Population of Fast Radio Bursts at Cosmological Distances. *Science* **2013**, *341*, 53–56. [CrossRef] [PubMed]
59. Katz, J.I. Fast radio bursts. *Prog. Part. Nucl. Phys.* **2018**, *103*, 1–18. [CrossRef]
60. MAGIC Collaboration; Acciari, V.A.; Ansoldi, S.; Antonelli, L.A.; Arbet Engels, A.; Arcaro, C.; Baack, D.; Babić, A.; Banerjee, B.; Bangale, P.; et al. Constraining very-high-energy and optical emission from FRB 121102 with the MAGIC telescopes. *Mon. Not. R. Astron. Soc.* **2018**, *481*, 2479–2486. [CrossRef]
61. Hardy, L.K.; Dhillon, V.S.; Spitler, L.G.; Littlefair, S.P.; Ashley, R.P.; De Cia, A.; Green, M.J.; Jaroenjittichai, P.; Keane, E.F.; Kerry, P.; et al. A search for optical bursts from the repeating fast radio burst FRB 121102. *Mon. Not. R. Astron. Soc.* **2017**, *472*, 2800–2807. [CrossRef]
62. Valyavin, G.; Beskin, G.; Valeev, A.; Galazutdinov, G.; Fabrika, S.; Romanyuk, I.; Aitov, V.; Yakovlev, O.; Ivanova, A.; Baluev, R.; et al. EXPLANATION: Exoplanet and Transient Event Investigation Project—Optical Facilities and Solutions. *Photonics* **2022**, *9*, 950. [CrossRef]
63. Nakar, E.; Gottlieb, O.; Piran, T.; Kasliwal, M.M.; Hallinan, G. From γ to Radio: The Electromagnetic Counterpart of GW170817. *Astrophys. J.* **2018**, *867*, 18. [CrossRef]
64. Ma, S.B.; Lei, W.H.; Gao, H.; Xie, W.; Chen, W.; Zhang, B.; Wang, D.X. Bright Merger-nova Emission Powered by Magnetic Wind from a Newborn Black Hole. *Astrophys. J. Lett.* **2018**, *852*, L5. [CrossRef]
65. Petroff, E.; Barr, E.D.; Jameson, A.; Keane, E.F.; Bailes, M.; Kramer, M.; Morello, V.; Tabbara, D.; van Straten, W. FRBCAT: The Fast Radio Burst Catalogue. *Publ. Astron. Soc. Aust.* **2016**, *33*, e045. [CrossRef]
66. Beskin, G.M.; Karpov, S.V.; Plokhotnichenko, V.L.; Bondar, S.F.; Perkov, A.V.; Ivanov, E.A.; Katkova, E.V.; Sasyuk, V.V.; Shearer, A. Wide-field subsecond temporal resolution optical monitoring systems for the detection and study of cosmic hazards. *Phys. Uspekhi* **2013**, *56*, 836–842. [CrossRef]
67. Beskin, G.; Biryukov, A.; Bondar, S.; Ivanov, E.; Karpov, S.; Katkova, E.; Orekhova, N.; Perkov, A.; Plokhotnichenko, V.; Sasyuk, V. SAINT: A Wide-Field Multichannel Optical Telescope Project. In Proceedings of the Stars: From Collapse to Collapse, Nizhny Arkhyz, Russia, 3–7 October 2016; Astronomical Society of the Pacific Conference Series; Balega, Y.Y., Kudryavtsev, D.O., Romanyuk, I.I., Yakunin, I.A., Eds.; Astronomical Society of the Pacific: San Francisco, CA, USA, 2017; Volume 510, p. 530.
68. Valyavin, G.; Beskin, G.; Valeev, A.; Galazutdinov, G.; Fabrika, S.; Aitov, V.; Yakovlev, O.; Ivanova, A.; Baluev, R.; Vlasyuk, V.; et al. EXPLANATION: Exoplanet and Transient Events Investigation Project. *Astrophys. Bull.* **2022**, *77*, 495–508. [CrossRef]
69. Shugarov, A.S. The prospects of modern CMOS and CCD detectors for wide field telescopes. *INASAN Sci. Rep.* **2020**, *5*, 236–242. [CrossRef]
70. Karpov, S.; Beskin, G.; Bondar, S.; Guarnieri, A.; Bartolini, C.; Greco, G.; Piccioni, A. Wide and Fast: Monitoring the Sky in Subsecond Domain with the FAVOR and TORTORA Cameras. *Adv. Astron.* **2010**, *2010*, 784141. [CrossRef]
71. Alard, C.; Lupton, R.H. A Method for Optimal Image Subtraction. *Astrophys. J.* **1998**, *503*, 325–331. [CrossRef]
72. Zackay, B.; Ofek, E.O.; Gal-Yam, A. Proper Image Subtraction—Optimal Transient Detection, Photometry, and Hypothesis Testing. *Astrophys. J.* **2016**, *830*, 27. [CrossRef]
73. Hu, L.; Wang, L.; Chen, X.; Yang, J. Image Subtraction in Fourier Space. *Astrophys. J.* **2022**, *936*, 157. [CrossRef]
74. Sedaghat, N.; Mahabal, A. Effective image differencing with convolutional neural networks for real-time transient hunting. *Mon. Not. R. Astron. Soc.* **2018**, *476*, 5365–5376. [CrossRef]
75. Acero-Cuellar, T.; Bianco, F.; Dobler, G.; Sako, M.; Qu, H.; LSST Dark Energy Science Collaboration. What's the Difference? The Potential for Convolutional Neural Networks for Transient Detection without Template Subtraction. *Astron. J.* **2023**, *166*, 115. [CrossRef]
76. Cabrera-Vives, G.; Reyes, I.; Förster, F.; Estévez, P.A.; Maureira, J.C. Deep-HiTS: Rotation Invariant Convolutional Neural Network for Transient Detection. *Astrophys. J.* **2017**, *836*, 97. [CrossRef]
77. Gieseke, F.; Bloemen, S.; van den Bogaard, C.; Heskes, T.; Kindler, J.; Scalzo, R.A.; Ribeiro, V.A.R.M.; van Roestel, J.; Groot, P.J.; Yuan, F.; et al. Convolutional neural networks for transient candidate vetting in large-scale surveys. *Mon. Not. R. Astron. Soc.* **2017**, *472*, 3101–3114. [CrossRef]
78. Carrasco-Davis, R.; Cabrera-Vives, G.; Förster, F.; Estévez, P.A.; Huijse, P.; Protopapas, P.; Reyes, I.; Martínez-Palomera, J.; Donoso, C. Deep Learning for Image Sequence Classification of Astronomical Events. *Publ. ASP* **2019**, *131*, 108006. [CrossRef]

79. Makhlouf, K.; Turpin, D.; Corre, D.; Karpov, S.; Kann, D.A.; Klotz, A. O'TRAIN: A robust and flexible 'real or bogus' classifier for the study of the optical transient sky. *Astron. Astrophys.* **2022**, *664*, A81. [CrossRef]
80. Davenport, J.R.A. The Kepler Catalog of Stellar Flares. *Astrophys. J.* **2016**, *829*, 23. [CrossRef]
81. Günther, M.N.; Zhan, Z.; Seager, S.; Rimmer, P.B.; Ranjan, S.; Stassun, K.G.; Oelkers, R.J.; Daylan, T.; Newton, E.; Kristiansen, M.H.; et al. Stellar Flares from the First TESS Data Release: Exploring a New Sample of M Dwarfs. *Astron. J.* **2020**, *159*, 60. [CrossRef]
82. Yang, Z.; Zhang, L.; Meng, G.; Han, X.L.; Misra, P.; Yang, J.; Pi, Q. Properties of flare events based on light curves from the TESS survey. *Astron. Astrophys.* **2023**, *669*, A15. [CrossRef]
83. Howard, W.S.; Corbett, H.; Law, N.M.; Ratzloff, J.K.; Glazier, A.; Fors, O.; del Ser, D.; Haislip, J. EvryFlare. I. Long-term Evryscope Monitoring of Flares from the Cool Stars across Half the Southern Sky. *Astrophys. J.* **2019**, *881*, 9. [CrossRef]
84. Jackman, J.A.G.; Wheatley, P.J.; West, R.G.; Gill, S.; Jenkins, J.S. Big flares from small stars: Detecting flares from faint low-mass stars with NGTS full-frame images. *Mon. Not. R. Astron. Soc.* **2023**, *525*, 1588–1600. [CrossRef]
85. Kupfer, T.; Prince, T.A.; van Roestel, J.; Bellm, E.C.; Bildsten, L.; Coughlin, M.W.; Drake, A.J.; Graham, M.J.; Klein, C.; Kulkarni, S.R.; et al. Year 1 of the ZTF high-cadence Galactic plane survey: Strategy, goals, and early results on new single-mode hot subdwarf B-star pulsators. *Mon. Not. R. Astron. Soc.* **2021**, *505*, 1254–1267. [CrossRef]
86. Aizawa, M.; Kawana, K.; Kashiyama, K.; Ohsawa, R.; Kawahara, H.; Naokawa, F.; Tajiri, T.; Arima, N.; Jiang, H.; Hartwig, T.; et al. Fast optical flares from M dwarfs detected by a one-second-cadence survey with Tomo-e Gozen. *Publ. Astron. Soc. Jpn.* **2022**, *74*, 1069–1094. [CrossRef]
87. Weryk, R.J.; Campbell-Brown, M.D.; Wiegert, P.A.; Brown, P.G.; Krzeminski, Z.; Musci, R. The Canadian Automated Meteor Observatory (CAMO): System overview. *Icarus* **2013**, *225*, 614–622. [CrossRef]
88. Jenniskens, P. Meteor showers in review. *Planet. Space Sci.* **2017**, *143*, 116–124. [CrossRef]
89. Sako, S.; Ohsawa, R.; Takahashi, H.; Kojima, Y.; Doi, M.; Kobayashi, N.; Aoki, T.; Arima, N.; Arimatsu, K.; Ichiki, M.; et al. The Tomo-e Gozen wide field CMOS camera for the Kiso Schmidt telescope. In Proceedings of the Ground-based and Airborne Instrumentation for Astronomy VII, Austin, TX, USA, 10–15 June 2018; Society of Photo-Optical Instrumentation Engineers (SPIE) Conference Series; Evans, C.J., Simard, L., Takami, H., Eds.; SPIE: Bellingham, WA, USA, 2018; Volume 10702, p. 107020J. [CrossRef]
90. Ceplecha, Z.; Borovička, J.; Elford, W.G.; Revelle, D.O.; Hawkes, R.L.; Porubčan, V.; Šimek, M. Meteor Phenomena and Bodies. *Space Sci. Rev.* **1998**, *84*, 327–471. [CrossRef]
91. Love, S.G.; Brownlee, D.E. A Direct Measurement of the Terrestrial Mass Accretion Rate of Cosmic Dust. *Science* **1993**, *262*, 550–553. [CrossRef] [PubMed]
92. Plane, J.M.C. Cosmic dust in the earth's atmosphere. *Chem. Soc. Rev.* **2012**, *41*, 6507–6518. [CrossRef] [PubMed]
93. Baggaley, W.J. Radar Observations. In *Meteors in the Earth's Atmosphere*; Murad, E., Williams, I.P., Eds.; Cambridge University Press: Cambridge, UK, 2002; pp. 123–147.
94. Reylé, C.; Jardine, K.; Fouqué, P.; Caballero, J.A.; Smart, R.L.; Sozzetti, A. The 10 parsec sample in the Gaia era. *Astron. Astrophys.* **2021**, *650*, A201. [CrossRef]
95. Choi, J.; Dotter, A.; Conroy, C.; Cantiello, M.; Paxton, B.; Johnson, B.D. Mesa Isochrones and Stellar Tracks (MIST). I. Solar-scaled Models. *Astrophys. J.* **2016**, *823*, 102. [CrossRef]
96. Engle, S.G.; Guinan, E.F. Living with a Red Dwarf: The Rotation-Age Relationships of M Dwarfs. *Astrophys. J. Lett.* **2023**, *954*, L50. [CrossRef]
97. Kasting, J.F.; Whitmire, D.P.; Reynolds, R.T. Habitable Zones around Main Sequence Stars. *Icarus* **1993**, *101*, 108–128. [CrossRef] [PubMed]
98. Dressing, C.D.; Charbonneau, D. The Occurrence of Potentially Habitable Planets Orbiting M Dwarfs Estimated from the Full Kepler Dataset and an Empirical Measurement of the Detection Sensitivity. *Astrophys. J.* **2015**, *807*, 45. [CrossRef]
99. Hsu, D.C.; Ford, E.B.; Terrien, R. Occurrence rates of planets orbiting M Stars: Applying ABC to Kepler DR25, Gaia DR2, and 2MASS data. *Mon. Not. R. Astron. Soc.* **2020**, *498*, 2249–2262. [CrossRef]
100. Cameron, A.C.; Extrasolar Planetary Transits. In *Methods of Detecting Exoplanets: 1st Advanced School on Exoplanetary Science*; Bozza, V., Mancini, L., Sozzetti, A., Eds.; Springer International Publishing: Cham, Swizerland, 2016; pp. 89–131. [CrossRef]
101. Giacobbe, P.; Damasso, M.; Sozzetti, A.; Toso, G.; Perdoncin, M.; Calcidese, P.; Bernagozzi, A.; Bertolini, E.; Lattanzi, M.G.; Smart, R.L. Photometric transit search for planets around cool stars from the western Italian Alps: A pilot study. *Mon. Not. R. Astron. Soc.* **2012**, *424*, 3101–3122. [CrossRef]
102. Berta, Z.K.; Irwin, J.; Charbonneau, D. Constraints on Planet Occurrence around Nearby Mid-to-late M Dwarfs from the MEARTH Project. *Astrophys. J.* **2013**, *775*, 91. [CrossRef]
103. Gibbs, A.; Bixel, A.; Rackham, B.V.; Apai, D.; Schlecker, M.; Espinoza, N.; Mancini, L.; Chen, W.P.; Henning, T.; Gabor, P.; et al. EDEN: Sensitivity Analysis and Transiting Planet Detection Limits for Nearby Late Red Dwarfs. *Astron. J.* **2020**, *159*, 169. [CrossRef]
104. Kovács, G.; Zucker, S.; Mazeh, T. A box-fitting algorithm in the search for periodic transits. *Astron. Astrophys.* **2002**, *391*, 369–377. [CrossRef]
105. Kovács, G.; Bakos, G.; Noyes, R.W. A trend filtering algorithm for wide-field variability surveys. *Mon. Not. R. Astron. Soc.* **2005**, *356*, 557–567. [CrossRef]

106. Hippke, M.; Heller, R. Optimized transit detection algorithm to search for periodic transits of small planets. *Astron. Astrophys.* **2019**, *623*, A39. [CrossRef]
107. Agol, E. Transit Surveys for Earths in the Habitable Zones of White Dwarfs. *Astrophys. J. Lett.* **2011**, *731*, L31. [CrossRef]
108. Agol, E. Finding habitable earths around white dwarfs with a robotic telescope transit survey. In Proceedings of the Telescopes from Afar, Waikoloa Beach Marriott, HI, USA, 28 February–3 March 2011; p. 25.
109. Farihi, J.; Parsons, S.G.; Gänsicke, B.T. A circumbinary debris disk in a polluted white dwarf system. *Nat. Astron.* **2017**, *1*, 0032. [CrossRef]
110. Stephan, A.P.; Naoz, S.; Gaudi, B.S. Giant Planets, Tiny Stars: Producing Short-period Planets around White Dwarfs with the Eccentric Kozai-Lidov Mechanism. *Astrophys. J.* **2021**, *922*, 4. [CrossRef]
111. Veras, D. Planetary Systems Around White Dwarfs. In *Oxford Research Encyclopedia of Planetary Science*; Oxford University Press: New York, NY, USA, 2021; p. 1. [CrossRef]
112. Vanderburg, A.; Rappaport, S.A.; Xu, S.; Crossfield, I.J.M.; Becker, J.C.; Gary, B.; Murgas, F.; Blouin, S.; Kaye, T.G.; Palle, E.; et al. A giant planet candidate transiting a white dwarf. *Nature* **2020**, *585*, 363–367. [CrossRef] [PubMed]
113. Faedi, F.; West, R.G.; Burleigh, M.R.; Goad, M.R.; Hebb, L. Detection limits for close eclipsing and transiting substellar and planetary companions to white dwarfs in the WASP survey. *Mon. Not. R. Astron. Soc.* **2011**, *410*, 899–911. [CrossRef]
114. Kilic, M.; Agol, E.; Loeb, A.; Maoz, D.; Munn, J.A.; Gianninas, A.; Canton, P.; Barber, S.D. Habitable Planets Around White Dwarfs: An Alternate Mission for the Kepler Spacecraft. *arXiv* **2013**, arXiv:1309.0009.
115. Granados, A.; Hermes, J. Using TESS Data to Search for Transiting Exoplanets Around White Dwarfs. In Proceedings of the 237th American Astronomical Society Meeting, Online, 11–15 January 2021; Bulletin of the American Astronomical Society; Volume 53, p. 134.05.
116. Krushevska, V.N.; Kuznyetsova, Y.G.; Veles, O.A.; Andreev, M.V.; Romanyuk, Y.O.; Garai, Z.; Pribulla, T.; Budaj, J.; Shugarov, S.; Kundra, E.; et al. Search for extrasolar planets around white dwarfs. *Contrib. Astron. Obs. Skaln. Pleso* **2019**, *49*, 380–383.
117. Carrazco-Gaxiola, J.S.; Gómez Maqueo Chew, Y.; Pereyra, M. Sample of White Dwarfs to Search for Transiting Exoplanets from OANSPM. *RMxAC* **2023**, *55*, 93. [CrossRef]
118. Aloisi, R.; Vanderburg, A.; Selep, P.; Popowicz, A.; Bernacki, K.; Hambsch, J.; Kaye, T.G.; Oksanen, A.; Rappaport, S. TESS Full Frame Image Search for White Dwarf Exoplanets. In Proceedings of the Posters from the TESS Science Conference II (TSC2), Online, 2–6 August 2021; p. 119. [CrossRef]
119. Kleinman, S.J.; Kepler, S.O.; Koester, D.; Pelisoli, I.; Peçanha, V.; Nitta, A.; Costa, J.E.S.; Krzesinski, J.; Dufour, P.; Lachapelle, F.R.; et al. SDSS DR7 White Dwarf Catalog. *Astrophys. J. Suppl.* **2013**, *204*, 5. [CrossRef]
120. Gentile Fusillo, N.P.; Tremblay, P.E.; Cukanovaite, E.; Vorontseva, A.; Lallement, R.; Hollands, M.; Gänsicke, B.T.; Burdge, K.B.; McCleery, J.; Jordan, S. A catalogue of white dwarfs in Gaia EDR3. *Mon. Not. R. Astron. Soc.* **2021**, *508*, 3877–3896. [CrossRef]
121. Piotrowski, L.; Małek, K.; Mankiewicz, L.; Sokołowski, M.; Wrochna, G.; Żadroźny, A.; Żarnecki, A.F. Limits on the flux of heavy compact objects from the the "Pi of the Sky" project. In Proceedings of the 37th International Cosmic Ray Conference, Berlin, Germany, 15–22 July 2021; p. 536. [CrossRef]
122. Ratzloff, J.K.; Law, N.M.; Fors, O.; Corbett, H.T.; Howard, W.S.; del Ser, D.; Haislip, J. Building the Evryscope: Hardware Design and Performance. *Publ. ASP* **2019**, *131*, 075001. [CrossRef]
123. Mankiewicz, L.; Batsch, T.; Castro-Tirado, A.; Czyrkowski, H.; Cwiek, A.; Cwiok, M.; Dabrowski, R.; Jelínek, M.; Kasprowicz, G.; Majcher, A.; et al. Pi of the Sky full system and the new telescope. *RMxAC* **2014**, *45*, 7. Available online: https://www.redalyc.org/articulo.oa?id=57132995003 (accessed on 15 March 2012).
124. Han, X.; Xiao, Y.; Zhang, P.; Turpin, D.; Xin, L.; Wu, C.; Cai, H.; Dong, W.; Huang, L.; Kang, Z.; et al. The Automatic Observation Management System of the GWAC Network. I. System Architecture and Workflow. *Publ. ASP* **2021**, *133*, 065001. [CrossRef]
125. Law, N.M.; Carlberg, R.; Salbi, P.; Ngan, W.H.W.; Ahmadi, A.; Steinbring, E.; Murowinski, R.; Sivanandam, S.; Kerzendorf, W. Exoplanets from the Arctic: The First Wide-field Survey at 80°N. *Astron. J.* **2013**, *145*, 58. [CrossRef]
126. Vestrand, W.T.; Borozdin, K.N.; Brumby, S.P.; Casperson, D.E.; Fenimore, E.E.; Galassi, M.C.; McGowan, K.; Perkins, S.J.; Priedhorsky, W.C.; Starr, D.; et al. The RAPTOR experiment: A system for monitoring the optical sky in real time. In Proceedings of the Advanced Global Communications Technologies for Astronomy II, Waikoloa, HI, USA, 22–28 August 2002; Society of Photo-Optical Instrumentation Engineers (SPIE) Conference Series; Kibrick, R.I., Ed.; SPIE: Bellingham, WA, USA, 2002; Volume 4845, pp. 126–136. [CrossRef]
127. Pepper, J.; Kuhn, R.B.; Siverd, R.; James, D.; Stassun, K. The KELT-South Telescope. *Publ. ASP* **2012**, *124*, 230. [CrossRef]
128. Pollacco, D.L.; Skillen, I.; Collier Cameron, A.; Christian, D.J.; Hellier, C.; Irwin, J.; Lister, T.A.; Street, R.A.; West, R.G.; Anderson, D.R.; et al. The WASP Project and the SuperWASP Cameras. *Publ. ASP* **2006**, *118*, 1407–1418. [CrossRef]
129. Ofek, E.O.; Ben-Ami, S.; Polishook, D.; Segre, E.; Blumenzweig, A.; Strotjohann, N.L.; Yaron, O.; Shani, Y.M.; Nachshon, S.; Shvartzvald, Y.; et al. The Large Array Survey Telescope-System Overview and Performances. *Publ. ASP* **2023**, *135*, 065001. [CrossRef]
130. Ben-Ami, S.; Ofek, E.O.; Polishook, D.; Franckowiak, A.; Hallakoun, N.; Segre, E.; Shvartzvald, Y.; Strotjohann, N.L.; Yaron, O.; Aharonson, O.; et al. The Large Array Survey Telescope—Science Goals. *arXiv* **2023**, arXiv:2304.02719.
131. Law, N.; Corbett, H.; Vasquez Soto, A.; Galliher, N.; Gonzalez, R.; Glazier, A.; Howard, W.; Argus Array Science Team. The Argus Array: The Deep Sky Every Minute. In Proceedings of the 237th American Astronomical Society Meeting, Online, 11–15 January 2021; Bulletin of the American Astronomical Society; Volume 53, p. 235.02.

132. Bakos, G.Á.; Csubry, Z.; Penev, K.; Bayliss, D.; Jordán, A.; Afonso, C.; Hartman, J.D.; Henning, T.; Kovács, G.; Noyes, R.W.; et al. HATSouth: A Global Network of Fully Automated Identical Wide-Field Telescopes. *Publ. ASP* **2013**, *125*, 154. [CrossRef]
133. Wheatley, P.J.; West, R.G.; Goad, M.R.; Jenkins, J.S.; Pollacco, D.L.; Queloz, D.; Rauer, H.; Udry, S.; Watson, C.A.; Chazelas, B.; et al. The Next Generation Transit Survey (NGTS). *Mon. Not. R. Astron. Soc.* **2018**, *475*, 4476–4493. [CrossRef]
134. Kornilov, V.G.; Lipunov, V.M.; Gorbovskoy, E.S.; Belinski, A.A.; Kuvshinov, D.A.; Tyurina, N.V.; Shatsky, N.I.; Sankovich, A.V.; Krylov, A.V.; Balanutsa, P.V.; et al. Robotic optical telescopes global network MASTER II. Equipment, structure, algorithms. *Exp. Astron.* **2012**, *33*, 173–196. [CrossRef]
135. Watson, A.M.; Lee, W.H.; Troja, E.; Román-Zúñiga, C.G.; Butler, N.R.; Kutyrev, A.S.; Gehrels, N.A.; Ángeles, F.; Basa, S.; Blanc, P.E.; et al. DDOTI: The deca-degree optical transient imager. In Proceedings of the Observatory Operations: Strategies, Processes, and Systems VI, Edinburgh, UK, 26 June–1 July 2016; Society of Photo-Optical Instrumentation Engineers (SPIE) Conference Series; Peck, A.B., Seaman, R.L., Benn, C.R., Eds.; SPIE: Bellingham, WA, USA, 2016; Volume 9910, p. 99100G. [CrossRef]
136. Dyer, M.J.; Steeghs, D.; Galloway, D.K.; Dhillon, V.S.; O'Brien, P.; Ramsay, G.; Noysena, K.; Pallé, E.; Kotak, R.; Breton, R.; et al. The Gravitational-wave Optical Transient Observer (GOTO). In Proceedings of the Society of Photo-Optical Instrumentation Engineers (SPIE) Conference Series, Online, 14–18 December 2020; Volume 11445, p. 114457G. [CrossRef]
137. Larson, S.; Beshore, E.; Hill, R.; Christensen, E.; McLean, D.; Kolar, S.; McNaught, R.; Garradd, G. The CSS and SSS NEO surveys. In Proceedings of the AAS/Division for Planetary Sciences Meeting Abstracts #35, Monterey, CA, USA, 4–7 May 2003; Volume 35, p. 36.04.
138. Tonry, J.L.; Denneau, L.; Heinze, A.N.; Stalder, B.; Smith, K.W.; Smartt, S.J.; Stubbs, C.W.; Weiland, H.J.; Rest, A. ATLAS: A High-cadence All-sky Survey System. *Publ. ASP* **2018**, *130*, 064505. [CrossRef]
139. Akerlof, C.W.; Kehoe, R.L.; McKay, T.A.; Rykoff, E.S.; Smith, D.A.; Casperson, D.E.; McGowan, K.E.; Vestrand, W.T.; Wozniak, P.R.; Wren, J.A.; et al. The ROTSE-III Robotic Telescope System. *Publ. ASP* **2003**, *115*, 132–140. [CrossRef]
140. Groot, P.J.; Bloemen, S.; Vreeswijk, P.M.; Jonker, P.G.; Pieterse, D.; Engels, A.; Michiels, J.; Bakker, R.; Hahn, F.; Raskin, G.; et al. BlackGEM: The wide-field multi-band optical telescope array. In Proceedings of the Ground-Based and Airborne Telescopes IX, Montreal, QC, Canada, 17–22 July 2022; Society of Photo-Optical Instrumentation Engineers (SPIE) Conference Series; Marshall, H.K., Spyromilio, J., Usuda, T., Eds.; SPIE: Bellingham, WA, USA, 2022; Volume 12182, p. 121821V. [CrossRef]
141. Nir, G.; Ofek, E.O.; Ben-Ami, S.; Segev, N.; Polishook, D.; Hershko, O.; Diner, O.; Manulis, I.; Zackay, B.; Gal-Yam, A.; et al. The Weizmann Fast Astronomical Survey Telescope (W-FAST): System Overview. *Publ. ASP* **2021**, *133*, 075002. [CrossRef]
142. Chambers, K.C.; Magnier, E.A.; Metcalfe, N.; Flewelling, H.A.; Huber, M.E.; Waters, C.Z.; Denneau, L.; Draper, P.W.; Farrow, D.; Finkbeiner, D.P.; et al. The Pan-STARRS1 Surveys. *arXiv* **2016**, arXiv:1612.05560.
143. Ruprecht, J.D.; Stuart, J.S.; Woods, D.F.; Shah, R.Y. Detecting small asteroids with the Space Surveillance Telescope. *Icarus* **2014**, *239*, 253–259. [CrossRef]
144. Becker, A.C.; Wittman, D.M.; Boeshaar, P.C.; Clocchiatti, A.; Dell'Antonio, I.P.; Frail, D.A.; Halpern, J.; Margoniner, V.E.; Norman, D.; Tyson, J.A.; et al. The Deep Lens Survey Transient Search. I. Short Timescale and Astrometric Variability. *Astrophys. J.* **2004**, *611*, 418–433. [CrossRef]

 *photonics*

Article

Cryogenic Systems for Astronomical Research in the Special Astrophysical Observatory of the Russian Academy of Sciences

Yurii Balega [1], Oleg Bolshakov [2], Aleksandr Chernikov [2,3], Valerian Edelman [4], Aleksandr Eliseev [2], Eduard Emelyanov [1], Aleksandra Gunbina [2], Artem Krasilnikov [1,2], Ilya Lesnov [2], Mariya Mansfeld [1,2], Sergey Markelov [1], Mariya Markina [4], Guram Mitiani [1], Evgenii Pevzner [2], Nickolay Tyatushkin [2], Gennady Valyavin [1], Anton Vdovin [1,2] and Vyacheslav Vdovin [1,2,*]

[1] Special Astrophysical Observatory Russian Academy of Sciences, Nizhnij Arkhyz 369167, Russia; balega@sao.ru (Y.B.); eddy@sao.ru (E.E.); gvalyavin@sao.ru (G.V.); vdovinav@ipfran.ru (A.V.)
[2] A.V. Gaponov-Grekhov Institute of Applied Physics Russian Academy of Sciences, Nizhniy Novgorod 603950, Russia; chern@nf.jinr.ru (A.C.); gunbina@ipfran.ru (A.G.); lesnov@ipfran.ru (I.L.); ttshkn3@bk.ru (N.T.)
[3] Joint Institute for Nuclear Research, Dubna 141980, Russia
[4] P.L. Kapitza Institute for Physical Problems Russian Academy of Sciences, Moscow 119334, Russia; edelman@kapitza.ras.ru (V.E.); mamarkina@edu.hse.ru (M.M.)
* Correspondence: vdovin@ipfran.ru

Abstract: This article presents the main results and new plans for the development of receivers which are cooled cryogenically to deep cryogenic temperatures and used in optical and radio astronomy research at the Special Astrophysical Observatory of the Russian Academy of Sciences (SAO RAS) on both the Big Telescope Alt-Azimuthal optical telescope (BTA) and the Radio Astronomical Telescope Academy of Sciences (RATAN-600) radio telescope, 600 m in diameter. These two instruments almost completely cover the frequency range from long radio waves to the IR and optical bands (0.25–8 mm on RATAN and 10–0.3 µm, on BTA) with a certain gap in the terahertz part (8–0.01 mm) of the spectrum. Today, this range is of the greatest interest for astronomers. In particular, the ALMA (Atacama Large Millimeter Array) observatory and the worldwide network of modern telescopes called the EVH (Event Horizon Telescope) operate in this range. New developments at SAO RAS are aimed at mastering this part of the spectrum. Cryogenic systems of receivers in these ranges are a key element of the system and differ markedly from the cooling systems of optical and radio receivers that ensure cooling of the receivers to sub-Kelvin temperatures.

Keywords: optical and radio astronomy; cryogenic receivers; cryogenic systems; radio waves; IR and optical ranges; terahertz range; superconductivity; BTA (Big Telescope Alt-Azimuthal); RATAN-600 (Radio Astronomical Telescope Academy of Sciences—600 m in diameter)

Citation: Balega, Y.; Bolshakov, O.; Chernikov, A.; Edelman, V.; Eliseev, A.; Emelyanov, E.; Gunbina, A.; Krasilnikov, A.; Lesnov, I.; Mansfeld, M.; et al. Cryogenic Systems for Astronomical Research in the Special Astrophysical Observatory of the Russian Academy of Sciences. *Photonics* **2023**, *10*, 1263. https://doi.org/10.3390/photonics10111263

Received: 31 July 2023
Revised: 12 November 2023
Accepted: 13 November 2023
Published: 15 November 2023

1. Introduction

The modern progress of astronomy, with its extremely weak signals and extremely remote objects of studies, imposes most ambitious requirements on receiving equipment. These requirements are several orders of magnitude higher than those for the ground-based receivers used in communications, radar, etc., and actually determine the level of development of modern instrumentation. The main characteristic of astronomical receivers is their sensitivity. Achieving extremely high sensitivity leads almost inevitably to the need for deep cryogenic cooling of the input elements of astronomical receivers.

Initially, astronomy developed in the optical range, and since the second quarter of the 20th century, in the radio range as well. The last century of the development of astronomy was characterized by the expansion of the spectrum and the filling of the space between the optical and radio ranges [1,2], as well as an advance towards shorter wavelengths than optics [3–5]. The critical limitation in this range was not only the technology of the equipment with its fundamental quantum limitations, but also the atmosphere, which affects the propagation of electromagnetic waves [6].

As a result of the development of the equipment, ground-based astronomy went partially into space [7–9], where there was no limitation due to the atmospheric influence. But astronomy is still impossible without ground-based research, not to mention the huge price and short life spans of space observatories. Therefore, there is a boom in the construction of new observatories around the world.

This work presents the development of cryogenic systems for radio astronomy research at telescopes of the SAO RAS. The following two subsections of the introduction define SAO's place among modern observatories, describe the main types of cryosystems for astronomy and some of SAO's instruments using cryogenics, and present the atmospheric conditions of the site for observations. The next section presents the original developments of cryosystems for telescopes of SAO RAS. In Section 3, we describe the latest developments of the authors related to the field of the sub-THz astronomy technology, which is being actively developed in the world recently. The results of creating superconducting radio detectors cooled to sub-K temperatures for use as elements of the BTA optical telescope are presented.

1.1. The Special Astrophysical Observatory of the Russian Academy of Sciences among Astronomical Observatories: Instrumentation and Astroclimate

SAO RAS [10] has two telescopes in its arsenal: the 600 m RATAN-600 radio telescope, which covers a wide range of radio waves up to millimeter waves; and the BTA (Big Alt-Azimuth) telescope, a 6 m optical reflector that successfully operates in the visible and infrared parts of the spectrum. In fact, a part of the spectrum called the terahertz gap remains unfilled. To fill this gap, specialized antennas, good astroclimate, and highly sensitive receivers are needed. Atmospheric absorption of the THz waves is so great that there are only limited opportunities for ground-based astronomy to work in its low-frequency part (0.1–1.0 THz), which is called the sub-THz range. At the lower boundary of the band (100 GHz), the atmosphere gives not very significant absorption almost everywhere, and observatories are located even on the seashore [11]. The upper edge of the sub-THz range is open to ground observations only in unique places on Earth, such as the Atacama Desert [1,2], Pamir [12], Tibet [13], Mauna Kea, Hawaii [14], and the Antarctic Plateau [15]. In the central part of the sub-THz range, the atmospheric transparency windows are 1.3 mm (~230 GHz) (the operating range of the current Event Horizon Telescope (EHT) [16]) and especially 0.8 mm (~375 GHz) (the EHT perspective window), which is highly critical for choosing the location of the observatory. Usually, these are mountains from 3 km above the sea level and above in dry climate and remote from large water bodies.

EHT is a global network of radio telescopes operating together on the Very Long Baseline Radio Interferometer (VLBI) principle. VLBI allows you to combine data from several remote receivers. Using VLBI, multiple independent radio antennas separated by hundreds or thousands of kilometers can act as a phased array, a virtual telescope that can be electronically aimed with an effective aperture equal to the diameter of the planet, greatly improving its angular resolution. Moreover, the wider and more complete the coverage of the UV plane of the network on the Earth's surface is, the higher is the quality of the final result. Today, the plans to expand the EHT VLBI base are aimed at filling the blank spot on the map of the Earth, specifically, the northeastern part of Eurasia. Our project will follow this trend. And the fact that this is a pilot project will determine the construction of a full-scale sub-THz telescope in this part of the planet.

In 2019 [16], one of the most impressive achievements in millimeter-scale astronomy was obtained: for the first time, an image of the shadow of a supermassive black hole in the M87 galaxy was obtained. Publications [17] provide an overview of the plans to expand EHT and highlight the North Caucasian region as promising and requiring a new telescope.

SAO RAS does not have an exceptional astroclimate [18,19], and the altitude above sea level is not very significant: BTA is at an altitude of 2100 m, and RATAN-600 is at an altitude of 970 m. The astroclimate is quite satisfactory for observations in the entire radio range of RATAN-600 observations down to millimeter waves. The location of BTA has a satisfactory astroclimate in the optical and near-infrared ranges. There is a significant level of cloudy days. Sub-THz waves have a noticeable feature, and even the upper site of SAO RAS, where BTA is located, is far from the best sub-THz observatory site in terms of atmosphere transparency. Moreover, even in the Caucasus, there are more promising sites in Dagestan, but not only the tools, but also the infrastructure are not there yet.

At the same time, the astroclimate studies [18,19] make it possible to work reliably in the two lower sub-THz transparency windows (3 mm and 2 mm) without any problem. There are reliable and rather long periods in winter when there is good atmospheric quality of around 0.1 Np absorption at the zenith in the 1.3 mm transparency window. Sub-THz transparency windows of 0.8 and shorter are practically closed for observations from the BTA site.

More than 50 years of development of astronomy and the equipment for it in SAO were accompanied by the development of cryogenic systems of various types and for different temperatures, providing cryogenic cooling and high sensitivity of receivers. The cryosystems at SAO operate in the temperature range of liquid nitrogen (~70 K), hydrogen (~20 K), and helium (~4 K). An extensive list of similar developments of cryosystems is presented in publications of well-known observatories, for example ALMA or Sardinia [20]. A more detailed overview of the achievements and comparison with our developments will be presented in [21].

The construction of a sub-THz telescope, starting with the selection of a site with low atmospheric absorption, is an expensive and time-consuming undertaking. At the same time, since the beginning of the 1990s, attempts have been made to use sub-THz receivers on the BTA optical telescope. The size of the mirror for a radio telescope is rather small, and the astroclimate is not the best for sub-THz waves, but the quality of the surface exceeds the requirements of THz waves by two orders of magnitude. An important feature of bolometric receivers in the sub-terahertz range is the need for cryogenics at the sub-Kelvin level. This article is concerned with the analysis of already completed and ongoing developments of various cryogenic systems for the RATAN-600 and BTA telescopes of the SAO RAS, as well as on their basis for other telescopes. These systems for heterodyne radio receivers and optical photodetectors had temperatures no lower than the temperature of liquid helium. The article also presents a new project being developed for the Special Astrophysical Observatory of the Russian Academy of Sciences, specifically a radio receiver as part of the instruments of an optical telescope. A matrix bolometric receiver was chosen, rather than a heterodyne. Therefore, a new challenge for the project, a cryogenics system at the sub-K level of temperature, is presented here.

1.2. Main Types of Cooling Systems for Astronomical Receivers

The both main types of cooling systems are used to cool astronomical radiation receivers in SAO RAS: accumulators (Dewar vessels) and refrigerators (containing a working refrigerator). However, astronomers have long preferred refrigerators that do not require monitoring and periodic refueling with a cryoagent. In addition to the principle of cooling, refrigeration systems differ in temperature levels. In the optical range, noncryogenic cooling systems (above 100 K) are widely used. Typically, this type of cooling uses thermoelectric coolers to reduce the dark current of the CCD receiver arrays. Nitrogen level cooling (~80 K) is also quite popular for cooling optical and infrared photodetectors (e.g., [22–24]). This is provided both by Dewars filled with liquid nitrogen and nitrogen level coolers, which are relatively cheap and commercially available [24], and usually built on the Stirling cycle.

Since the 1970s, refrigerators with a closed cycle of the hydrogen temperature level (~20 K), built on the basis of the Gifford–McMahon thermodynamic cycle, have been widely used in astronomy [25,26]. In wide practice in astronomy, there is no information on the use of a liquid hydrogen cryostat filler. However, such cryostats were developed for other purposes, and liquid hydrogen is much cheaper than helium used for cooling systems.

Moreover, according to experts, liquid hydrogen is a much calmer and more convenient gas to work with [27], but the prejudice of its being explosive does not allow them to be used in astronomy still. Refrigerators of the hydrogen temperature level, but operating with helium, have so far reached the almost-helium temperatures (5–6 K) and are actively used in astronomy for cooling nonsuperconducting microwave receivers based on Schottky barrier diode (SBD) detectors and high electron moving transistor (HEMT) amplifiers. Classical helium systems (operating at a temperature level of about 4 K) with cooling both in the filler configuration and in the refrigerator have found wide application in ground-based astronomy. This temperature level makes it possible to ensure reliably the transition to the superconducting state of the main structural materials (Nb, Pb, NbN, etc.), on the basis of which various types of modern superconducting heterodyne receivers are developed. Along with the Gifford–McMahon cycle, pulsed tubular coolers that do not contain moving parts in the cooling head have been used actively in recent decades. This has radically reduced the level of vibration being the main disadvantage of refrigeration cooling systems as compared to cryoaccumulators, i.e., Dewars.

Helium systems, both filled by pumping and cooled, can reach temperatures below 4 K (about 1.6–2.0 K). However, in order to overcome the milestone at 1 K, other approaches and other technologies are required. Today, sub-Kelvin (<1 K) temperature levels in astronomy are provided by cryosorption systems (0.3 K) and dilution cryostats (0.01 K). These cycles are implemented based on precooling systems up to 4 K using a refrigerator or liquid helium. Sub-K temperatures are necessary to cool the most sensitive receiving systems, e.g., direct bolometric detectors and their arrays.

Almost all of these systems have been implemented over the past 50 years of development of cooled receivers for the BTA and RATAN-600 telescopes in various configurations and are presented in the following sections. Some of them are currently under development and will be used for the BTA observations in 2025. In this work, we focused on examples of the development of cryogenic receivers for telescopes of the SAO RAS. A more complete list of our results, including equipment for other devices, as well as a more detailed overview of the world practice of creating such systems, is presented in our second publication [21], which will be published after this one.

2. Materials and Methods: Development of Cryogenic Systems for Optical and Radio Astronomy at SAO RAS

2.1. Cooling Systems for CCD Arrays

CCD arrays (charge-coupled devices) are currently the main working tool in optical astronomy, including the visible, IR, and UV ranges. One of the most effective ways to improve noise characteristics and eliminate the dark current, and hence increase the sensitivity of the receiving device, is to cool it down. Below is a line of various cooling systems operated at different temperature levels and based on different principles of cooling CCD arrays for BTA and other instruments, created in cooperation with SAO RAS and IAP RAS for more than 30 years. Some of the first results are presented in review [22].

2.1.1. Nitrogen-Cooled Optical Cryostat

The main advantages of the presented product are the vibration-free system and the liquid-free cryostat. See Figure 1 for photos.

The main characteristics of the product are as follows:

- T = 80 K;
- Vacuum level is 10^{-4} mbar (at least 3 months);
- Overall dimensions: diameter 224 × 464 mm;
- Vacuum camera dimensions: 160 (diameter) × 80 mm (height);
- Optical window: window with a diameter of 109 mm, window material is quartz.

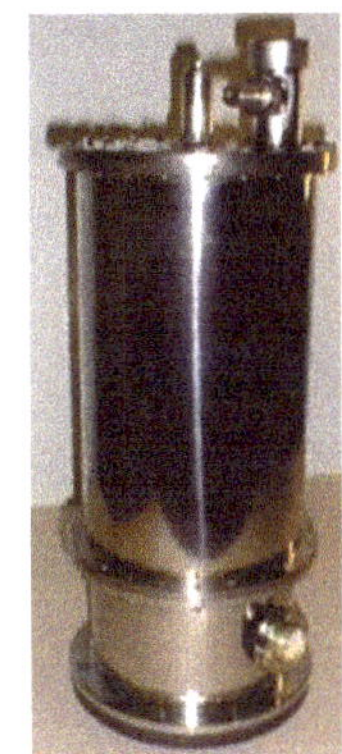

Figure 1. Photos of an optical vacuum chamber with nitrogen cooling. Operating temperature T = 80 K.

2.1.2. Optical Vacuum Chambers with Cooling by the CryoTiger Cryogenic System

The created optical vacuum chambers are intended for cooling the receivers used on the BTA telescope [10]. Two chambers were developed, which are cooled by one and two microcryogenic systems (the photos are shown in Figure 2; the main characteristics and advantages are given in Table 1). The main feature of the presented devices is the maintenance of a stable operating temperature of 80 K with extra-large optical windows with a diameter of 220 mm and 326 mm. The presence of optical windows of this diameter makes it possible to accommodate a sufficiently large number of photodetectors (a very large matrix).

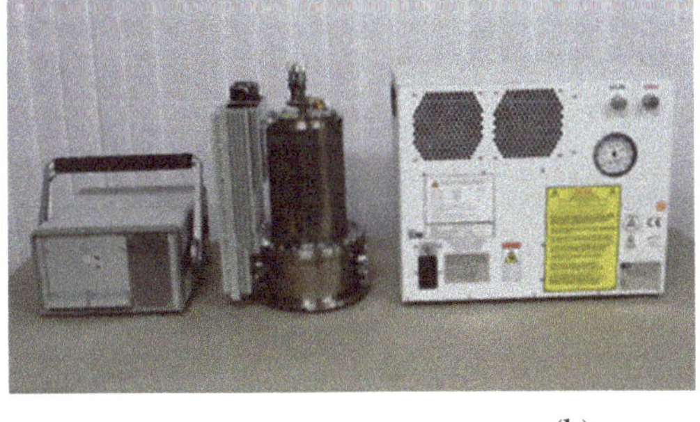

(a) (b)

Figure 2. Photos of optical vacuum chambers with CryoTiger microcryogenic cooling system: (a) cooling by the CryoTiger; (b) cooled plate cooled by two CryoTigers.

The uniqueness of the design with two CryoTiger [28] units is the fact that two identical cryogenic machines work in parallel for one cooling unit. Usually, this problem does not have an accurate solution due to the incomplete identity of the refrigerators and their operating conditions. The authors managed to solve this problem and almost double the cooling capacity of the system by selecting identical refrigerators and controlling their modes using adjustable thermal switches on cryogenic interfaces. An additional benefit is the increased reliability. In case of failure of one of the machines, the second one will provide cooling at a slightly higher level, which will further reduce the dark current of the photodetector, compared to the uncooled version.

Table 1. Main characteristics and advantages of optical vacuum chambers with the CryoTiger microcryogenic cooling system.

	Cooling with One CryoTiger	Cooling with Two CryoTiger Units
Main Features	- Operating temperature: T = 80 K; - Vacuum level is 10^{-4} mbar; - Overall dimensions: - Diameter—300 × 430 mm; - Dimensions of the cavity: - Diameter is 258 × 100 mm; - Optical window: Diameter—220 mm; Material—quartz.	- Operating temperature: T = 80 K; - Vacuum level is 10^{-4} mbar; - Overall dimensions: - Diameter—388 × 533 mm; - Dimensions of the cavity: - Diameter is 344 × 140 mm; - Optical window: Diameter—326 mm; Material—quartz.
Advantages	- Large optical window; - The presence of a detector suspension unit made of hard-to-process material (superinvar); - The possibility of placing and cooling a large number of photodetector devices (CCD arrays).	- Extra-large optical window with a diameter of 326 mm; - The possibility of placing and cooling a large number of photodetector devices (CCD arrays).

2.1.3. Chambers with Thermoelectric Modules for Cooling Large-Format CCD Arrays

Cooling without the use of liquid or compressed gases seems very attractive. This opportunity is provided by thermoelectric, thermomagnetic and laser cooling systems. In practice, there is only experience in using Peltier elements. And there are two radical limitations here. The temperature was supposed to be no lower than 150 K. Moreover, it is achieved by cascading a fairly significant number of Peltier elements (up to 7–8) and extremely low efficiency, the need to remove a significant heat flow from the hot junction of the thermoelectric battery. The photos of the created chamber cooled by Peltier batteries are presented in Figure 3.

Figure 3. Photos of the created chamber with cooling by thermoelectric modules.

The main characteristics of the product are as follows:

- T = 150 K;
- Vacuum level is 10^{-4} mbar;
- Dimensions of the vacuum cavity: $174 \times 176 \times 89$ mm;
- Optical window:
- Diameter—90 mm;
- Material—quartz.

2.1.4. Nitrogen Optical Cryostat for Cooling the IR Spectrophotometer

To expand the range of BTA observations into the near-IR range (wavelengths of 0.8–2.5 microns), a unique nitrogen optical cryostat was developed for cooling the IR spectrophotometer. In this article, we will only briefly present the description and main characteristics of the cooling system. The receiving device is placed in a cylindrical Dewar, Figure 4a.

A part of the cryostat is occupied by a no-spill container with liquid nitrogen intended for operation at the primary focus of BTA. Similar designs were implemented for other BTA tasks. The optical circuit contains only two areas that require deep cooling, specifically, the light receiver and the output iris with diffraction gratings. The high stability of the maintained temperature is needed only for the light receiver. The nitrogen tank is located in the upper part of the cryostat and is a cylinder with an asymmetric recess. The shell and top covers of the nitrogen tank are made of stainless steel. The bottom ring of the container with heat-removing contact pads is made of copper. The nitrogen tank is fixed inside the cryostat body by welding an inner filling tube to it (which in turn is welded to an external nitrogen tube, which is also welded to the outer flange of the cryostat). In addition, the lower part of the shell of the nitrogen tank is stretched with Kevlar threads to ensure the desired rigidity of the structure. The cooling of the main elements of the optical circuit is carried out by means of flexible copper cooling wires attached to the copper rim of the nitrogen tank. To ensure the operation of the cryostat at any angle of inclination, the tank covers are connected inside it by copper cooling wires. All optical and electromechanical equipment is placed on a welded stainless steel frame attached to the lower flange of the device through heat-insulating gaskets (Figure 4b).

Main characteristics of the cryogenic system:

- Minimal temperature is T = 80 K;
- Filler nitrogen cryostat with a nitrogen tank of complex shape;
- All elements of the optical scheme of the IR spectrophotometer are cooled to different temperatures inside the cryostat, including moving parts;
- System dimensions: diameter 572 mm, height 660 mm;
- Weight without cryoagent (nitrogen) 30 kg;
- Volume of the cryoagent tank22 L;
- Optical window: diameter 90 mm, material is quartz;
- Electrical connector SNC13-102—3 pieces;
- Pumping flange—KF16.

The main advantages of the system:

- Large-volume nitrogen optical no-spill Dewar;
- Ensures different temperatures of parts of the optical path;
- Minimal temperature is T = 80 K $\pm$ 2 K;
- It is possible to place the cryostat in the primary focus of the BTA at SAO RAS.

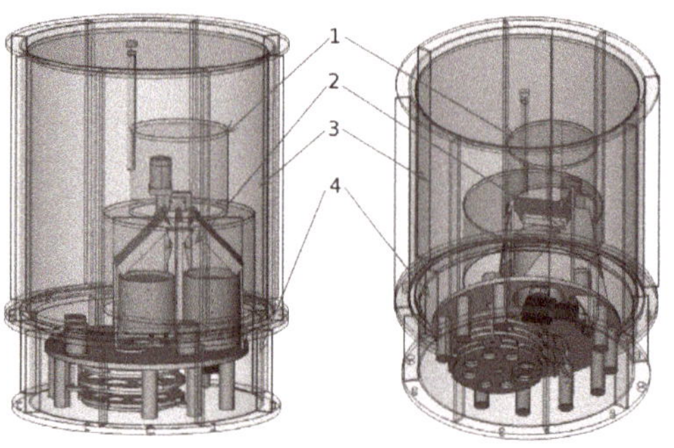

1—nitrogen tank, 2—main frame, 3—upper part of the cryostat body, 4—lower part of the cryostat body

(**a**)

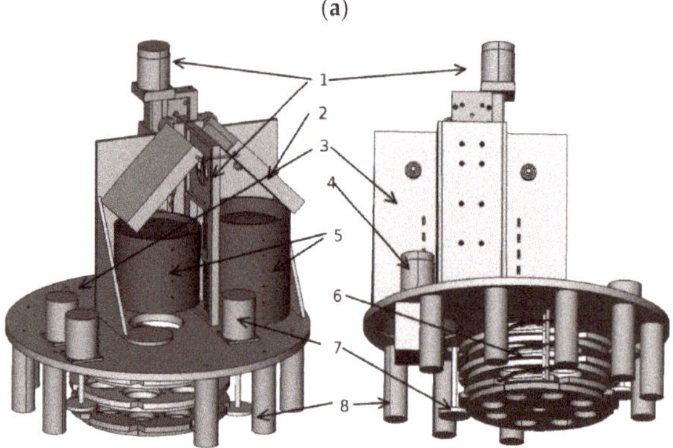

1—movement with a block of slits and an iris mask, 2—diagonal mirrors, 3—the frame of the device, 4—focus movement, 5—collimator and camera, 6—turret block, 7—turret drives, 8—thermal insulation supports

(**b**)

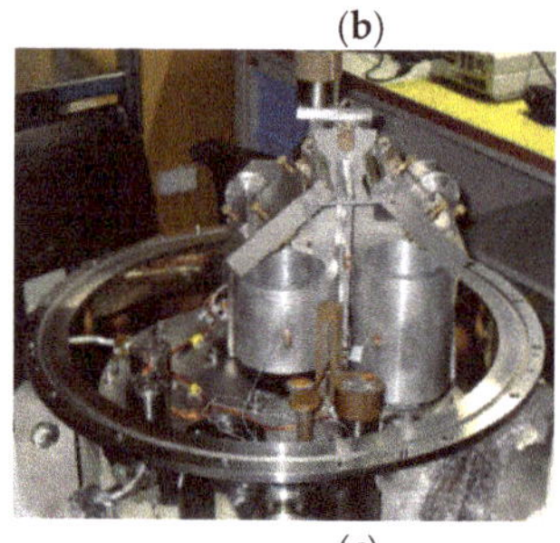

(**c**)

Figure 4. Nitrogen optical cryostat: (**a**) 3D model and photograph of the general view of the cryostat; (**b**) general view of the cryostated IR spectrometer; (**c**) photographs of the elements of a cooled IR spectrometer.

2.1.5. Systems with Remote Cooling

According to the authors, the most interesting design of a cryostat with a unique set of thermal and optical interfaces was created in SAO RAS and IAP RAS for a two-frequency array receiver with nitrogen cooling and a head with a CCD matrix carried outside of the filler nitrogen cryostat (remote cooling) [29]. The task was to minimize the shading of the cryogenic system by the elements of the radiation beam in the two-mirror system of the telescope. The solution was to remove the Dewar with nitrogen from the beam zone. At the same time, the head with the matrix had a ratio of the effective area with the matrix to the total area of the head and the cryochannel—cooling wire about 90%. An additional difficulty consisted in sealing the window on an extremely small landing flange. A photo of the system is shown in Figure 5.

Figure 5. The photo of the system with remote cooling.

2.2. Cooling Systems for Radioreceivers of RATAN-600

The RATAN-600 radio telescope from the first years of operation on the initiative of D.V. Korolkov [30] implied the use of radio receivers with cryogenically cooled components and the cryogenic department existed as part of the observatory until recently. The ideology of cryogenic refrigerators was chosen as the basis of cryogenic cooling systems, and the cryogenic department was engaged in the operation, maintenance, and repair of refrigerators. Basically, these were hydrogen-level refrigerators manufactured by JSC Cryogenic Equipment (formerly Sibkriotechnika) [31], since the semiconductor receivers, which were cooled mainly, did not require deeper cooling, as compared with the superconducting ones. The vast majority of the SAO RAS receivers are presented in detail in the extensive bibliography and will not be considered here. The three cryogenic radiometers, developed on the initiative with the participation or by order of SAO RAS, which did not work for a single day as part of the RATAN-600 telescope, but largely determined the development of the culture of cryogenic receivers for astronomy, will be presented below.

2.2.1. The 20 K Cryogenic Radiometer of 4 mm Band

At the end of the 70 s, on the initiative of D.V. Korolkov, with the participation of the collaboration of SAO RAS, IAP RAS, and Saturn Research Institute (Kiev), a cryogenic radiometer of the 4 mm wavelength range with a cooled Schottky diode mixer and an intermediate-frequency parametric amplifier was fabricated [26]. The cryostat was made of thick (10–12 mm) stainless steel and was cooled by a Gifford–McMahon refrigerator at a hydrogen temperature level (~12 K); see Figure 6. It demonstrated excellent performance at that time, but the untimely death of D.V. Korolkov prevented it from being installed on RATAN-600. At the same time, the demand for such an instrument was high, and after small modifications, first as a 4 mm and then a 3 mm device, it was used for radioastronomical observations on the RT-22 KrAO radio telescope [25,26]. The complexity and necessity of periodic tuning of the parametric amplifier and the progress in the development of HEMT (high electron mobility transistor) [32] amplifier technologies soon led to the replacement of PA with HEMT [33], and for a long time determined the architecture of the radio receiver for astronomy up to short-mm or sub-THz waves. At this stage, the world leader in the development of cryogenic receivers for astronomy was Virginia Diodes. In Europe, such receivers were successfully developed by Antti Raisanen [34,35]. This ideology and the components of these SAO receivers were actively, and for a long time, used for other telescopes and applications before the era of superconducting receivers [36] in the sub-THz range began to dominate.

Figure 6. Photo of the 20 K cryogenic radiometer of 4 mm band.

The first cryostat (Figure 6) was designed as a vacuum cryostat in the form of a parallelepiped with a volume of about 20 L with thick walls (10 mm) made of stainless steel. A nickel-plated copper radiation shield is visible inside. The shield cover and the cover of the cryostat have been removed. Inside the shield, there is a cooled DBS mixer and a parametric intermediate frequency amplifier. Quasi-optical vacuum windows with antireflective quarter-wave coating are directed opposite to the reader, and their edges are visible at the left edge of the cryostat. The Gifford–McMahon cooler head is visible from above, and the shield and receiver elements are attached to it.

2.2.2. The Last Cryogenic Receiver of RATAN-600

The choice of the receiver's physical temperature is a complex multifactorial process. It would seem that the goal in astronomical receivers is the same: to achieve extremely high sensitivity or extremely low noise of receiver. However, in reality, the choice of temperature and cooling system is determined by a compromise between:

- Available cryogenic systems and their maximum cooling levels and cooling capacity, as well as being ready to pay big money for unique systems;
- The development of semiconductor and superconducting technologies of detectors and amplifiers of the corresponding frequency range of observations within a total spectrum from RF to UV waves for SAO RAS;
- Atmospheric limitations of the location of the observatory also determining the effectiveness of the telescope in a particular frequency range.

The first two factors determined the course of the development of cryogenic systems at SAO for many years.

The combination of the second and third factors led about 10 years ago to the liquidation of the cryogenic department in SAO [21]. The fact is that the noise temperatures of uncooled receivers up to the MM range have become very low: units and tens of Kelvin depending on the frequency. Of course, cryogenic receivers are still much better than uncooled ones and practically reach the quantum limit of sensitivity. But the sensitivity of the receiver [37] is determined not by the noise of the receiver, but (1) by the noise of the system, including the noise of the antenna and the atmosphere above it:

$$\delta T = k \times T_{sys} \times (\delta f \times \tau)^{-1/2} \tag{1}$$

where δT is the fluctuation limit of sensitivity; $T_{sys} = T_r + T_a$ is the noise temperature of the system determined as the sum of the receiver noise and the antenna noise, including the atmospheric noise; k is the Boltzmann constant; δf is the band; and τ is the time integration constant.

In fact, with an extremely low receiver noise, the atmosphere can already make a dominant contribution, and the effect of cooling on sensitivity drops sharply. Moreover, the presence of vacuum interfaces in cryogenic receivers, i.e., windows that contribute their losses and noise, negates the feasibility of cryogenics on RATAN-600, located at a low altitude above the sea level (980 m) and having a significant contribution of the atmosphere at 3 mm waves. This fact was indeed the reason for the closure of the cryogenic department and the lack of demand for the last of the mm wave band radiometers developed for SAO RAS, presented below in Figure 7. The last RATAN cryostat, shown in Figure 7, is made of thin steel (1.5 mm) in the original strong and shape-stable topology with approximately the same volume as the first, while containing four receiver channels inside for two frequency ranges and two polarizations. The shape of the cryostat is oval in the cross section, with shortened narrow edges. The cold head of the CryoTiger refrigerator is shown at the top. The horns that feed the RATAN-600 antenna are located on the left, and the blue and orange boxes on the right are the containers for the uncooled components of the receivers.

Figure 7. The last cryogenic receiver of RATAN-600 at SAO RAS.

Device characteristics:
- Frequency range 18.5–21.5 GHz (center frequency 20 GHz) and 27.5–32.5 GHz;
- The possibility of receiving two mutually perpendicular polarizations of electromagnetic radiation;
- Working temperature is 80 K;
- Construction scheme—modulating with a waveguide ferrite switch and a cooled matched load;
- Fluctuation sensitivity is not worse than 5 mK per 1 s of the time constant of the output filter.

Consideration of the sub-THz range for the RATAN-600 has limited prospects not only because of the small height and large absorption, but also because of the quality of the antenna. It meets the needs of observations of the cm and long mm waves.

However, following the logic of analyzing the ratio of atmospheric noise and receivers, we can confidently conclude that it is advisable to cool the sub-THz radiometers even for low altitudes. In more promising and high-altitude areas, cryogenics will remain a key tool of astronomical observatories for many years to come. The receiver presented here operates in the centimeter range, rather than in the mm or sub-THz one. And the atmosphere affects sensitivity much less than at a wavelength of 3 mm, where it is still rational to use cryogenic receivers.

2.2.3. Is Cryogenics Required for Optical Receivers?

The sad conclusion of the previous paragraph that cryogenics is no longer needed for cooling radio receivers for the RATAN-600 up to the mm range raises the question: is cryogenics needed for optical receivers on the BTA? For many years, electric vacuum photomultiplier tubes (PMTs) on optical telescopes successfully solved the problem of astronomical observations in the optical range. Here, the radiation quantum is energetic enough to be reliably detected. The PMT era ended with the introduction of CCD matrices. However, now there are [38] solid-state multipliers that solve a similar problem, not for a single pixel, but for matrices. The signal level after the single-photon receiver and solid-state multiplier would seem to be sufficient even without cooling. However, this turned out to be not entirely true. Cooling, although not cryogenic, is still necessary to minimize dark current. Minus 20 C is still required. And this is not difficult to provide with a simple Peltier cooling machine.

3. Results and Discussions

3.1. Results of the Development of Cooling of Optical Receivers for an Optical Telescope and Radio Receivers for a Radio Telescope

Results have been achieved in the development of cryogenic systems for cooling photodetectors for the BTA optical telescope and radio receivers for the RATAN-600 radiotelescope. A total of over 20 [22] cryosystems of various temperature levels from nitrogen to helium were created and successfully used for astronomical observations on the both telescopes in the radio range up to 3 mm and in optics, including IR. There are lots of new astrophysical results obtained due to the use of the developed receivers:

- Direct evidence of the fossil origin of large-scale magnetic fields of chemically peculiar stars obtained [39];
- The discovery of new LBVs in the Local Volume galaxy NGC 1156 [40];
- Planet TOI1408.01: a grazing transit and probably a highly eccentric orbit [41];
- Collected RATAN-600 multifrequency data for the BL Lacertae objects [42];
- The RATAN-600 telescope helps to understand the origin of cosmic neutrinos [43];
- Identified Quasi-periodic Pulsations in a Solar Microflare [44].

3.2. Cooling Systems of the Sub-THz Range for Radioreceivers as a Part of the BTA Telescope

Within the framework of Project 23-62-10013 of the Russian Science Foundation "Development of Russian sub-terahertz observatory prototype as part of an optical telescope", the authors of the article set themselves the task of expanding the capabilities of the optical Big Telescope Alt-Azimuth (BTA) of SAO RAS in the sub-terahertz range. In this article, we will not present the entire project in detail, but the issue of creating one of the key nodes, specifically, a specialized sub-Kelvin cryogenic system and a set of its interfaces with elements of the receiving complex and the telescope itself will be considered in detail. The key motivation of this project and this article was the intention to place a sub-THz receiver at the focus of the BTA optical telescope. Highly sensitive bolometric receivers of sub-THz waves require deep sub-Kelvin cooling.

The first experience of placing a sub-THz receiver on a cooled Shottky barrier diode mixer that was developed at IAP RAS for BTA dates back to 1991. The experience was not successful, but led to the intensification of cooperation between SAO RAS and IAP RAS, which gave rise to a series of successful projects of cooled photodetectors presented in Section 3.1.

We returned to this idea [45]. Two samples of a sorption cryostat with an operating temperature of 300 mK were manufactured (photo in Figure 8).

Figure 8. Sorption cryostat with an operating temperature of 300 mK.

The main characteristics:

- Operating temperature: 300 mK $\pm$ 0.1 K;
- Heat load: 1 mW;
- Vacuum level: 10^{-4} mbar;
- Two flanges for the installation of optical windows with a diameter of 25 mm;
- Flange KF D25 for pumping;
- Input of electrical and RF connectors;
- The size of the working cavity: diameter 185 mm, height—70 mm, dimensions—not more than 1600 mm;
- Diameter: 700 mm.

The main advantages:
Advantages of sorption pumping:

- Compact placement inside the cryostat;
- Possibility of automation;
- Low cost compared to mechanical pumping means.

Advantages:

- A two-stage sorption pumping system has been developed;
- The gas heat key is integrated with the sorption pump into a single unit;
- The technology of manufacturing sorption pumps makes it possible to obtain a high degree of pumping compared to Western analogues;
- Two models were developed: with filler and using a closed-loop cryogenic system.

The second sample of a similar cryosorption system contained a precooling system based on a 4 K closed-cycle refrigerator, and not in the form of a filler helium cryostat. However, it did not go beyond a series of laboratory tests. The problem was the lack of progress in developing reliable, reproducible, and capable of folding into matrices of identical pixels high sensitive detectors. At the moment, the problem has been solved and a series of both single detectors, such as in [45], and matrices, which are not difficult to modify for the BTA telescope irradiation system, has been created. Over the next two years of the project, the detectors will be finalized, the receiving path will be built and integrated into the BTA optical system, the cryogenic system and the necessary interfaces will be created. In particular, one of the 0.3 K variants of the BTA receiver cooler is built on the basis of the refrigerator shown in 8, but with the replacement of the Dewar with a 4 K refrigerator head.

3.3. Sorption Cryostat

One of the first mentions of the creation of a 0.3 K cryogenic system for installing of a sub-THz receiver on BTA is given in publication [45], and the system was at the design stage. He3-He4 cryogenic sorption and dilution systems give unique possibilities to cool down astronomical receivers to the temperature level below 1 K [46–50]. In this article, a cryogenic system was proposed that provides the operating temperature of the receiving antenna arrays T~0.3 K. It is a closed-cycle system, i.e., it does not require the use of liquid gases (nitrogen and helium), based on a refrigerator on pulse pipes and two sorption pumps on ^{3}He and ^{4}He. The system includes a refrigerator based on pulse tubes. Despite the lack of funding for this project at that time, work on the creation of such a cryogenic system was continued, and the next stage of development was already presented in publication [51].

The project proposes the development and installation of a cooled receiver based on SINIS [52] (superconductor–insulator–normal metal–insulator–superconductor) detectors at the BTA site of SAO RAS.

Preliminary assessments of astroclimatic studies at the BTA site showed that the most successful range for test studies of the receiving system is 3 mm, since observations can be carried out year–round in this window of transparency. Also, in the future, transparency windows of 1.3 mm and possibly 0.8 mm will be considered (however, in this case, favorable conditions account for only 2% of the time a year in winter, or observations should be directed to a powerful radiation source). The optimal operating temperature of the designed SINIS detectors is 300 mK or lower (to improve sensitivity), which requires a complex cryogenic system. It should be noted that over the past fifteen years, cryosorption cryostats and dissolution cryostats have become a common commercial product. Oxford Instruments, Bluforce, and others offer [53] convenient commercial products to achieve these temperatures with reasonable cooling capacity of units and tenths of a microwatt. However, these are large and extremely inconvenient machines for mounting in the focus of the telescope, and in fact it is only a tool for laboratory experiments. For installation on a telescope, it is necessary to develop a specialized cryostat and, most importantly, a set of highly efficient interfaces, in particular mechanical, vacuum, cryogenic, optical, electrical ones, etc., that are capable of providing effective cooling of the detector cooling object, as well as the supply/removal of signals to it without loss and with maximum thermal insulation from the environment. At the moment, three variants of cryogenic systems are being considered, the descriptions of which are given in this section. The first one is based on the 2008 groundwork described above (and presented in Figure 9).

3.4. Cryogenic System for Sub-THz Detectors for the BTA Based on a Dilution Microcryostat–Deep Stick

The dilution microcryostat–deep stick [54] (Figure 10) was developed by V.S. Edelman at the Kapitsa Institute of Physical Problems of the Russian Academy of Sciences. The minimum achievable operating temperature in such a cryostat can reach 50 mK (without a background load in the absence of an optical window) and nanowatt capacity. In addition to the achievability of such low temperatures, another advantage of this device is the absence of pulse tubes (which contribute to the total noise of the receiving system) because pumping of ^{3}He and ^{4}He is carried out using built-in sorption pumps. But there is also a certain drawback of such a cryogenic system, specifically that the duty cycle is no more than 8 h. This generally fits into the ideology of using time unoccupied by night optical observations for sub-THz observations on an optical telescope. At night, when observations are made in optics, ^{3}He will be recondensed. The dilution cryostat is controlled by sorption heaters, sorption gas heat key heaters and evaporator heater.

Figure 9. The 0.3 K sorption cryostat.

Such equipment is promising for conducting short-term observations, but requires high sensitivity of the receiving system (due to the absence of external vibrations from pulsating pipes and low operating temperature).

3.5. A Continuously Operating Dilution Microcryostat

To ensure long-term temperature maintenance of the operating temperature of the detector, a modification of the cryogenic system presented in the previous section was carried out [54]. The principle of operation is based on the technology of sorption pumping ^{3}He and ^{4}He and condensation pumping of the mixture ^{3}He–^{4}He in the dissolution cycle. The continuous maintenance of a low temperature of the order of 0.1 K during the regeneration of sorbents is carried out due to the large heat capacity of the block filled with holmium plates, in which the condensation of circulating helium occurs.

The structurally presented cryostat is in many ways similar to the designs given in the works [54,55]. The cryostat is made in the form of an insert (deep stick) into an industrial 35 L transport liquid helium Dewar. The cold sample holder is located at the top of the device, which facilitates access to it. A total of 35 L of liquid helium is enough to precool the device from nitrogen temperature and for 6–7 days of operation. During this time, a low temperature of the sample holder can be maintained with accuracy of the order of a millikelvin.

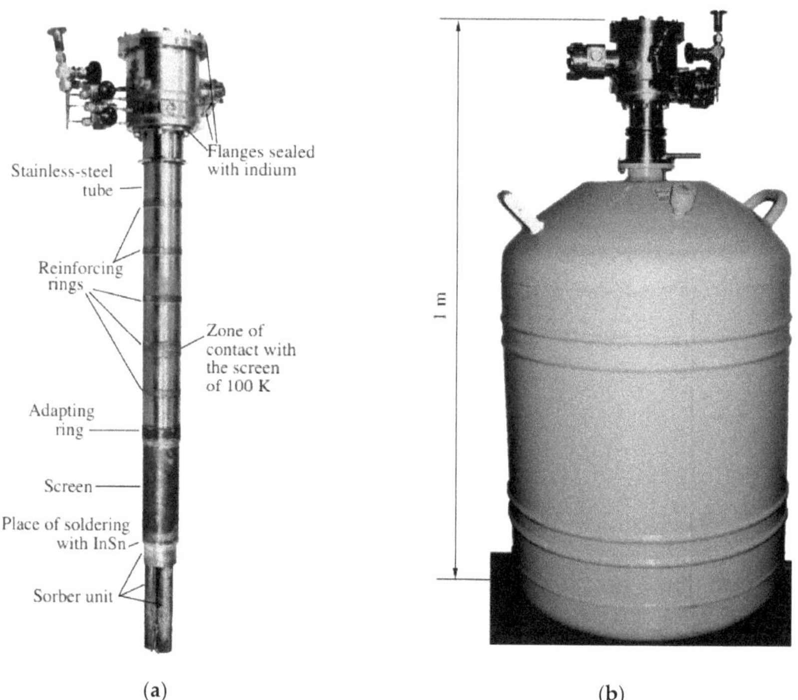

(a)   (b)

Figure 10. (**a**) Appearance of the cryostat–deep stick mounted in (**b**) a portable helium cryostat with a volume of 35 L.

A simplified schematic representation of the structure is shown in Figure 11. The microcryostat contains the following functional units located in a common vacuum volume:

- The ^{4}He unit that includes a 1 K chamber and a ^{3}He condensation volume, which constitute a common unit, and a sorption pump that evacuates helium vapors. The ^{4}He sorber (as well the ^{3}He sorber) is above the ^{4}He bath and both ^{3}He baths. Heat is withdrawn from the sorber to liquid helium in the portable Dewar flask through copper heat conductors. The sorber on which a heater is wound is manufactured from stainless steel and placed in a sealed copper container. The evacuation channel consists of a stainless steel tube into which a copper tube about 5 mm long is soldered. The latter passes through the container cover and is soldered to it. In this tube, ^{4}He condenses and trickles down to the 1 K chamber. A thermal valve, during whose heating/cooling the heat exchanging gas (^{3}He) is desorbed/sorbed, serves for controlling the heat exchange between the sorber and the container walls. This is necessary when changing from the ^{4}He desorption regime to its evacuation.
- The upper ^{3}He unit that contains the ^{3}He sorber and the condenser of ^{3}He vapors, which is soldered into it and serves for the lower ^{3}He bath. The ^{3}He bath is cooled by the evacuation of vapors of liquid ^{3}He by its own sorption pump, which is analogous to the ^{3}He pump. When the pump is regenerated, the bath is filled with ^{3}He, which is liquefied when being in contact with the 1 K chamber.

- The lower ^{3}He unit that contains the second ^{3}He bath, which is under the first bath and is connected to the condenser of ^{3}He vapors of the upper circuit with a stainless steel tube. The both circuits are sealed with respect to each other. If the temperature of the upper bath is lower than that of the lower bath, a good thermal contact establishes between them; otherwise, the thermal interaction is substantially weakened.
- The dilution unit that contains a dilution bath (mixer), a heat exchanger, a still, and a condenser of vapors of a ^{3}He–^{4}He mixture. The condenser is filled with plates of holmium that has the high specific heat at T < 1 K. For a He mass of ~50 g, its specific heat at T = 0.4 K is ~2 J/K. This allowed us not to use a large amount of ^{3}He for maintaining the working conditions during the regeneration of ^{3}He in the upper circuit, and to restrict ourselves to a rather small amount of ^{3}He, only 0.015 mol in the lower circuit for establishing thermal coupling. The dilution circuit is placed above the sorption pumps, and the condenser included in it is connected to the lower ^{3}He bath via a copper heat conductor. The temperature difference between the upper ^{3}He bath and the condenser at T = 0.4 K is approximately 0.02 K/100 µW.

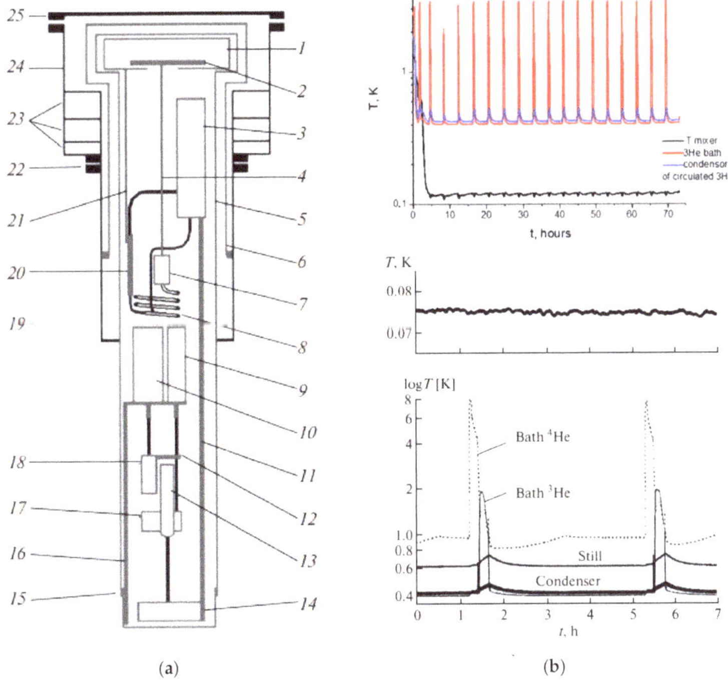

Figure 11. A continuously operating dilution microcryostat (**54**). (**a**) Simplified diagram of the microcryostat that shows the arrangement of the main units: (1, 5, 6) 0.6, 4.2, and 100 K shields, respectively; (2) platform at 0.06–0.1 K; (3) holmium-filled condenser of the mixture vapors; (4, 11, 16, 21) 0.1, 0.4, 4.2, and 0.6 K copper heat conductors, respectively; (7) mixer; (8) tubular heat exchanger; (9) ^{3}He sorber; (10) ^{4}He sorber; (12) condenser of ^{3}He arriving at the upper bath; (13) ^{3}He condenser of the lower circuit; (14) lower ^{3}He bath; (15) vacuum-tight soldered spots of shield 5; (17) upper ^{3}He bath; (18) ^{4}He bath (1 K chamber); (19) outer stainless steel tube with a 56 mm diameter; (20) still; (22) flange of the outer tube; (23) vessels for collecting the working gases that are stored in them at a pressure of 25–50 atm at room temperature of the instrument; (24) housing; and (25) detachable cap. Stainless steel tubes and parts are black-colored; copper parts are gray-colored. (**b**) Time dependences of the temperatures of the mixer (upper curve) and the ^{4}He bath, upper ^{3}He bath, still, and condenser of the mixture vapors (lower curves).

3.6. Dilution Microcryostat with Cooling by a Refrigerator with a Pulse Tube

More promising for autonomous operation on the BTA telescope is the dilution microcryostat designed for scientific research at temperatures up to 0.1 K and below. To ensure its operation, a refrigerator with a pulse tube is used. It is possible to maintain the temperature at 0.1 K for 5 h with the refrigerator not working. The device is used to study low-temperature radiation receivers. The photo of dilution microcryostat with cooling by a refrigerator with a pulse tube is presented in Figure 12.

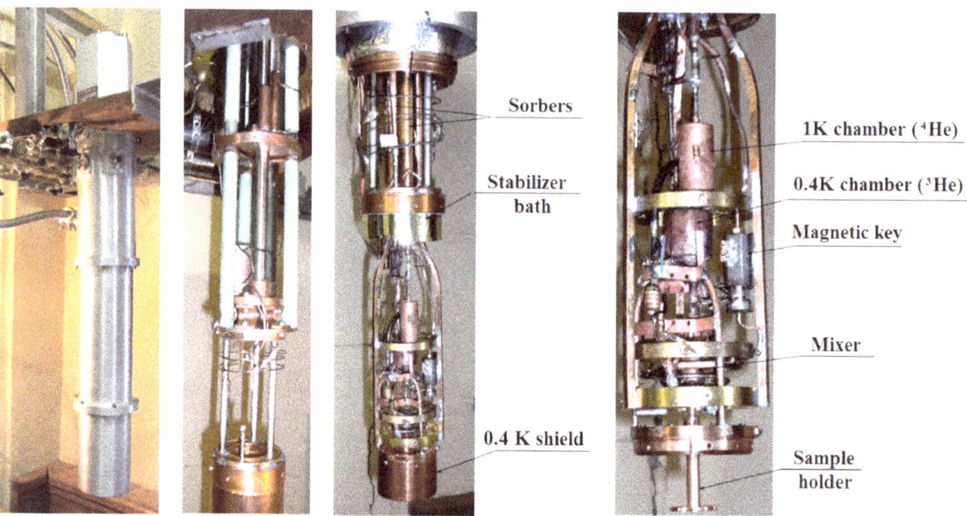

Figure 12. The photo of dilution microcryostat with cooling by a refrigerator with a pulse tube.

The main characteristics of the system:

- Minimum fixed temperature of the sample holder: 0.07 K.
- Performance at 0.1 K: about 1–3 μW.
- Cooling time to 0.1 K (from the start of pumping ^{3}He): 30–40 min.
- Time of regeneration of sorbers, condensation, and cooling ^{3}He up to 1 K: about 1.5 h.
- Time of temperature maintenance below 0.1 K: 4–6 h.
- Time to maintain working conditions when PT is turned off: up to 8–9 h.
- The time to restore the initial conditions after a new start PT: approximately 1 h.
- Cooling time from room temperature to 0.1 K: 10–12 h.
- Quantity of gases used: 4.5 n.l. ^{4}He; 3 n.l ^{3}He; 1 n.l of mixture 30% ^{3}Ha + 70% ^{4}He.
- Energy and water consumption during PT operation according to the manufacture's passport: 6–8 kWh and 8–10 L/min.

In essence, the latter option, called the "liquefaction circuit", is the most promising, because it combines the advantages of a refrigerated circuit and requires filling of a cryogent. But at the same time, for a long time (up to 8–9 h), it is able to maintain the operating temperature with the compressor and cooler turned off, i.e., absolutely without vibrations and noises and temperature fluctuations caused by the frequency of operation of the cooler with a frequency of about 1 s. Refrigeration systems without a liquefaction circuit, operating continuously generate noise and temperature fluctuations that dramatically affect the characteristics of the receiver and require special measures.

The main goals of this work have been achieved. At the same time, fixing certain specific temperatures is not our scientific achievement. The temperatures are just normal and usual for cryogenics. Today, cooling records are below nanokelvin. Our achievements lie in the fact that a sufficient temperature has been reached to achieve a certain level of sensitivity of receivers or a state of reliable superconductivity of superconducting detectors, that is, to ensure their operability. For cooled photonics, this means the disappearance of dark current and improved image quality. We would like to emphasize that an important feature of the cryogenic systems being created is their high level of functionality and availability. In particular, reaching a given temperature cooling regime takes about one hour for cryosystems to the hydrogen level and several (2–4) hours for helium and subhelium machines. This ensures ease of use of telescopes and does not take up a significant part of the time of astronomical observations. Another important result is the successful spatial, thermal, and optical matching of the developed cryogenic systems with built-in receivers and optical-mechanical structures of telescopes.

4. Conclusions

This paper provides a brief overview of the development of cryogenic systems operated at different temperature levels and the cooling principles for cooling radio and photo receivers of various frequency ranges for astronomical research using the main instruments of the Special Astrophysical Observatory of the Russian Academy of Sciences, specifically the BTA optical telescope and the RATAN-600 radio telescope.

The main results of the presented work are as follows:

- A successful line of cryogenic systems actually working at various temperature levels has been developed actively used over the years, and is now used for cooling photodetectors in the BTA telescope and radio receivers for the RATAN radio telescope;
- This work presents a new unique project of a cryosystem for the ultradeep cooling of a radio receiver for the BTA optical telescope;
- Using these instruments, new unique astronomical results were obtained and presented in a wide list of highly rated publications reviewed in [39–44,56];
- Dozens of highly efficient and reliable cryogenic systems have been developed that cool highly sensitive radio and photodetectors of the two telescopes at the Special Astrophysical Observatory of the Russian Academy of Sciences. Using these instruments, new unique astronomical results were obtained, presented in a wide list of highly rated publications reviewed in [39–44,56].

This is the first presentation of the new project connected with the development of a sub-THz radio receiver for the optical BTA telescope. Several variants of cryogenic systems of the sub-Kelvin level for operation of the sub-THz receiver of the BTA optical telescope at SAO RAS have been proposed for the new project.

The proposed approach will make it possible to fill part of the gap in the spectrum of astronomical research between the RATAN-600 observation windows in the radio range and the operating range of the receivers of the BTA optical telescope, of course, taking into account the limited transmission of THz waves shorter than 1 mm at the location of SAO RAS. There is a real astronomical science for such combination of optical telescope and radio receiver [56].

Author Contributions: Conceptualization, Y.B. and G.V.; methodology, S.M., N.T. and V.E.; software, O.B. and I.L.; formal analysis, V.V.; investigation, A.G., A.V., M.M. (Maria Markina), A.E., E.E., G.M. and A.K.; design and hardware, M.M. (Maria Mansfeld), E.P. and A.C.; writing—original draft preparation, A.G. and V.V.; writing—review and editing, G.V. and V.V.; supervision, Y.B.; project administration, Y.B. and V.V.; funding acquisition, Y.B. All authors have read and agreed to the published version of the manuscript.

Funding: The development of cryogenic systems for cooling of the sub-THz receiving device on the BTA telescope of SAO RAS is supported by the grant of the Russian Science Foundation # 23-62-10013 "Development of Russian subterahertz observatory prototype as part of an optical telescope".

Institutional Review Board Statement: Not applicable.

Informed Consent Statement: Not applicable.

Data Availability Statement: Data are contained within the article.

Acknowledgments: The authors are grateful to the employees of SAO RAS, IRE RAS, and IAP RAS who are currently working and have left us for the persistent development of cooled receivers for radio astronomy, as well as colleagues from the Cryogenic Technology Research Center (Omsk) and the Saturn Research Institute (Kiev) who actively contributed to these works. Cryogenic equipment was developed and manufactured on the basis of Large-scale research facilities «Center of Microwave Research of Materials and Substances» (CKP-7, UNU # 3589084).

Conflicts of Interest: The authors declare no conflict of interest.

References

1. Tychoniec, Ł.; Guglielmetti, F.; Arras, P.; Enßlin, T.; Villard, E. Bayesian Statistics Approach to Imaging of Aperture Synthesis Data: RESOLVE Meets ALMA. *Phys. Sci. Forum* **2022**, *5*, 52.
2. ALMA Observatory. About ALMA, at First Glance | ALMA. Available online: https://www.almaobservatory.org/en/about-alma (accessed on 10 November 2023).
3. Sachkov, M.; de Castro, A.I.G.; Shustov, B.; Sichevsky, S.; Shugarov, A. World Space Observatory-Ultraviolet mission: Status 2022. In *Space Telescopes and Instrumentation 2022: Ultraviolet to Gamma Ray*; SPIE: Bellingham, DC, USA, 2022.
4. Chen, Z.-R.; Tsai, W.-C.; Huang, S.-F.; Li, T.-Y.; Song, C.-Y. Classification of Plank Techniques Using Wearable Sensors. *Sensors* **2022**, *22*, 4510. [CrossRef] [PubMed]
5. Khokhriakova, A.D.; Chugunov, A.I.; Popov, S.B.; Gusakov, M.E.; Kantor, E.M. Observability of HOFNARs at SRG/eROSITA. *Universe* **2022**, *8*, 354. [CrossRef]
6. Lamb, J.W. Miscellaneous data on materials for millimetre and submillimetre optics. *Int. J. Infrared Millim. Waves* **1996**, *17*, 1997–2034. [CrossRef]
7. Lovyagin, N.; Raikov, A.; Yershov, V.; Lovyagin, Y. Cosmological Model Tests with JWST. *Galaxies* **2022**, *10*, 108. [CrossRef]
8. NASA. Webb Image Release- Webb Space Telescope GSFC/NASA. Available online: https://webb.nasa.gov/ (accessed on 10 November 2023).
9. NSF. The South Pole Telescope. Available online: https://pole.uchicago.edu/public/Home.html (accessed on 10 November 2023).
10. SAO RAS. SAO RAS Home Page. Available online: https://www.sao.ru/Doc-en/index.html (accessed on 10 November 2023).
11. Shulga, V.M.; Zinchenko, I.I.; Nesterov, N.S.; Myshenko, V.V.; Andriyanov, A.F.; Isaev, V.F.; Knyazkov, L.B.; Lapinov, A.V.; Litvenenko, L.N.; Maltsev, V.A.; et al. Molecular Line Observations in the 85-GHZ to 90-GHZ Band Using a Maser Receiver at the Crimean Astrophysical Observatory 22-METER Radio Telescope. *Sov. Astron. Lett.* **1991**, *17*, 448.
12. Lapinov, A.V.; Lapinova, S.A.; Petrov, L.Y.; Ferrusca, D. On the benefits of the Eastern Pamirs for sub-mm astronomy. In *Millimeter, Submillimeter, and Far-Infrared Detectors and Instrumentation for Astronomy X*; SPIE: Bellingham, DC, USA, 2020.
13. Khaikin, V.B.; Khaikin, V.B.; Shikhovtsev, A.Y.; Mironov, A.P.; Qian, X. A study of the astroclimate in the Dagestan mountains Agul region and at the Ali Observatory in Tibet as possible locations for the Eurasian SubMM Telescopes (ESMT). *Proc. Sci.* **2002**, *425*, 72–79.
14. Shikhovtsev, A.Y.; Kovadlo, P.G.; Khaikin, V.B.; Kiselev, A.V. Precipitable Water Vapor and Fractional Clear Sky Statistics within the Big Telescope Alt-Azimuthal Region. *Remote Sens.* **2022**, *14*, 6221. [CrossRef]
15. Battistelli, E.S.; Capalbo, V.; Isopi, G.; Radiconi, F. Status of Cosmic Microwave Background Observations for the Search of Primordial Gravitational Waves. *Universe* **2022**, *8*, 489. [CrossRef]
16. Collaboration, The Event Horizon Telescope; Akiyama, K.; Alberdi, A.; Alef, W.; Asada, K.; Azulay, R.; Baczko, A.K.; Ball, D.; Baloković, M.; Barrett, J.; et al. First M87 Event Horizon Telescope Results. I. The Shadow of the Supermassive Black Hole. *Astrophys. J. Lett.* **2019**, *875*, L1.
17. Raymond, A.W.; Palumbo, D.; Paine, S.N.; Blackburn, L.; Rosado, R.C.; Doeleman, S.S.; Farah, J.R.; Johnson, M.D.; Roelofs, F.; Tilanus, R.P.J.; et al. Evaluation of New Submillimeter VLBI Sites for the Event Horizon Telescope. *Astrophys. J. Suppl. Ser.* **2021**, *253*, 5. [CrossRef]
18. Bubnov, G.M.; Abashin, E.B.; Balega, Y.Y.; Bolshakov, O.S.; Dryagin, S.Y.; Dubrovich, V.K.; Marukhno, A.S.; Nosov, V.I.; Vdovin, V.F.; Zinchenko, I.I.; et al. Searching for New Sites for THz Observations in Eurasia. *IEEE Trans. Terahertz Sci. Technol.* **2015**, *5*, 64–72. [CrossRef]
19. Zinchenko, I.I.; Lapinov, A.V.; Vdovin, V.F.; Zemlyanukha, P.M.; Khabarova, T.A. Measurements and Evaluations of the Atmospheric Transparency at Short Millimeter Wavelengths at the Candidate Sites for Millimeter and Submillimeter Wave Telescopes. *Appl. Sci.* **2023**, *13*, 11706. [CrossRef]
20. Ladu, A.; Schirru, L.; Ortu, P.; Saba, A.; Pili, M.; Navarrini, A.; Gaudiomonte, F.; Marongiu, P.; Pisanu, T. Adaptation of an IRAM W-Band SIS Receiver to the INAF Sardinia Radio Telescope: A Feasibility Study and Preliminary Tests. *Sensors* **2023**, *23*, 7414. [CrossRef] [PubMed]
21. Baryshev, A.; Bolshakov, O.; Chernikov, A.; Gunbina, A.; Efimova, M.; Eliseev, A.; Krasilnikov, A.; Lapkin, I.; Lesnov, I.; Mansfeld, M.; et al. Development of cryogenic systems for astronomical research. *Photonics* 2023, *to be submitted*.
22. Murzin, V.A.; Markelov, S.V.; Ardilanov, V.I.; Afanaseva, T.; Borisenko, A.; Ivashchenko, N.; Pritychenko, M.; Mitiani, G.; Vdovin, V. Astronomical CCD systems for the 6-meter telescop BTA (a review). *Appl. Phys.* **2016**, *4*, 500–506. (In Russian)

23. Gardner, D.V. Does Your CCD Camera Need Cooling? Available online: https://www.photonics.com/Articles/Does_Your_CCD_Camera_Need_Cooling/a12810 (accessed on 10 November 2023).
24. Cease, H.; DePoy, D.; Derylo, G.; Diehl, H.T.; Estrada, J.; Flaugher, B.; Kuk, K.; Kuhlmann, S.; Lathrop, A.; Schultz, K.; et al. Cooling the dark energy camera CCD array using a closed-loop two-phase liquid nitrogen system. In *Moden Technologies in Space and Ground-Base Telescopes and Instrumentation*; SPIE: Bellingham, DC, USA, 2010; p. 77393N.
25. Malkov, M.P.; Danilov, I.B.; Zeldovich, A.G.; Fradkov, A.B. *Handbook on the Physical and Technical Foundations of Cryogenics*, 3rd ed.; Energoatomizdat: Moscow, Russia, 1985; p. 432.
26. Vdovin, V.F. Problems of cryogenic cooling of superconductor and semiconductor receivers in the range 0.1–1 THz. *Radiophys. Quantum Electron.* **2005**, *48*, 779–791. [CrossRef]
27. Belyaeva, A.I.; Silaev, V.I.; Stelmakhov, Y.N.; Stetsenko, Y.E. Continuous flow cryostats for experiments in the presence of appreciable heat influx to the specimen. *Cryogenics* **1983**, *23*, 303–308. [CrossRef]
28. Cryotiger Coolers. Available online: https://www.ing.iac.es/~eng/detectors/CRYOTIGER_man.pdf (accessed on 10 November 2023).
29. Borisenko, A.N.; Vdovin, V.F.; Eliseev, A.I.; Lapkin, I.V.; Markelov, S.V. Cryostating of large CCD matrices. *Petersburg J. Electron.* **2001**, *28*, 39–43. (In Russian)
30. Esepkina, N.A.; Korolkov, D.V.; Pariysky, Y.N. *Radio Telescopes and Radiometers*; Nauka: Moscow, Russia, 1972; p. 416.
31. Cryogenic Technique. STC Cryogenic Technique—Home. Available online: https://cryontk.ru/en/ (accessed on 10 November 2023).
32. Mimura, T. The early history of the high electron mobility transistor (HEMT). *IEEE Trans. Microw. Theory Tech.* **2002**, *50*, 780–782. [CrossRef]
33. Vdovin, V.F.; Grachev, V.G.; Dryagin, S.Y.; Eliseev, A.I.; Kamaletdinov, R.K.; Korotaev, D.V.; Lesnov, I.V.; Mansfeld, M.A.; Pevzner, E.L.; Perminov, V.G.; et al. Cryogenically cooled low-noise amplifier for radio-astronomical observations and centimeter-wave deep-space communications systems. *Astrophys. Bull.* **2016**, *71*, 125–128. [CrossRef]
34. Räisänen, A.A. *THz Receivers for Ground and Space*; International Seminar on Terahertz Electronics: Darmstadt, Germany, 1994; pp. s.2–s.33.
35. Virginia Diodes. Virginia Diodes, Inc.—Your Source for Terahertz and mm-Wave Products. Available online: https://www.vadiodes.com/en/ (accessed on 10 November 2023).
36. Vdovin, V.F.; Eliseev, A.I.; Zinchenko, I.I.; Koshelets, V.P.; Kuznetsov, I.V.; Lapkin, I.V.; Mujunen, A.; Oinaskallio, E.; Peltonen, J.; Pilipenko, O.M.; et al. A Two-Frequency Two-Polarization Superconducting Receiver for Radio-Astronomical Investigations in the Millimeter Wave Band. *J. Commun. Technol. Electron.* **2005**, *50*, 1118–1122.
37. Kraus, J.D. *Radio Astronomy*; Cygnus-Quasar Books: Powell, OH, USA, 1986; p. 719, ISBN 1882484002.
38. Masuda, T.; Hiramoto, A.; Ang, D.G.; Meisenhelder, C.; Panda, C.D.; Sasao, N.; Uetake, S.; Wu, X.; DeMille, D.P.; Gabrielse, D.G.; et al. High-sensitivity low-noise photodetector using a large-area silicon photomultiplier. *Opt. Express* **2023**, *31*, 1943–1957. [CrossRef]
39. Semenko, E.; Romanyuk, I.; Yakunin, I.; Kudryavtsev, D.; Moiseeva, A. Spectropolarimetry of magnetic Chemically Peculiar stars in the Orion OB1 association. *Mon. Not. R. Astron. Soc.* **2022**, *515*, 998–1011. [CrossRef]
40. Solovyeva, Y.; Vinokurov, A.; Tikhonov, N.; Kostenkov, A.; Atapin, K.; Sarkisyan, A.; Moiseev, A.; Fabrika, S.; Oparin, D.; Valeev, A. Search for LBVs in the Local Volume galaxies: Study of two stars in NGC 1156. *Mon. Not. R. Astron. Soc.* **2022**, *518*, 4345–4356. [CrossRef]
41. Galazutdinov, G.A.; Baluev, R.V.; Valyavin, G.; Aitov, V.; Gadelshin, D.; Valeev, A.; Sendzikas, E.; Sokov, E.; Mitiani, G.; Burlakova, T.; et al. Doppler confirmation of TESS planet candidate TOI−1408.01: Grazing transit and likely eccentric orbit. *Mon. Not. R. Astron. Soc. Lett.* **2023**, *526*, L111–L115. [CrossRef]
42. Mingaliev, M.G.; Sotnikova, Y.V.; Udovitskiy, R.Y.; Mufakharov, T.V.; Nieppola, E.; Erkenov, A.K. RATAN-600 multi-frequency data for the BL Lacertae objects. *Astronomy* **2014**, *572*, A59. [CrossRef]
43. Plavin, A.; Kovalev, Y.Y.; Kovalev, Y.A.; Troitsky, S. Observational Evidence for the Origin of High-energy Neutrinos in Parsec-scale Nuclei of Radio-bright Active Galaxies. *Astrophys. J.* **2020**, *894*, 101. [CrossRef]
44. Nakariakov, V.M.; Anfinogentov, S.; Storozhenko, A.A.; Kurochkin, E.A.; Bogod, V.M.; Sharykin, I.N.; Kaltman, T.I. Quasi-periodic Pulsations in a Solar Microflare. *Astrophys. J.* **2018**, *859*, 154. [CrossRef]
45. Vystavkin, A.N.; Shitov, S.V.; Bankov, S.E.; Kovalenko, A.G.; Pestriakov, A.V.; Cohn, I.A.; Uvarov, A.V.; Vdovin, V.F.; Perminov, V.G.; Trofimov, V.N.; et al. High sensitive 0.13–0.38 THz TES array radiometer for the big telescope azimuthal of Special Astrophysical Observatory of Russian Academy of Sciences. In Proceedings of the 2007 Joint 32nd International Conference on Infrared and Millimeter Waves and the 15th International Conference on Terahertz Electronics, Cardiff, UK, 2–9 September 2007.
46. Camus, P.; Vermeulen, G.; Volpe, A.; Triqueneaux, S.; Benoit, A.; Butterworth, J.; d'Escrivan, S.; Tirolien, T. Status of the Closed-Cycle Dilution Refrigerator Development for Space Astrophysics. *J. Low Temp. Phys.* **2013**, *176*, 1069–1074. [CrossRef]
47. Frossati, G. Experimental techniques: Methods for cooling below 300 mK. *J. Low Temp. Phys.* **1992**, *87*, 595–633. [CrossRef]
48. Benoit, A.; Pujol, S. A dilution refrigerator insensitive to gravity. *Phys. B Condens. Matter* **1991**, *169*, 457–458. [CrossRef]
49. Benoît, A.; Caussignac, M.; Pujol, S. New types of dilution refrigerator and space applications. *Phys. B Condens. Matter* **1994**, *197*, 48–53. [CrossRef]

50. Sirbi, A.; Pouilloux, B.; Benoit, A.; Lamarre, J.-M. Influence of the astrophysical requirements on dilution refrigerator design. *Cryogenics* **1999**, *39*, 665–669. [CrossRef]
51. Chernikov, A.N.; Trofimov, V.N. Helium-3 adsorption refrigerator cooled with a closed-cycle cryocooler. *J. Surf. Investigation. X-ray Synchrotron Neutron Tech.* **2014**, *8*, 956–960. [CrossRef]
52. Tarasov, M.; Gunbina, A.; Chekushkin, A.; Yusupov, R.; Edelman, V.; Koshelets, V. Microwave SINIS Detectors. *Appl. Sci.* **2022**, *12*, 10525. [CrossRef]
53. Cryogenic Society. Cold Facts. Available online: https://www.cryogenicsociety.org/cold-facts (accessed on 10 November 2023).
54. Edelman, V.S. A continuously operating dilution microcryostat. *Instrum. Exp. Tech.* **2012**, *55*, 145–148. [CrossRef]
55. Herrmann, R.; Ofitserov, A.V.; Khlyustikov, I.N.; Edel\textquotesingleman, V.S. A Portable Dilution Refrigerator. *Instrum. Exp. Tech.* **2005**, *48*, 693–702. [CrossRef]
56. Stolyarov, V.; Balega, Y.; Mingaliev, M.; Sotnikova, Y.; Vdovin, V.; Gunbina, A.; Tarasov, M.; Fominskii, M.; Chekushkin, A.; Edelman, V.; et al. Prospects for the Development of Millimeter Astronomy in SAO RAS. *Astrophys. Bull.* **2023**, *under review*.

Article

A Tutorial on Retroreflectors and Arrays Used in Satellite and Lunar Laser Ranging

John J. Degnan [†]

Independent Researcher, Odenton, MD 21113, USA; jjdegnan3rd@gmail.com

[†] Semi-Retired Technical Consultant (2018-present) formerly held various research and supervisory positions at NASA Goddard Space Flight Center (1964–2002), Chief Scientist at Sigma Space Corporation, (2002–2018), and Distinguished Adjunct Professor of Physics at The American University (1988–1993). Cofounder of International Laser Ranging Service (1998) and 1st ILRS Governing Board Chairman (1998–2002), Optica Fellow. IEEE Senior Life Member.

Abstract: The present paper discusses the basics of retroreflector theory and the manner in which they are combined in arrays to service the laser tracking of artificial satellites and the Moon. We begin with a discussion of the relative advantages and disadvantages of solid versus hollow cube corners and the functional dependence of their optical cross-sections and far-field patterns on cube diameter. Because of velocity aberration effects, the design of an array for a particular space mission depends on many factors, including the desired range accuracy and the satellite's orbital altitude, velocity, and pass geometry relative to the tracking station. This generally requires the individual retroreflectors in the array to be "spoiled" by perturbing one or more of the 90-degree angles that define a perfect cube corner, or alternatively, by adding a curved surface to a hollow cube. In order to obtain adequate return signal strengths from all points along the satellite path, the rotational orientation of the retroreflectors within the array may need to be varied or "clocked". Possible approaches to developing millimeter-accuracy arrays with both large cross-sections and ultrashort satellite signatures are discussed, as are new designs proposed to replace aging reflectors on the Moon. Finally, we briefly discuss methods for laser ranging beyond the Moon.

Keywords: retroreflectors; cube corners; Satellite Laser Ranging; Lunar Laser Ranging; velocity aberration; spoiling; clocking; cross-section

Citation: Degnan, J.J. A Tutorial on Retroreflectors and Arrays Used in Satellite and Lunar Laser Ranging. *Photonics* **2023**, *10*, 1215. https://doi.org/10.3390/photonics10111215

Received: 13 September 2023
Revised: 23 October 2023
Accepted: 26 October 2023
Published: 31 October 2023

1. Introduction

The first successful Satellite Laser Ranging (SLR) experiment was carried out on 31 October 1964 by Dr. Henry H. Plotkin and his team at the NASA Goddard Space Flight Center in Greenbelt, Maryland, USA [1]. In this first experiment, the NASA Explorer 22B satellite was equipped with an array of retroreflectors designed to reflect laser pulses from a ground station back to their point of origin. The roundtrip distance to the satellite can then be determined by multiplying the roundtrip transit time by the speed of light. In 1969, a team led by Prof. Carroll Alley of the University of Maryland succeeded in ranging to a flat panel of 100 retroreflectors left on the lunar surface by NASA's Apollo 11 crew [2]. In the almost six decades since these early experiments, dozens of Earth-orbiting satellites carrying retroreflector arrays have been launched to support a wide variety of scientific investigations, and five arrays have been placed on the lunar surface by three manned US Apollo missions (11, 14, and 15) and two unmanned Soviet Lunakhod (17 and 21) missions carrying arrays provided by France. The present author has previously published a comprehensive review of the active ground-based SLR hardware and operations [3], whereas this paper will concentrate on the requirements placed on the passive spaceborne segment, i.e., the retroreflector arrays. For a more complete discussion of recent SLR capabilities and trends, the reader is referred to the following reviews [4,5].

2. SLR Link Equation

During the first four decades of SLR (1964 to 2004), ground stations were characterized by a moderate-to-large (submeter-to-meter-class) receiver telescope, and the laser beam was transmitted via a parallel, small-diameter tube mounted on the side of the telescope. Since the laser beam divergence is proportional to the ratio of the laser wavelength to the beam radius (λ/ω), this approach resulted in an unnecessarily high divergence beam that required much higher-energy lasers to obtain an adequate satellite return. These high-energy beams posed a potential eye safety hazard to personnel in low-flying aircraft and to ground personnel during ground calibrations. As a result, all early SLR stations were equipped with dedicated aircraft radars and safety personnel to man them. However, in 1994, the present author proposed a new system, SLR2000, to NASA management [6], which optimally sized and transmitted a low-divergence laser beam via a common transmit–receive telescope to achieve maximum illumination of the satellite array while simultaneously reducing the laser energy and the threat to eye safety. The optimum SLR2000 design drew heavily from earlier studies on optical antenna gain carried out in support of early NASA laser communications studies [7,8]. The NASA SLR2000 funding started slowly in 1998, but by 2003, the newly named Next-Generation Satellite Laser Ranging (NGSLR) system had tracked its first satellite and continued tracking until 2007, when it was sadly destroyed by a lightning bolt. NASA's replacement system, Space Geodesy Satellite Laser Ranging (SGSLR), adds additional automation and is currently being installed at three sites—(1) the NASA Goddard Space Flight Center in Greenbelt, MD, USA, (2) the McDonald Observatory at the University of Texas, USA, and (3) Ny Alesund, Norway [9].

The SLR link equation defines the number of photons detected by the ground station and is given by

$$n_s = \frac{E}{h\tau}\eta_t\,\frac{2}{\pi(\theta_d R)^2}exp\left[-2\left(\frac{\Delta\theta_p}{\theta_d}\right)^2\right]\left[\frac{1}{1+\left(\frac{\Delta\theta_j}{\theta_d}\right)^2}\right]\left(\frac{\sigma A_r}{4\pi R^2}\right)\eta_r\eta_d T_a^2 T_c^2 \tag{1}$$

where

n_s = the number of detected satellite photoelectrons per laser pulse
E_t = the laser pulse energy
$h\nu$ = the laser photon energy = 3.73×10^{-19} J @ 532 nm (Doubled Nd:YAG)
η_t = the transmitter optical throughput efficiency
θ_d = the Gaussian beam divergence half angle
R = the slant range between the station and satellite
$\Delta\theta_p$ = the laser beam pointing error
$\Delta\theta_j$ = the RMS tracking mount jitter
σ = the satellite optical cross-section
A_r = the telescope receiving area.
η_r = the receiver optical throughput efficiency
η_c = the detector counting efficiency
T_a = the one-way atmospheric transmission
T_c = the one-way cirrus cloud transmission

We note from Equation (1) that the received signal strength varies as $1/R^4$ and that the satellite optical cross-section, σ, is the sole contribution of the space segment to the link equation. Retroreflectors, also referred to as "cube corners", can be either hollow or solid, and the solid retros can be coated or uncoated. Table 1 provides a summary of the various retro types, their unique characteristics, and their historical usage. Note also that the satellite optical array cross-section, σ, is the sole link contribution of the space segment.

Table 1. Three types of cube corners and their properties.

Type	Al Back-Coated Solid	Uncoated Solid (TIR)	Hollow
Frequency of Use	Most Common	Occasional Use	Not currently used in the visible range
Satellite Examples	Most satellites	Apollo, LAGEOS, AJISAI, ETS-VIII	ADEOS RIS, REM, TES
Reflectivity, ρ	0.78	0.93	Can approach 1.0
Polarization Sensitive	No	Yes	No—metal coating Yes—dielectric coating
Weight	Heavy	Heavy	Light
Far-Field Pattern	Wide	Wide	Narrow
Issues	Metal coatings absorb sunlight and create thermal gradients. Not as well shielded at high orbital altitudes.	Fewer thermal problems, but TIR "leaks" at incidence angles > 17°. Polarization effects reduce cross-section by a factor of 4.	Thermal heating and gradient effects on joints

Retroreflectors (or "cube corners") are designed to reflect light back to the point of origin in a narrow beam. Increasing the size or number of retroreflectors increases the return signal strength. Figure 1 illustrates the two basic types of cube corner, i.e., solid or hollow. Table 1 lists the relevant properties and/or advantages of (1) aluminum back-coated solid retroreflectors, (2) uncoated solid Total Internal Reflection (TIR) retroreflectors, and (3) hollow retroreflectors. For normally incident light, a single unspoiled retroreflector (cube corner) has a peak, on-axis, optical cross-section defined by

$$\sigma_{cc} = \rho A_{cc}\left(\frac{4\pi}{\Omega}\right) = \rho\frac{4\pi A_{cc}^2}{\lambda^2} = \frac{\pi^3 \rho D^4}{4\lambda^2} \tag{2}$$

where the reflectivity of the cube corner, ρ, is typically equal to 0.78 or 0.93 for aluminum-coated back faces and uncoated Total Internal Reflection (TIR) surfaces, respectively, A_{cc} is the collecting aperture of the corner cube, D is the cube diameter, $4\pi/\Omega$ is the on-axis reflector gain, and Ω is the effective solid angle occupied by the far-field diffraction pattern (FFDP) of the retroreflector.

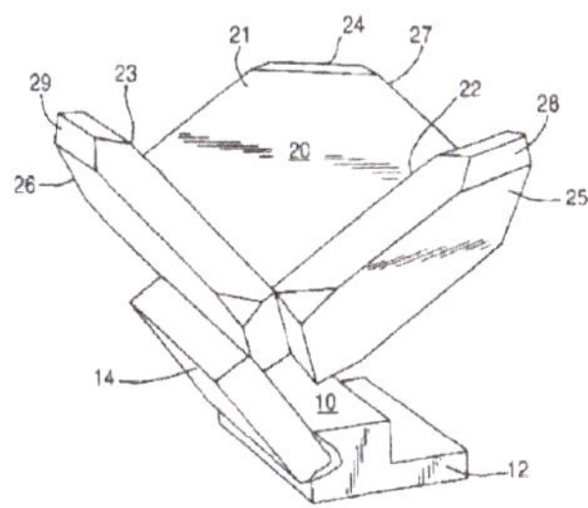

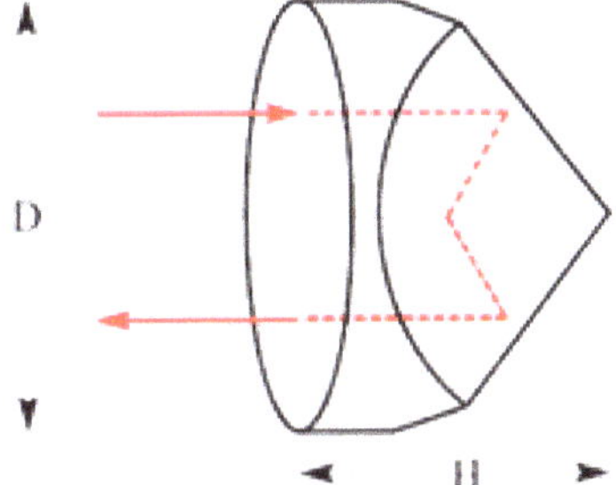

Figure 1. *Cont.*

Hollow Cube Corner   **Solid Cube Corner**

Figure 1. Two types of retroreflector/cube corner—hollow and solid.

As illustrated in Figure 2, the peak optical cross-section rises rapidly as the retroreflector diameter to the fourth power. For the popular 1.5 in (38 mm)-diameter cube with a physical cross-section of 0.001 m^2, the peak optical cross-section is about 5.8×10^7 m^2, an increase of over ten orders of magnitude.

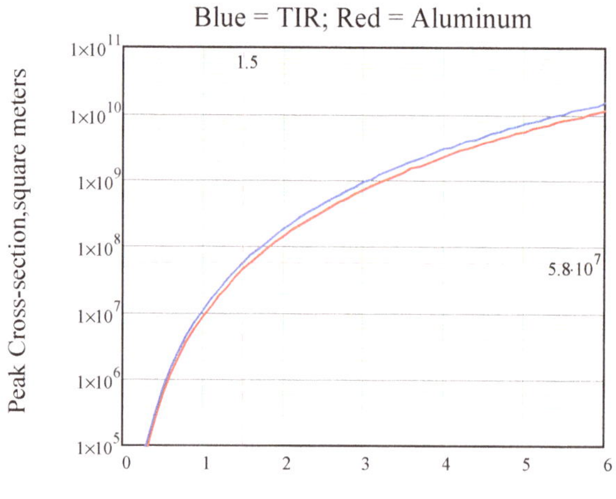

Figure 2. The peak cross-section in square meters as a function of the solid retroreflector diameter in inches for both Total Internal Reflection (blue curve) and aluminum back-coating (red curve) surfaces. The peak optical cross-section rises rapidly as the retroreflector diameter to the fourth power. For the popular 1.5 in (38 mm)-diameter cube with a physical cross-section of 0.001 m^2, the peak optical cross-section is about 5.8×10^7 m^2, an increase of over ten orders of magnitude.

3. Retroreflector Far-Field Diffraction Pattern (FFDP)

For a uniformly illuminated circular aperture, the FFDP of the reflected wave, as plotted in Figure 3, is the familiar Airy function, given by

$$\sigma(x) = \sigma_{cc} \left[\frac{2J_1(x)}{x} \right]^2 \tag{3}$$

$$x = \frac{\pi D}{\lambda} \sin\theta \tag{4}$$

where the frequency-doubled Nd:YAG laser wavelength, $\lambda = 532$ nm, is the most widely used in the SLR ground network, and D is the cube aperture diameter.

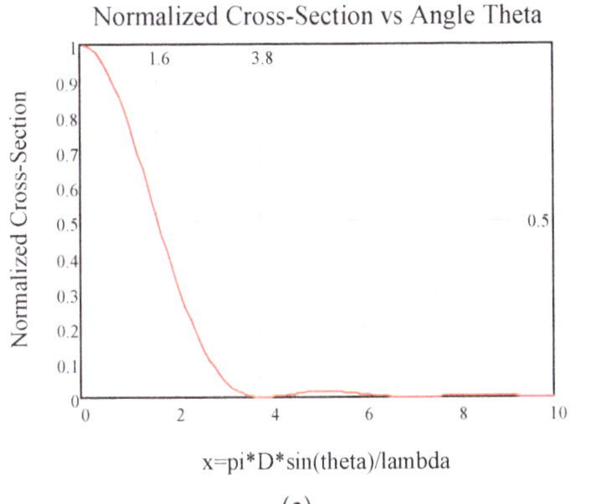

(**a**)

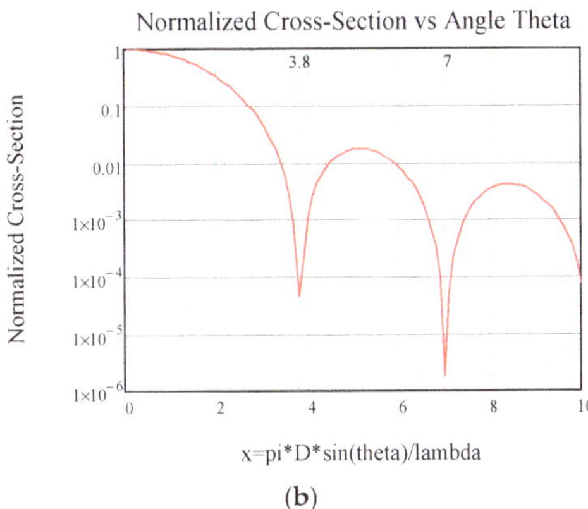

(**b**)

Figure 3. The FFDP of an unspoiled circular aperture retroreflector is plotted on a linear scale and a logarithmic scale in parts (**a**,**b**) of the figure, respectively. The half-power and first null occur at $x = 1.6$ and 3.8, respectively. For the popular 1.5 in (38 mm)-diameter cube at 532 nm, this corresponds to $\theta = 7.1$ and 16.9 microradians (1.5 and 3.5 arcsec), respectively.

At an arbitrary incidence angle, θ_{inc}, the effective area of the cube is reduced by the factor

$$\eta(\theta_{inc}) = \frac{2}{\pi}\left(\sin^{-1}\mu - \sqrt{2}\tan\theta_{ref}\right)\cos\theta_{inc} \tag{5}$$

where θ_{inc} is the incidence angle and θ_{ref} is the internal refracted angle, as determined by Snell's Law, i.e.,

$$\theta_{ref} = \sin^{-1}\left(\frac{\sin\theta_{inc}}{n}\right) \tag{6}$$

where n is the cube index of refraction, and the quantity μ is given by the formula

$$\mu = \sqrt{1 - \tan^2\theta_{ref}} \tag{7}$$

Thus, as the incidence angle, θ_{inc}, increases, the peak optical cross-section in the center of the reflected lobe falls off as

$$\sigma_{eff}(\theta_{inc}) = \eta^2(\theta_{inc})\sigma_{cc} \tag{8}$$

and is plotted in Figure 4.

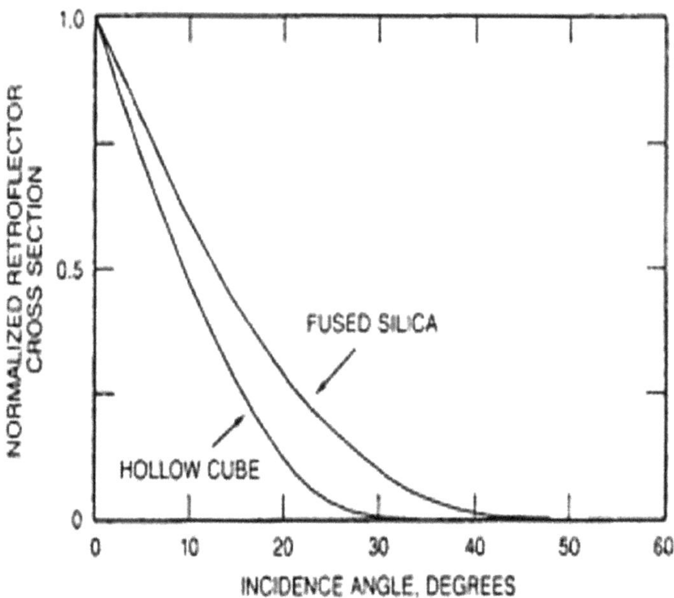

Figure 4. The normalized cross-section as a function of beam incidence angle for hollow and solid fused silica retroreflectors. The 50% and 0% efficiency points for fused silica ($n = 1.455$) are 13° and 45°, respectively, whereas the 50% and 0% efficiency points for a hollow cube ($n = 1$) are 9° and 31°, respectively, i.e., hollow cubes have a narrower angular response range than solid cubes.

4. Retroreflectors as Ground Calibration Targets

All SLR stations make use of a calibration scheme to initially determine and then monitor changes in optical and/or electronic system delays that might be caused by changes in hardware or environmental conditions (e.g., temperature). The most common approach is to place a single retroreflector at some carefully surveyed distance from the ground station "invariant point", defined as the intersection of the elevation and azimuth axes of the telescope assembly. The retro acts as a point source with a delta function response. This calibration distance is usually measured at the 1 or 2 mm level using accurate ground surveying techniques. Subtracting the known target range from the measured range provides a "range correction" which is then applied to all future satellite measurements. For maximum accuracy, calibrations are typically performed hourly.

5. Science Applications of SLR Impact the Satellite Array Design

The current SLR constellation spans a wide range of altitudes (500 km to 36,000 km) and inclinations depending on mission requirements, and each retroreflector array has to be designed accordingly based on orbital altitude, mission goals, desired signal strength, etc. The artificial satellites typically fall into four altitude realms with very different scientific goals, e.g.,

1. **Low-Earth Orbiting (LEO: h < 1500 km):** Earth gravity field studies and/or orbital support for spaceborne radars or lidars observing terrain heights, sea/ice levels, tree mass, etc.

2. **Medium Earth Orbiting (MEO: h~6000 km):** Low-drag MEO satellites such as LAGEOS 1 and 2 are ideal for accurately determining relative station positions, tectonic plate motion, regional crustal deformation, etc.

3. **Global Navigation System Satellites (GNSS. h~20,000 km):** SLR provides precise orbital support to International Navigation Constellations, such as GLONASS (Rus-

sia), GALILEO (EU), and COMPASS/BeiDou (China). All of the satellites in the aforementioned constellations are, or will be, equipped with laser retroreflectors, thereby enhancing the accuracy of GNSS orbits and their associated ground networks and transportation guidance. Only two of the US GPS satellites carry retroreflectors.

4. **Geosynchronous satellites (GEO: h~36,000 km):** in equatorial orbits provide continuous monitoring of roughly a third of the Earth's surface. Besides obvious defense and military applications, civilian applications include global communications, weather forecasting, television broadcasting, as well as communications between ground stations and/or relaying information between other satellites or spacecraft.

For maximum accuracy in orbit determination, the distance of the effective light reflection point from the satellite center of mass is ideally independent of the viewing angle. Thus, passive LEO and GEO satellites are typically spheres embedded with retros. Furthermore, since the signal strength decreases with satellite range as $1/R^4$, the sphere diameter typically increases with altitude to accommodate more retros in order to meet cross-section (signal strength) requirements. To improve detection probability and range precision, modern SLR stations typically employ fast, single-photon-sensitive, ultralow-timewalk detectors such as Micro Channel Plate Photo Multiplier Tubes (MCP/PMTs) or Single Photon Avalanche Diode (SPAD) arrays. When operated at or near single-photon-return levels, these detectors have allowed satellite detection at few kHz rates with low laser pulse energies as opposed to the historical few pulses per second with high pulse energies [6]. In the kHz low-signal scenario, each retroreflector in the array has a diminishing probability of providing the detected photon in a given measurement as it is further displaced from the laser line-of-sight (retro #7) and the laser beam incidence angle increases (Figure 5).

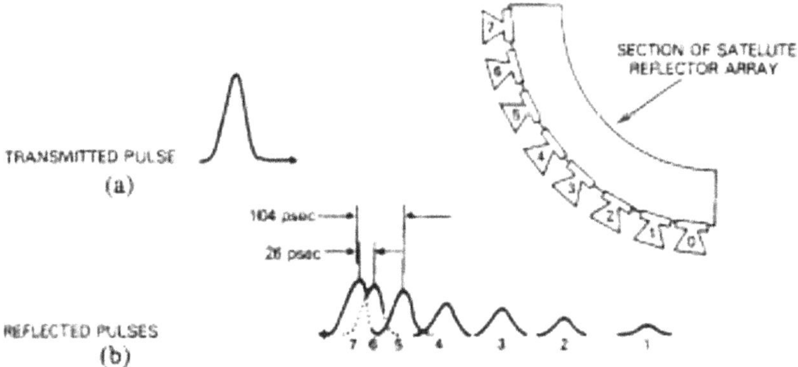

Figure 5. Pulse spreading induced by spherical LEO and GEO satellites (**a**). In multiphoton detection mode, the returned light pulse is the sum of the ultrashort (~100 psec) light pulses generated by the individual retroreflectors, which are reduced in intensity at the higher beam incidence angles (**b**). In multiphoton mode, the detector in the ground station sums over the individual retro returns before sending them to the timing electronics, thereby degrading the range precision. In single-photon mode, returns from individual reflectors are received at kHz rates and averaged over time to form "Normal Points".

In Figure 6, τ is a time normalized to the time it takes a light pulse to travel the diameter of the satellite, i.e., $2R_s/c$, where R_s is the radius of the sphere. Increasing the radius of the satellite will increase the cross-section by allowing the impinging laser beam to illuminate more cubes at low incidence angles. At the same time, however, the larger satellite will also broaden the impulse response, thereby degrading range precision. However, one can narrow the temporal response of the larger-diameter satellite while maintaining a high cross-section at low incidence angles by either using hollow rather than solid cube corners (as illustrated in Figure 6) or by recessing solid cubes inside hollow tubes to limit the response to cubes within a narrower range of incidence angles. The latter approaches effectively eliminate the weaker contributions of the pulses to the right in Figure 5b.

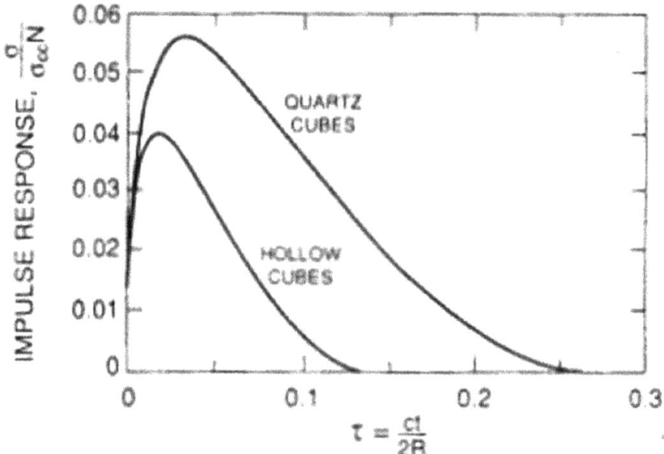

Figure 6. Impulse response in the large spherical satellite limit ($R_s \gg nL$) for both hollow and solid quartz cube corners. The difference is due to the smaller acceptance angle in hollow cubes resulting in fewer cubes contributing to the return.

Finally, because GNSSs and geosynchronous satellites generally perform a utilitarian function (e.g., Earth observation, communications, navigation, etc.), the nadir side of the satellite approximately faces the Earth Center of Mass (CoM). For the typical maximum zenith tracking angle of 70°, the beam incidence angles from the array normal can vary from 0 to β, where

$$\beta = asin\left[\frac{R_E}{R_E + h} sin110°\right] \tag{9}$$

For GNSS satellites at h = 20,000 km, β = 13.1 deg, and for GEO satellites at h = 36,000km, β = 8.2 deg. The smaller range of incidence angles ensures: (1) near-maximum-strength returns from a planar array; and (2) limited pulse spreading, especially if the array is compact in size and the retros are densely packed together to achieve the required cross-section. Nevertheless, the maximum flat-panel-induced laser pulse spreading per linear foot of array due to the zenith tracking angle is still 474 psec (7 cm) and 292 psec (4.4 cm) for GNSS and GEO satellites, respectively. Thus, typical rectangular or square planar arrays can introduce an angle-dependent broadening of the pulse return, resulting in lower-ranging precision. Although a planar circular array with the same area would result in more uniform pulse spreading, the use of a segmented sphere mounted on the nadir face of the satellite would largely eliminate angle-dependent pulse spreading and range biases.

6. Velocity Aberration and "Spoiled" Retroreflectors

If there is no relative velocity between the station and satellite, the beam reflected by the retroreflector will fall directly back onto the station. However, as illustrated in Figure 7, a relative velocity, v, between the satellite and station causes the reflected beam to be angularly deflected from the station in the forward direction of the satellite motion by an angle $\alpha = 2v/c$. This phenomenon is referred to as "velocity aberration". Since small-diameter cubes have small optical cross-sections but large-angle far-field diffraction patterns (FFDPs), the signal at the station is not significantly reduced by velocity aberration. On the other hand, large-diameter cubes with high cross-sections have small-angle FFDPs, and the signal at the station is therefore substantially reduced by velocity aberration. In general, the signal is reduced by half or more if the cube diameter, D_{cc}, satisfies the inequality

$$D_{cc} > D_{1/2} = \frac{1.6\,\lambda}{\pi\alpha} = \frac{0.8\,\lambda c}{\pi v} \tag{10}$$

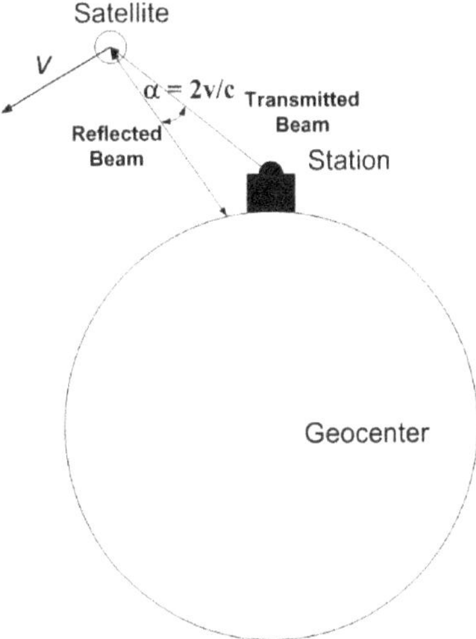

Figure 7. Velocity aberration causes the reflected beam to be angularly displaced from the ground station in the direction of the satellite velocity.

If there is a relative velocity between the satellite and the station, the coordinates of the FFDP are translated in the direction of the velocity vector. The magnitude of the angular displacement in the FFDP is given by

$$\alpha(h_s, \theta_{zen},\, \omega) = \alpha_{max}(h_s)\sqrt{cos^2\omega + \Gamma^2(h_s,\, \theta_{zen})sin^2\omega} \tag{11}$$

where the maximum and minimum angular displacement values are given by

$$\alpha_{max}(h_s) = \alpha(h_s, 0, 0) = \frac{2v_s}{c} = \frac{2}{c}\sqrt{\frac{gR_E^2}{R_E + h_s}}$$

$$\alpha_{min}(h_s) = \alpha(h_s,\, 70^\circ, 90^\circ) = \alpha_{max}(h_s)\,\Gamma(h_s,\, 70^\circ)$$

$$\Gamma(h_s, \theta_{zen}) = \sqrt{1 - \left(\frac{R_E\,sin\theta_{zen}}{R_E + h_s}\right)^2} \tag{12}$$

$$\omega = cos^{-1}\left[\left(\hat{r}\; x\; \hat{p}\right) * \hat{v}\right]$$

where v = the satellite velocity at an altitude h_s above sea level, R_E = the Earth's radius = 6378 km, g = the surface gravity acceleration = 9.8 m/s², c = the velocity of light = 3×10^8 m/s, θ_{zen} = the largest satellite zenith angle for tracking = 70°, $\hat{r}$ = the unit vector to the satellite from the geocenter, $\hat{p}$ = the unit vector from the station to the satellite, and $\hat{v}$ = unit vector in the direction of the satellite velocity. The maximum and minimum values of the parameter α are plotted in Figure 8 as a function of the satellite altitude above sea level, h_s.

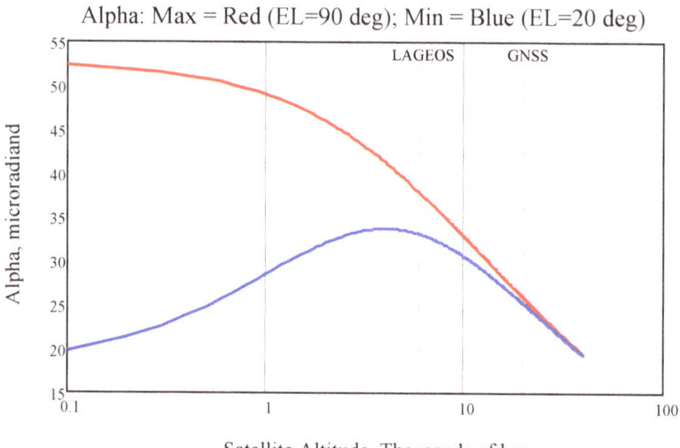

Figure 8. The maximum (red) and minimum (blue) values of the parameter α as a function of the satellite altitude, h_s. By the time one reaches geosynchronous altitude of 36,000 km, the two curves converge to roughly the same value of α.

"Spoiling" of the retroreflector is used to compensate for velocity aberration and improve the signal return from the satellite over the full range of observation angles. If we offset one or more (N = 1 to 3) of the cube dihedral angles from 90° by an amount δ, the central lobe of the FFDP splits into $2N$ spots, as in Figure 9. If n is the cube index of refraction, the mean angular distance of the lobe from the center of the original Airy pattern increases linearly with the dihedral angle offset, δ, according to

$$\gamma = \frac{4}{3}\sqrt{6}n\delta = 3.27\, n\, \delta \tag{13}$$

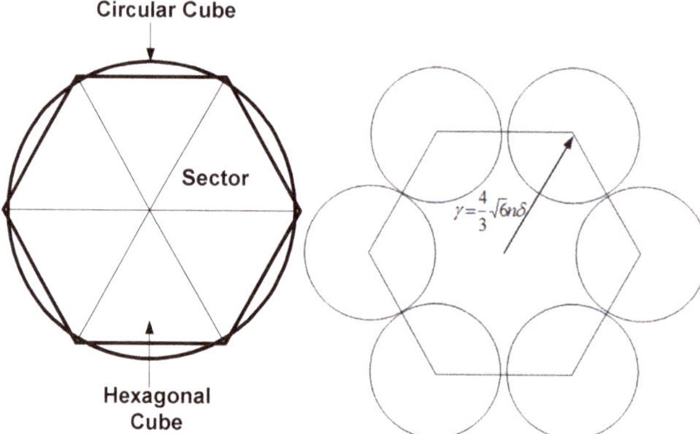

Figure 9. Far-field diffraction pattern (FFDP) of a retroreflector "spoiled" in all three axes. To simplify the drawing, the six resulting lobes are shown as circles, but, in reality, they are better described by the two-dimensional Fourier transform of either a triangular aperture or a 60-degree sector of a circle.

As before, the angular size of any given lobe decreases as the cube diameter gets larger, and the FFDP of each lobe is the 2D Fourier transform of an individual 60° sector. The energy distribution is complex but has a hexagonal symmetry if all three δs are equal. Furthermore, the effective area and peak cross-section of each lobe is reduced to

$$A_{eff} = \eta(\theta_{inc}) \frac{A_{cc}}{2N} \tag{14a}$$

and

$$\sigma_{peak} = \eta^2(\theta_{inc}) \frac{\sigma_{cc}}{(2N)^2} \tag{14b}$$

Since the return signal is weakest at the lowest elevation tracking angle, choosing $\gamma = \alpha_{min}$ places the peak of the lobe there, but any lobe energy inside the dark inner ring or outside the dark outer ring is wasted. Filling in the circumferential gaps between the lobes produced by 'spoiling" can be accomplished by rotating cubes by an angle equal to $60°$ divided by an integer greater than one, generally referred to as "clocking". A larger-diameter cube will result in smaller lobe dimensions and will reduce the spillover into the region outside the active angular ring, but it may also create a larger angular spacing between lobes, which in turn may require more clocking positions, as shown in Figure 10.

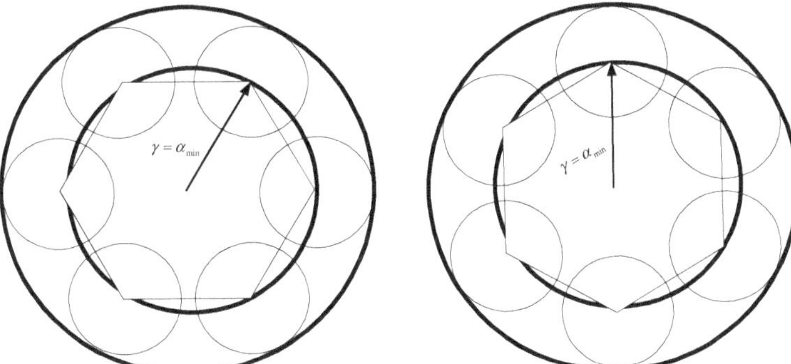

Figure 10. Example of "clocking" where two adjacent "spoiled" retroreflectors are rotated by 30 degrees with respect to one another in order to fill in potential gaps in the array FFDP due to "spoiling". The angular radius to the inner and outer dark circles in the figure correspond to the minimum and maximum values of α plotted in Figure 8.

7. Lunar Laser Ranging

Lunar Laser Ranging (LLR: h~384,000 km) tracks the distance and angular position of the surface retroreflector array relative to the Earth-based ground station and contributes to a wide range of applications, including astronomy, lunar science, gravitational physics, geodesy, and geodynamics [10]. From Equation (1), the return signal strength varies as $1/R^4$, and typically meter-class telescopes, combined with relatively high-energy lasers, were required to achieve detectable signals.

The earliest lunar-range measurements were conducted in 1962 by MIT Lincoln Laboratories by reflecting high-energy laser pulses (50 Joules) off the lunar surface and recording the received photons [11]. Subsequently, in 1969, in order to reduce the ground station telescope and laser energy requirements and simultaneously improve range accuracy to a specific lunar location, an array of one hundred 38 mm-diameter retroreflectors was placed on the lunar surface by the manned NASA Apollo 11 mission. In subsequent years, two other manned Apollo missions (14 and 15) also placed retroreflector panels on the Moon. Of these, the Apollo 15 array was the largest (300 vs. 100 cubes), thereby increasing the effective array cross-section (and return signal strength) by a factor of three relative to the earlier Apollo 11 and 14 arrays. Two unmanned Soviet Lunakhod (17 and 21) missions also landed arrays provided by France. Images of the current arrays and their locations on the lunar surface are provided in Figure 11.

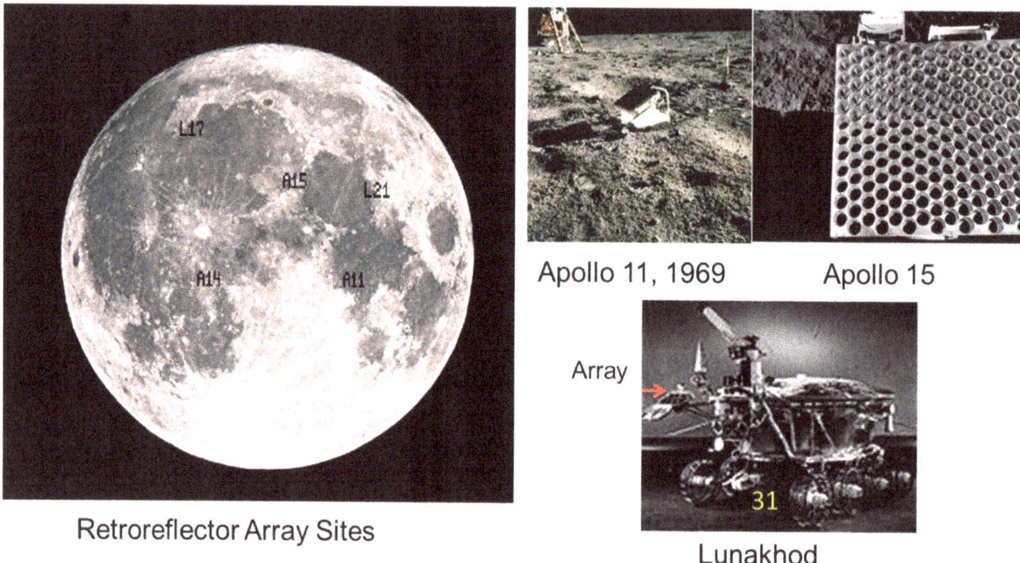

Figure 11. Images and locations of current lunar retroreflector arrays.

Although the Moon presents the same face to the Earth at all times, the angular velocity of the arrays as viewed by a particular Earth station varies slightly on a monthly basis due to a varying lunar velocity resulting from the Moon's slightly elliptical orbit about the Earth and, on a daily basis, due to the Earth's rotation about its polar axis. The Earth station velocity decreases with station latitude and also with time as the Moon approaches the station horizon. Fortunately, the beams reflected from the small-diameter (D_{cc} = 38 mm) cubes used in the current lunar arrays are highly divergent (on the order of $4\lambda/D_{cc}$ = 4 (532 nm/38 mm) = 56 μrad for the widely used 532 nm wavelength), and therefore velocity aberration has minimal effect on the cube performance. On the negative side, the high divergence of the retros also meant that many reflectors would be required to achieve measurable return signals, and, as discussed previously in Section 5 of this paper, the larger panels produce more pulse spreading at increasingly large incidence angles, resulting in diminished range precision.

Using the Earth–moon distance, $R_{EM} = h + R_E = 384.4 \times 10^6$ m, in Equation (12), we obtain α_{max} = 6.74 μrad or 1.40 arcsec and α_{min} = 6.68 μrad or 1.39 arcsec at an elevation angle of 20 degrees, where the relative velocity between the lunar target and the Earth ground station due to lunar orbital motion v = 1 km/s.

However, the latter equations ignore the smaller contribution of the station's motion due to the Earth's rotation (~0.46 km/s) to the relative velocity, which typically reduces α to 4 or 5 μrad for LLR but is negligible for LEO to GEO satellites. If the Apollo reflector arrays are pointed at the center of the Earth, the maximum beam incidence angle on the array from any Earth station (ignoring lunar libration) is

$$\theta_{inc} = atan\left(\frac{R_E}{R_{EM}}\right) = 0.95 \; deg \tag{15}$$

and the unspoiled cube diameter for which the cross-section falls to half its peak value is $D_{1/2}$ = 40.6 mm = 1.6 inches, but the typical manufacturing tolerances are 0.5 arcsec for dihedral angles and $\lambda/10$ for surface flatness. Apollo 15 has a flat array of three hundred 38 mm fused quartz cubes, each with an unspoiled peak cross-section of 5.8×10^7 m^2. Thus, the theoretical array cross-section, ignoring manufacturing tolerances and local environmental effects, is $\sigma \cong 300(0.5)(5.8 \times 10^7 \text{ m}^2) = 8.7 \times 10^9 \text{ m}^2$. According to Arnold [12], polarization losses due to the uncoated TIR faces reduce the cross-section by a factor of four, resulting in $\sigma = 2.2 \times 10^9 \text{ m}^2$, but the International Laser Ranging Service (ILRS) lists a slightly lower Apollo 15 array cross-section of $\sigma = 1.4 \times 10^9 \text{ m}^2$.

For the first 30 years following the Apollo 11 landing, the vast majority of the LLR data set was provided by three stations equipped with meter class telescopes, i.e., (1) the 2.7 m MLRS telescope at the University of Texas McDonald Observatory, (2) the 1.5 m CERGA LLR telescope in Grasse, France, and (3) the 1.8 m University of Hawaii telescope on Mt. Haleakala. The latter site was decommissioned by NASA in 1992, but, since that time, two additional LLR sites have been added: (4) MLRO, Matera, Italy, and (5) the 3.5 m Apollo telescope located at Apache Point, New Mexico, USA. The large 3.5 m telescope at the Apache Point site resulted in multiphoton returns and unprecedented lunar-range precisions at the 1 to 3 mm level [13].

Since the turn of the century, more efficient, single, large retroreflectors have been proposed by various international teams to replace the large Apollo 15 array on future lunar-landing missions [14–19]. Whereas each of the 300 38 mm-diameter retroreflectors in the Apollo 15 array currently contributes about 0.33% to the overall signal seen by the ground stations due to the wide divergence angles of their reflected beams, no dihedral angle is required for small-diameter reflectors (<150 mm for coated and <100 mm for uncoated and hollow reflectors). However, since larger-diameter retroreflectors produce low-divergence reflected beams, they required dihedral angles of 0.20, 0.25, and 0.35 arcsec for coated, uncoated, and hollow reflectors, respectively. More recent publications provide updated analyses of these larger lunar reflectors [16,17].

8. Concluding Remarks

A common transmit–receive telescope maximizes the observed signal strength, allowing for the use of lower-power lasers and significantly higher measurement rates. Early SLR systems typically transmitted the quasi-Gaussian laser beam through a separate narrow parallel tube mounted on the receiving telescope. This resulted in a highly divergent transmitter beam with a full angular width equal to $4\lambda/D_b$ in our link equation, Equation (1), where D_b is the exiting laser beam diameter. Using the full aperture of the ground telescope to transmit the beam increased the received signal strength by minimizing the beam divergence of the outgoing beam, thereby concentrating a much greater fraction of the laser pulse energy on the satellite. Adopting this approach in NASA's photon-counting SLR2000 system [6] increased measurement rates from under 10 Hz to 2 kHz and, due to significantly lower transmitted pulse energies and large transmitted beamwidths, greatly reduced eye safety hazards to aircraft. In this instance, using Equation (2), we can rewrite the original link equation, Equation (1), as [7,8]

$$
\begin{aligned}
n_s &= n_t(\eta_t\,\eta_r\eta_d)T_{atm}^2\left(\frac{4\pi A_p}{\lambda^2}\right)g_T\,\frac{\sigma A_p}{4\pi R^2}g_R \\[2mm]
&= n_t(\eta_t\,\eta_r\eta_d)T_{atm}^2\left(\frac{\pi D_p^2}{\lambda^2}\right)g_T\,\frac{\pi^3\,\rho\,D_{cc}{}^4}{4\lambda^2}\frac{D_p^2}{4R^2}g_R \\[2mm]
&= n_t(\eta_t\,\eta_r\eta_d)T_{atm}^2\left(\tfrac{\pi}{2}\right)^4\left(\frac{D_{cc}D_p}{\lambda R}\right)^4\rho g_T g_R \leq 0.034 n_t T_{atm}^2\left(\frac{D_{cc}D_p}{\lambda R}\right)^4
\end{aligned}
\tag{16}
$$

where ρ is the reflectivity of the retro surfaces discussed in Section 2, n_t is the number of transmitted photons per laser pulse, $\eta_t \cong \eta_r \cong 0.66$ are the nominal optical throughput efficiencies of a telescope with aluminum- or silver-coated primary and secondary mirrors (plus a few additional low-loss optical elements) in the transmit and receive paths [20], η_d (<0.70) is the detector's counting efficiency, T_{atm} is the one-way transmission of the atmosphere, D_{cc} is the retro diameter, D_p is the diameter of the telescope's primary mirror, λ is the laser wavelength, R is the target range, and g_T ($\leq$0.81) and g_R ($\leq$1) are the reductions in the optimized transmitter and receiver gains due to the truncation of the outgoing quasi-

Gaussian beam by the primary mirror perimeter and/or the presence of a secondary mirror that blocks the central portion of the outgoing and incoming radiation [7,8].

Solving for n_t in Equation (16) and multiplying by the photon energy, hc/λ, the laser pulse energy required to achieve the desired number of signal photons, n_s, is

$$E_t = n_t \frac{hc}{\lambda} \geq \frac{30}{T_{atm}^2} n_s \frac{hc}{\lambda} \left(\frac{\lambda R}{D_{cc} D_p} \right)^4 \tag{17}$$

From Equation (17), the advantages of utilizing shorter laser wavelengths (λ) and the full primary diameter (D_p) of the ground-based telescope to transmit the laser beam is immediately apparent [6]. This is especially true for LLR, where the product $\lambda R = (532 \times 10^{-9}$ m$)(3.844 \times 10^8$ m$) = 204.5$ m^2 at the widely used 532 nm green wavelength of the frequency-doubled Nd:YAG laser. If we assume a lunar retro diameter $D_{cc} = 100$ mm and the Apollo telescope primary mirror diameter of $D_p = 3.5$ m used in both transmit and receive modes, then $D_{cc} D_p = 0.35$m^2. Ignoring atmospheric transmission losses, the minimum number of transmitted photons required to achieve a single photon return is $30 \, (\lambda R/D_{cc}D_p)^4 = 30(204.5/0.35)^4 = 3.5 \times 10^{12}$ photons, corresponding to a minimum laser pulse energy of 1.3 µJ at 532 nm. A one-meter telescope, on the other hand, would require a minimum of 195 µJ to realize a single photon return.

It should be noted that the use of a 100 mm-diameter cube, as proposed by several international LLR investigators, increases the signal from a single 38 mm Apollo cube by a factor of $(100 \text{ mm}/38 \text{ mm})^4 = 48$, or roughly a factor of 6 lower than the 300-retro Apollo 15 array response. It would take a single 158 mm cube to roughly match the Apollo 15 array cross-section. This value is in excellent agreement with Table 1 in [15]. Based on their simulations, the latter authors also concluded that only the y-axis of the hollow retro's three retro axes would have to be corrected by between 0.65 and 0.8 arcsec offset for velocity aberration. Chinese researchers, on the other hand, describe the design, manufacture, and testing of a larger 170 mm hollow retroreflector [17] with all three axes deviating from $90°$.

Millimeter-Accuracy LEO-to-MEO satellites use large-radius spherical satellites to better match the incoming plane wave, minimize pulse spreading, and allow for more reflectors within the active area to increase the cross-section. One can minimize satellite pulse broadening by reducing the range of accepted incidence angles through the use of hollow cubes or by recessing the cubes (hollow or solid) in tubes drilled into the satellite. It should also be noted that incidence angles $< 17°$ do not leak light in solid TIR reflectors. Finally, one must select cube diameters and clocking angles that best match the "α annulus"while favoring the response at high zenith (low elevation) angles, which is key to efficient array design.

GNSS and GEO satellites typically have a nadir face pointed near the Earth's center due to their other operational functions, e.g., Earth observation, communications, navigation, etc. Therefore, flat panels perform reasonably well but still exhibit reduced cross-sections and several hundred picoseconds of temporal spread at lower satellite elevation angles. Flat circular (rather than square or rectangular) arrays would largely eliminate azimuthal range biases, but replacing current flat panels by a sphere segment of comparable cross-section would further improve range accuracy at all observation angles.

Lunar Laser Ranging (LLR) The single large cube concept put forward by various international teams of investigators clearly has the beneficial effects of (1) better defining the point of reference on the lunar surface and (2) reducing the target-induced, angular-dependent, temporal spread of the return pulse. However, achieving the same (or larger) target cross-sections than are currently available with the Apollo 15 array would require single-retro diameters in excess of 160 mm. Furthermore, as the size of the single retroreflector grows, the divergence of the reflected beam decreases, and careful attention must be paid to velocity aberration effects due to lunar orbital motion and Earth rotation. Furthermore, Earth's rotation effects decrease with the increasing latitude of the ground station and peak at the point of closest approach to the lunar target.

Finally, unlike Earth-orbiting satellites, where the target of interest is surrounded by a vacuum, lunar arrays are surrounded by a lunar surface, which can potentially introduce additional unwanted laser or solar photons to the overall return signal and thereby potentially degrade the range measurement accuracy. Even if one uses the entire 3.5 m aperture of the largest Apollo Observatorytelescope to achieve the least divergent laser beam, the central lobe of the transmitted beam will span a diameter of roughly $2\lambda R/D_p = 2$ (532 nm) (384.4 $\times$ 10^6 m)/3.5 m = 117 m, corresponding to an angular FOV of 0.3 μrad. We can estimate the number of detected photons reflected by the lunar surface by applying the lidar equation (Equation (1) in [20]), which applies to a Lambertian-diffuse-reflecting surface that fills the receiver Field of View (FOV). Under the latter assumption, the number of laser photons received from the lunar surface is estimated at

$$
\begin{aligned}
n_{LS} &= nt(\eta_t \ \eta_r \eta_d)T^2_{atm}\left(\tfrac{4\pi A_p}{\lambda^2}\right)g\tau\rho_{LS}\tfrac{cos\theta}{\pi R^2}A_p g_R \\
&= n_t(\eta_t \ \eta_r \eta_d)T^2_{atm}g\tau g_R\rho_{LS}cos\theta\left(\tfrac{\pi^2}{4}\right)\left(\tfrac{D_p^4}{\lambda^2 R^2}\right)
\end{aligned}
\tag{18}
$$

where $\rho_{LS} = 0.12$ is the nominal reflectivity of the lunar surface at visible and near-infrared wavelengths and θ is the nominal surface slope as viewed by the Earth station's line of sight. Dividing Equation (18) by Equation (16), we obtain an estimated ratio of lunar surface returns vs. retro returns, i.e.,

$$
\frac{n_{LS}}{n_s} = \frac{\rho_{LS}}{\rho} \ cos\theta\left(\frac{4}{\pi^2}\right)\frac{\lambda^2 R^2}{D_{cc}^4}
\tag{19}
$$

The latter equation indicates that the overall lunar surface return greatly exceeds that of the reflector since $(\lambda R)^2 >> D_{cc}{}^4$ because, even when one uses the full aperture of the ground telescope to minimize laser beam divergence, the diameter of the illuminated surface typically spans hundreds of meters. However, the retro produces a "hot spot" in the return, which can be isolated spatially via a pinhole in the receiver FOV or, better yet, a multipixel SPAD (Single-Photon Avalanche Diode) array, which can detect the "hot spot" during target acquisition and drive it to the central pixel of the array in order to maximize the retro signal strength while simultaneously assigning the vast majority of the surface photons to other pixels in the array. Locating the Earth-facing retro near the perimeter of the lunar disk would drive the $cos\theta$ term in Equation (19) close to zero. The Southern lunar pole region might be a good choice since it is believed to be a potential source of water for future manned missions. In order to further minimize the impact of lunar surface returns on range accuracy to the target retro, the pixel FOV should be chosen as small as is practically possible for rapid target acquisition. It may even be advantageous to elevate the retro a known distance above the lunar surface to more clearly separate the retro returns from local surface returns that fall within the common pixel FOV.

Finally, a recent article by the LLR staff in Grasse, France, provides an interesting history of their evolving technology over decades of operations and their experiences with near-infrared wavelengths [21]. It should also be mentioned in closing that the LLR station in Grasse successfully tracked the Lunar Reconnaissance Orbiter (LRO) satellite using a 4 $\times$ 3 retroreflector array mounted on the anti-nadir side of the spacecraft [22].

Laser ranging throughout the solar system using conventional single-ended ranging to passive reflectors is unrealistic due to the R^{-4} signal loss in the link equation. However, precise ranging (as well as time transfer and wideband communications) can be accomplished using two-way asynchronous laser transponders equipped with transmitters and receivers on both ends of the link [23,24]. In this instance, the signal strength at either terminal varies only as $1/R^2$ as opposed to $1/R^4$. Furthermore, since most of the link burden (laser power or telescope aperture) can be carried by the Earth station, the space terminal can be relatively modest in size (submeter), weight, and power consumption [25]. The feasibility of the interplanetary transponder approach has already been demonstrated in experiments conducted at the 1.2 m telescope at NASA Goddard Space Flight Center with a satellite enroute to Mercury at a distance of 22 million km [24], which produced a range accuracy estimated at 20 cm [25]. Three months later, a second one-way experiment with the Mars Orbiter Laser Altimeter (MOLA) at a distance of 80 million km [26] resulted in roughly 500 laser pulses being detected by the Mars orbiter lidar receiver. In the latter case, the laser in orbit about Mars was sadly no longer operational when the interplanetary transponder experiment was conducted.

Funding: No funding was provided in support of this paper.

Institutional Review Board Statement: Not Applicable.

Informed Consent Statement: Not Applicable.

Data Availability Statement: Not applicable.

Conflicts of Interest: The author declares no conflict of interest.

References

1. Plotkin, H.H.; Johnson, T.S.; Spadin, P.; Moye, J. Reflection of Ruby Laser Radiation from Explorer 22. *Proc. IEEE* **1965**, *53*, 301–302. [CrossRef]
2. Alley, C.O.; Chang, R.; Curry, D.; Mullendore, J.; Poultney, S.; Rayner, J.D.; Silverberg, E.; Steggerda, C.; Plotkin, H.; Williams, W.; et al. Apollo 11 Laser Ranging Retro-Reflector: Initial Measurements from the McDonald Observatory. *Science* **1970**, *167*, 368–370. [CrossRef] [PubMed]
3. Degnan, J.J. Millimeter Accuracy Satellite Laser Ranging: A Review. *Contrib. Space Geod. Geodyn. Technol.* **1993**, *25*, 133–162.
4. Wilkinson, M.; Schreiber, U.; Procházka, I.; Moore, C.; Degnan, J.; Kirchner, G.; Zhongping, Z.; Dunn, P.; Shargorodskiy, V.; Sadovnikov, M.; et al. The next generation of satellite laser ranging systems. *J. Geodesy.* **2019**, *93*, 2227–2247. [CrossRef]
5. Pearlman, M.R.; Noll, C.E.; Pavlis, E.C.; Lemoine, F.G.; Combrink, L.; Degnan, J.J.; Kirchner, G.; Schreiber, U. The ILRS: Approaching 20 years and planning for the future. *Proc. SPIE* **1997**, *93*, 2161–2180. [CrossRef]
6. Degnan, J.J.; McGarry, J.F. SLR2000 Eyesafe and autonomous single photoelectron satellite laser ranging at kilohertz rates. *Proc. SPIE* **1997**, *93*, 2161–2180.
7. Klein, B.J.; Degnan, J.J. Optical Antenna Gain 1. Transmitting Antennas. *Appl. Opt.* **1974**, *13*, 2134–2141. [CrossRef]
8. Degnan, J.J.; Klein, B.J. Optical Antenna Gain 2. Receiving Antennas. *Appl. Opt.* **1974**, *13*, 2397–2401. [CrossRef]
9. McGarry, J.F.; Hoffman, E.D.; Degnan, J.J.; Cheek, J.W.; Clarke, C.B.; Diegel, I.F.; Donovan, H.L.; Horvath, J.E.; Marzouk, M.; Nelso, A.R.; et al. NASA's Satellite Laser Ranging Systems for the 21st Century. *J. Geodesy* **2018**, *93*, 2249–2282. [CrossRef]
10. Dickey, J.O.; Bender, P.L.; Faller, J.E.; Newhall, X.X.; Ricklefs, R.L.; Ries, J.G.; Shelus, P.J.; Veillet, C.; Whipple, A.L.; Wiant, J.R.; et al. Lunar Laser Ranging: A Continuing Legacy of the Apollo Program. *Science* **1994**, *265*, 482–490. [CrossRef]
11. Smullin, L.D.; Fioco, G. Optical Echoes from the Moon. *Nature* **1962**, *194*, 1267. [CrossRef]
12. Arnold, D.; (Harvard University, Boston, MA, USA). Personal Communication, 2013.
13. James, B.; Battat, R.; Murphy, T.W.; Adelberger, E.G.; Gillespie, B.A.; Boyle, C.D.; McMillan, R.J.; Michelson, E.I.; Nordtvedt, K.; Orin, A.E.; et al. The Apache Point Observatory Lunar Laser-ranging Operation (APOLLO): Two Years of Millimeter-Precision Measurements of the Earth-Moon Range. *Publ. Astron. Soc. Pac.* **2009**, *121*, 29–40.
14. Otsubo, T.; Kunimori, H.; Noda, H.; Hanada, H. Simulation of optical response of retroreflectors for future lunar laser ranging. *Adv. Space Res.* **2010**, *45*, 733–740. [CrossRef]
15. Otsubo, T.; Kunimori, H.; Noda, H.; Hanada, H.; Araki, H.; Katayama, M. Asymmetric Dihedral Angle Offsets for Large-Size Lunar Laser Ranging Reflectors. *Earth Planets Space* **2011**, *63*, 13–16. [CrossRef]
16. Currie, D.; Dell Agnello, S.; Delle Monache, G.O.; Behr, B.; Williams, J.G. Lunar Laser Ranging Reflector Array for the 21st Century. *Nucl. Phys. B* **2013**, 217–228.
17. He, Y.; Liu, Q.; He, J.; Li, M.; Duan, H.; Ye, H.; Luo, J. Development of a 170-mm hollow corner cube retroreflector for the future lunar laser ranging. *Chin. Phys. B* **2018**, *27*, 100701. [CrossRef]
18. Williams, J.G.; Porcelli, L.; Dell'Agnello, S.; Mauro, L.; Muccino, M.; Currie, D.G.; Wellnitz, D.; Wu, C.; Boggs, D.H.; Johnson, N.H. Lunar Laser Ranging Retroreflectors: Velocity Aberration and Diffraction Pattern. *Planet. Sci. J.* **2023**, *4*, 89. [CrossRef]
19. Garattinia, M.; Dell'Agnelloa, S.; Currie, D.; Delle Monachea, G.O.; Tibuzzia, M.; Patrizia, G.; Berardia, A.; Bonia, A.; Cantonea, C.; Itagliettaa, N.; et al. Moonlight: A New Lunar Laser Ranging Retroreflector Instrument. *Acta Polytech.* **2013**, *53*, 821. [CrossRef]
20. Degnan, J.J. Photon-counting multi-kilohertz microlaser altimeters for airborne and spaceborne topographic measurements. *J. Geodyn.* **2002**, *34*, 503–549. [CrossRef]
21. Chabe, J.; Corde, C.; Torre, J.M.; Bouquillon, S.; Borgin, A.; Aimar, M.; Albanese, D.; Chavineau, B.; Mariey, H.; Martinot-Lagarde, G.; et al. Recent Progress in Lunar Laser Ranging at Grasse Laser Ranging Station. *Earth Apace Sci.* **2019**, *7*, 3. [CrossRef]

22.   Mazarico, E.; Sun, X.; Torre, J.M.; Courde, C.; Chabe, J.; Aimar, M.; Mariey, H.; Maurice, N.; Barker, M.K.; Mao, D.; et al. First two-way laser ranging to a lunar orbiter: Infrared observations from the Grasse station to LRO's retroreflector array. *Earth Planets Space* **2020**, *72*, 113. [CrossRef]
23.   Degnan, J.J. Asynchronous Laser Transponders for Precise Interplanetary Ranging and Time Transfer. *J. Geodyn.* **2002**, *34*, 551–594. [CrossRef]
24.   Degnan, J.J. Multipurpose Laser Instrument for Interplanetary Ranging, Time Transfer, and Wideband Communications. *Photonics* **2023**, *10*, 98. [CrossRef]
25.   Smith, D.E.; Zuber, M.T.; Sun, X.; Neumann, G.A.; Cavanaugh, J.F.; McGarry, J.F.; Zagwodzki, T.W. Two Way Laser Link over Interplanetary Distance. *Science* **2006**, *311*, 53. [CrossRef]
26.   Abshire, J.B.; Sun, X.; Neumann, G.; McGarry, J.F.; Zagwodzki, T.; Jester, P.; Riris, H.; Zuber, M.; Smith, D. Laser pulses from Earth detected at Mars. In Proceedings of the Conference on Lasers and Electro-Optics and Quantum Electronics and Laser Science Conference, Beach, CA, USA, 21–26 May 2006; Optica Publishing Group: Washington, DC, USA, 2006; pp. 1–2.

Article

Atmospheric Turbulence with Kolmogorov Spectra: Software Simulation, Real-Time Reconstruction and Compensation by Means of Adaptive Optical System with Bimorph and Stacked-Actuator Deformable Mirrors

Ilya Galaktionov *, Julia Sheldakova, Vadim Samarkin, Vladimir Toporovsky and Alexis Kudryashov *

Sadovsky Institute of Geosphere Dynamics, Russian Academy of Sciences, Leninskiy Pr. 38/1, 119334 Moscow, Russia
* Correspondence: galaktionov@activeoptics.ru (I.G.); kud@activeoptics.ru (A.K.)

Abstract: Atmospheric turbulence causes refractive index fluctuations, which in turn introduce extra distortions to the wavefront of the propagated radiation. It ultimately degrades telescope resolution (in imaging applications) and reduces radiation power density (in focusing applications). One of the possible ways of researching the impact of turbulence is to numerically simulate the spectrum of refractive index fluctuations, to reproduce it using a wavefront corrector and to measure the resultant wavefront using, for example, a Shack–Hartmann sensor. In this paper, we developed turbulence simulator software that generates phase screens with Kolmogorov spectra. We reconstructed the generated set of phase screens using a stacked-actuator deformable mirror and then compensated for the introduced wavefront distortions using a bimorph deformable mirror. The residual amplitude of the wavefront reconstructed by the 19-channel stacked-actuator mirror was 0.26 λ, while the residual amplitude of the wavefront compensated for by the 32-channel bimorph mirror was 0.08 λ.

Keywords: atmospheric turbulence; Kolmogorov spectra; turbulence simulation; wavefront correction; bimorph deformable mirror; stacked-actuator deformable mirror; Shack–Hartmann wavefront sensor; adaptive optics

Citation: Galaktionov, I.; Sheldakova, J.; Samarkin, V.; Toporovsky, V.; Kudryashov, A. Atmospheric Turbulence with Kolmogorov Spectra: Software Simulation, Real-Time Reconstruction and Compensation by Means of Adaptive Optical System with Bimorph and Stacked-Actuator Deformable Mirrors. *Photonics* **2023**, *10*, 1147. https://doi.org/10.3390/photonics10101147

Received: 28 July 2023
Revised: 28 September 2023
Accepted: 2 October 2023
Published: 12 October 2023

Correction Statement: This article has been republished with a minor change. The change does not affect the scientific content of the article and further details are available within the backmatter of the website version of this article.

1. Introduction

It is well known that turbulence leads to refractive index fluctuations, which in turn lead to extra distortions of the wavefront of the radiation that propagates through the atmosphere. Atmospheric turbulence limits the resolution of telescopes and decreases the coherence of laser radiation [1–9]. As a result, the quality of the image of the objects observed by the telescopes degrades. Other tasks that turbulence affects are the wireless transmission of energy and information with the help of optical radiation [10–13]. This is important, in particular, for recharging batteries at remote sites; the organization of optical communication channels in free space [14]; wireless optical communication, wireless power delivery to flying objects [15,16] and low-earth-orbit satellites [11]; providing a backup power supply and wireless power supply for household and industrial devices; destroying unmanned aerial vehicles or space debris [17–19]; providing communication with aircraft (in order to monitor their condition and obtain information); creating a beam of the desired shape [20]; focusing a beam inside a limited-aperture pinhole for laser communication tasks [21]; increasing the radiation power density on a target (laser cutting); and improving the accuracy of beam positioning (optical recording of information on media).

As is known, the low efficiency of systems for the wireless transmission of energy and information using optical-range radiation is due to light beam diffraction, radiation scattering by atmospheric aerosol [22], and the influence of atmospheric turbulence [23–35]. When a laser beam passes through the turbulent atmosphere of the Earth, the wavefront becomes distorted, which limits the operational range of such systems [36]. This problem has been studied for more than a dozen years, yet has not lost its relevance. The technology for increasing the efficiency of laser radiation transmission to the receiver includes two main components: high-quality efficient single-mode laser radiation sources and adaptive optical systems for correcting for wavefront aberrations [37,38]. In this research, we will concentrate on the second component.

The problems of increasing the range of the propagation of laser radiation through the atmosphere using adaptive optics methods are being solved by research teams from Russia, Germany, Italy, the USA, and the Netherlands [39–41]. In terms of adaptive optics for astronomical applications, we begin with the Air Force Maui Optical Station (AMOS) [42] in Hawaii, USA, where the ADONIS (Daylight Optical Near-Infrared System) system was built in 1993–1995. This system was placed in a 1.2 m telescope and was used to increase the quality of the obtained images of the astronomical objects. There were no adaptive optics inside this system since it used postprocessing. After that, the ADONIS system was moved to the 3.6 m EOAR telescope and was equipped with an adaptive optical system.

In [43,44], the authors describe a vision system that uses a conventional adaptive optical system and works on the 2.5 km horizontal atmospheric path. Another vision system with adaptive optics included, produced in the Fraunhofer Institute [45,46], was developed to run in urban conditions. The adaptive system contained a conventional deformable mirror and works with a frequency of 800 Hz. In [47], the authors describe the use of an adaptive optical system within an 0.35 m telescope on the 20 km slant atmospheric path. In [48], the authors describe an adaptive system with a deformable mirror and tip-tilt corrector for a 0.12 m telescope on the 3 km horizontal atmospheric path.

Most of the papers described above are devoted to the compensation of turbulence phase fluctuations. In this paper, we will first numerically simulate the effect of the specific turbulent atmosphere conditions on the wavefront of laser radiation. The second step is to reconstruct the simulated distortions using the stacked actuator deformable mirror in laboratory conditions, followed by compensating for introduced distortions using the bimorph deformable mirror. Such a numerical–experimental simulation could allow us to estimate the influence of different turbulence conditions on the radiation wavefront and estimate the efficiency of the wavefront corrector.

2. Materials and Methods

2.1. Experimental Setup

The principal optical scheme of the adaptive system for atmospheric turbulence simulation and compensation is presented in Figure 1.

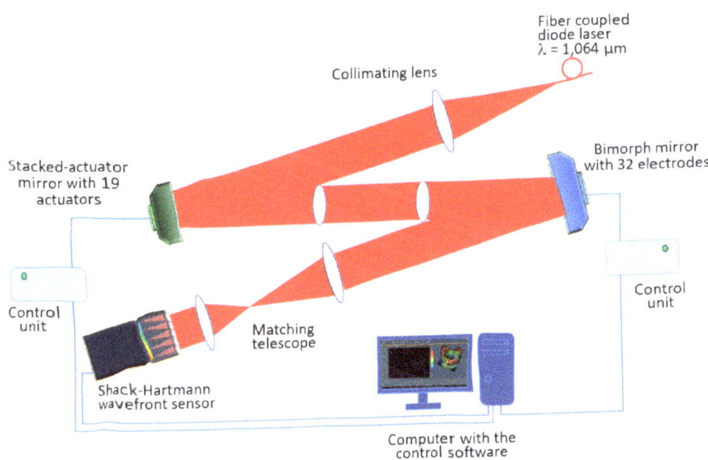

Figure 1. Optical scheme of the adaptive system for atmospheric turbulence simulation and compensation.

A point source (a fiber-coupled diode laser with a wavelength of 1.064 µm) is collimated with an achromatic lens to increase the beam diameter. The collimated laser beam is then incident on the 35 mm stacked-actuator deformable mirror with 19 control actuators, which introduces the phase delay in order to simulate the atmospheric turbulence. The distorted beam passes through the telescope (2 lenses), which optically conjugates the plane of the reflective surface of the stacked-actuator mirror with the plane of the bimorph mirror, and falls on the 30 mm bimorph deformable mirror with 32 control electrodes, which compensates for the introduced distortions. The corrected beam passes through the matching telescope and hits the Shack–Hartmann wavefront sensor.

2.2. Stacked-Actuator Deformable Mirror

The scheme of the actuator's layout and the principal scheme of the stacked-actuator mirror are presented in Figure 2, while the main mirror parameters are given in Table 1.

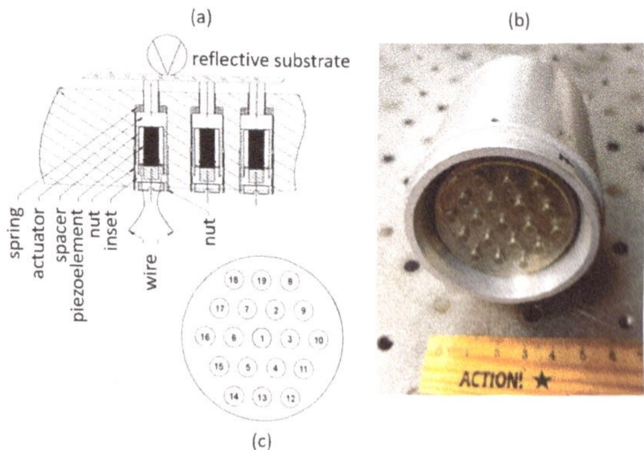

Figure 2. (**a**) The drawing of the stacked-actuator mirror, (**b**) the photo of the manufactured stacked-actuator mirror and (**c**) actuators' layout scheme.

Table 1. Main parameters of the stacked-actuator deformable mirror.

Parameter	Value
Substrate aperture	40 mm
Clear aperture	35 mm
Substrate material	glass
No. of control actuators	19
Type of actuators	PZT
Actuators geometry	Hexagonal
Maximum input voltage	−30…+130 V

There are a number of advantages of the stacked-actuator mirror. The first is the ability to replace individual actuators. If one actuator fails, there is no need to remove the substrate with the reflecting coating from all the other actuators—it is only necessary to unscrew the body of the broken actuator, keeping the tip of the actuator glued to the substrate, and then replace it with the working one. Secondly, the stacked-actuator mirror has the ability to polish the substrate in order to flatten the initial surface of the mirror in the assembled state, i.e., when all the actuator tips are glued to the substrate. The advantage of this mirror is that it requires a rather small initial adjustment of the surface in the assembled state, which preserves the dynamic range of the actuators for correcting the laser beam aberrations rather than the self-aberrations.

The bimorph mirror could also be used as a turbulence simulator. How~ are a few reasons why the stacked-actuator is a better choice. The fi~ stacked-actuator mirror was used instead of a bimorph one fo~

phase screens is that the bimorph mirror has a limited capacity to reproduce tilt aberrations, which have a big impact on the turbulent wavefront distortions. The stacked-actuator mirror, on the contrary, can reproduce the tilt aberrations without any significant sacrifice of dynamic range. The second reason is that it was interesting for us to test the efficiency of the newly developed stacked-actuator mirror.

2.3. Bimorph Deformable Mirror

A bimorph mirror consists of a passive glass substrate with a reflective coating and two piezoceramic disks glued to it [49,50]. A common electrode is applied to the internal piezoceramic disk, which is designed to change the curvature of the reflecting surface of the mirror. A grid of control electrodes is applied to the external piezoceramic disk. The number of control electrodes depends on the type of aberrations to be compensated [51,52]. The scheme of the electrode dislocation and the principal scheme of the bimorph mirror are presented in Figure 3, while the main mirror parameters are given in Table 2.

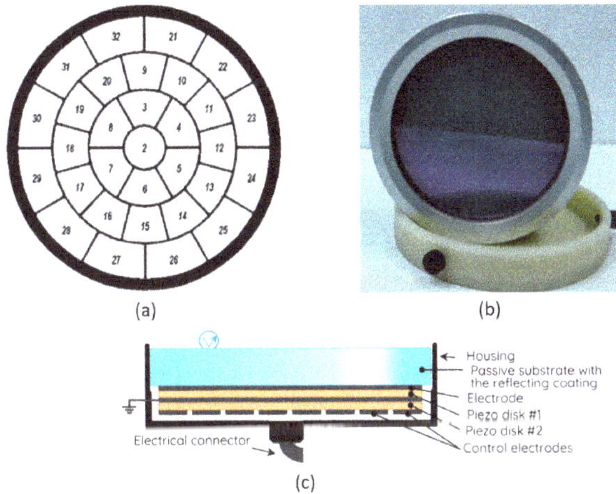

Figure 3. (**a**) Electrodes' layout scheme, (**b**) the photo of the manufactured bimorph mirror and (**c**) the principal scheme of the bimorph mirror construction.

Table 2. Parameters of the bimorph deformable mirror.

Parameter	Value
Substrate aperture	35 mm
Clear aperture	30 mm
Substrate material	glass
No. of PZT	2
No. of control electrodes	32
Type of actuators	PZT discs
Actuators geometry	sectorial
Maximum input voltage	$-200\ldots+300$ V

2.4. Shack–Hartmann Wavefront Sensor

In order to estimate the wavefront aberrations introduced by either the stacked-actuator or the bimorph deformable mirror, the Shack–Hartmann wavefront sensor [53–58] was used. The Shack–Hartmann wavefront sensor is a well-known device. It is quite simple to operate and can be easily calibrated for proper use [59]. Shack–Hartmann sensor is widely used in a large and diverse sets of applications, primarily to measure distortions of the wavefront of the radiation passing through different media, i.e., turbulent or scattering atmosphere [60], biological tissues [61], etc.

The principle of a conventional Shack–Hartmann sensor can be described as follows. The wavefront of the incident light is divided into a number of sub-apertures with the micro lens array. The micro lens array is a thin, flat base on which a grid of micro lenses is etched. Each micro lens has a diameter of 100 to 300 μm and a focal length f of 3 mm to 8 mm. Light passes through these micro lenses and creates a set of focal spots at the measurement (sensor) plane.

Since the diameter of each micro lens is small, the wavefront W is assumed to be flat and to have only tip-tilt aberration within a single micro lens. In the case of no aberrations (i.e., a wavefront is flat and parallel to the plane of the micro lens), the radiation is focused at the center of the corresponding sub-aperture of the sensor. If the wavefront in a micro lens has a non-zero tip-tilt, then the focal spot is displaced (S_x and S_y) from the center of the sub-aperture in proportion to the tip-tilt value. In other words, if we measure these displacements, S_x and S_y, of the focal spot per the X and Y axis, we will correspondingly obtain the values of the partial derivatives $\partial W/\partial x$ and $\partial W/\partial y$ of the wavefront W within each sub-aperture. On the other hand, to describe and visualize the wavefront surface analytically, one can use the polynomial approximation, for example, B-Splines [62] or Zernike polynomials [63–66], which are commonly used in optics. Thus, the partial wavefront derivatives $\partial W/\partial x$ and $\partial W/\partial y$ can be defined analytically using Zernike polynomials. They can also be calculated from the measured displacements S_x and S_y of the focal spots on the Shack–Hartmann sensor. Finally, we determined the overdetermined system of linear equations with the unknown coefficient's a_i. By solving the least squares problem [67], we obtain the coefficient's a_i. From here on, the wavefront can be analytically described and analyzed.

In addition to the wavefront measurements, the Shack–Hartmann sensor was used in order to measure the response functions of the actuators of our deformable mirrors (bimorph and stacked actuators one). Each control electrode or actuator is described by its own response function. The response function is a change in the profile of the mirror surface in response to the application of an electrical voltage to the electrode/actuator, while the remaining electrodes/actuators have a voltage under zero. The response function is presented as a set of focal spot displacements registered on the Shack–Hartmann wavefront sensor. Such a notation allows us to approximate the arbitrary wavefront measured by the Shack–Hartmann sensor (also presented as a set of focal spot displacements) using the response functions of the mirror.

2.5. Algorithm of Phase Screens Simulation

The main idea of this research is to simulate the set of phase screens with Kolmogorov spectra, to approximate and reproduce these phase screens using the 19-element stacked-actuator deformable mirror, and then to compensate for the introduced wavefront distortions using the 32-electrode bimorph deformable mirror.

In essence, the disturbance of an optical wave by a thin phase screen is the simplest model of the propagation in a turbulent atmosphere [68]. The fluctuations of the wave passing through a phase screen are similar to the fluctuations of the light field in a continuous randomly inhomogeneous medium. The thin phase screen most closely reproduces the influence of large-scale atmospheric inhomogeneities on the characteristics of the light field. In addition, the advantage of the phase screen method is that it allows the adaptive optical system to be analyzed to compensate for low-order phase aberrations in the atmosphere [69]. In other words, the phase screen approach is simple and presents a good approximation, and in most cases, it can reproduce the effect of turbulence on the wavefront with acceptable accuracy.

As a first step, we simulated the set of phase screens. To do this, we applied the Fast Fourier transform to the Kolmogorov spectrum of the phase fluctuations [70]:

$$p(u,v,t+\Delta t) = \iint_{-\infty}^{\infty} \sqrt{K(x,y)} \cdot f(x,y,t+\Delta t) \cdot e^{i \cdot \sqrt{x^2+y^2} \cdot V \cdot \Delta t} dx dy, \tag{1}$$

where $p(u, v, t + \Delta t)$ is the phase screen at moment $t + \Delta t$, (x, y) is a spectrum point, (u, v) is a phase screen point, V is the wind velocity, m/s, t is the moment of the previous phase screen generation, $t + \Delta t$ is the time moment of the new phase screen generation, Δt is the time interval between two phase screens and $K(x, y)$ is the spectrum of the phase fluctuations.

The spectrum of the phase fluctuations $K(x, y)$ is calculated as follows:

$$K(x, y) = 0.023 \cdot \left(\frac{2D}{r_0}\right)^{\frac{5}{3}} \cdot \left(x^2 + y^2\right)^{\frac{11}{3}}, \tag{2}$$

where D is the receiving aperture of the telescope, r_0 is the Fried radius and $f(x, y, t + \Delta t)$ is the function, defined as follows:

$$f(x, y, t + \Delta t) = p \cdot f(x, y, t) + \sqrt{1 - p^2} \cdot e^{i \cdot \varphi(x, y, t)}, \tag{3}$$

where $p = e^{-\Delta t / \tau}$, τ is the coherence (or freezing) time of the atmosphere, $\varphi(x, y, t)$ is the random delta-correlated value in the range of $[0; 2\pi]$.

When $t = 0$, the function f is expressed as:

$$f(x, y, t = 0) = e^{i \cdot \varphi(x, y, t = 0)}. \tag{4}$$

The calculated phase values should be normalized in accordance with the relation $\frac{D}{r_0}$ [47]. We used the phase structure function to do this:

$$DN = 6.88 \cdot \left(\frac{\sqrt{x^2 + y^2}}{r_0}\right)^{\frac{5}{3}}. \tag{5}$$

Each calculated phase screen was then approximated using the response functions of the stacked-actuator mirror and reproduced by this mirror. The response function of the individual actuator of the stacked-actuator deformable mirror represented the deformation of the surface of the mirror influenced by the voltage applied to this particular actuator, while other actuators still had a voltage under 0. In our case, the response function of each actuator of the mirror was measured by the Shack–Hartmann sensor and expressed as the vector of the displacements of the focal spots on the sensor.

Once the response functions of the mirror were measured, we could run the procedure of the reproduction of the particular phase screen by the mirror, described as follows:

1. As we had the Zernike approximation of the simulated phase screen, we could calculate the values of the wavefront derivatives in each sub-aperture of the wavefront sensor.
2. Knowing the values of the wavefront derivatives, we could calculate the displacements of the focal spots corresponding to these derivatives.
3. Thus, knowing the focal spot shifts associated with the mirror response functions and the focal spot displacements corresponding to the wavefront to be reproduced, we could solve the overdetermined system of linear equations using the least squares method and calculate the vector of voltages that had to be applied to the mirror actuators.

After applying the calculated set of voltages to the mirror actuators, we reconstructed the desired phase screen. The accuracy of the reconstruction is provided in the Results section.

2.6. Algorithm of Phase Screen Compensation

Once the phase screen was reconstructed by the stacked-actuator deformable mirror, the correction procedure ran. It includes the following steps:

1. The wavefront of the laser beam reflected from the stacked-actuator and bimorph mirrors was analyzed on the Shack–Hartmann wavefront sensor.

2. Having, on the one hand, the matrix of displacements of the focal spots on the Shack–Hartmann sensor $\begin{Bmatrix} S_x^k \\ S_y^k \end{Bmatrix}$, corresponding to the wavefront of the reconstructed phase screen, and, on the other hand, the matrix of values of the bimorph mirror response functions RF, also consisting of the focal spots shifts, we obtained an overdetermined system of linear equations for unknown coefficients, which were the values of the voltages at the mirror electrodes. To solve this system of equations, the least squares method was used.

3. The calculated voltages were applied to the electrodes of the bimorph mirror.

4. The residual wavefront was measured by means of the Shack–Hartmann sensor.

After the bimorph mirror compensated for the induced wavefront distortions, the phase screen simulation algorithm proceeded, and the described procedures were repeated.

3. Results and Discussion

The experiment was conducted as follows:

1. Simulation of the set of phase screens using the Fast Fourier transform for the Kolmogorov spectrum of phase fluctuations [71,72].

2. Calculation of the control voltages to be applied to the actuators of the stacked-actuator deformable mirror [73].

3. Reconstruction of the simulated phase screens using the stacked-actuator mirror.

4. Measurement of the introduced wavefront distortions using the Shack–Hartmann wavefront sensor.

5. Calculation of the control voltages to be applied to the electrodes of the bimorph deformable mirror in order to compensate for the wavefront distortions.

6. Compensation of the reconstructed phase screens by the bimorph mirror.

It is well known that the main parameter characterizing the strength of atmospheric turbulence is the refractive index structure parameter C_n^2. It can vary from 10^{-17} m$^{-\frac{2}{3}}$ for weak turbulence to 10^{-12} m$^{-\frac{2}{3}}$ for very strong atmospheric turbulence. For example, in Hefei city, China, C_n^2 varies from 6.69×10^{-16} m$^{-\frac{2}{3}}$ to 9.87×10^{-14} m$^{-\frac{2}{3}}$ for measurements performed in the summer month at 1 km horizontal atmospheric path [74]. For maritime atmospheric turbulence C_n^2 is equal to 10^{-15} m$^{-\frac{2}{3}}$ for the 10 km path, coherence radius $r_0 = 3.8$ cm for laser wavelength $\lambda = 0.85$ μm [75]. For ground-to-space communication between the International Space Station and the Optical Communications Telescope Laboratory (OCTL) in Wrightwood, California, a coherence radius of $r_0 = 4.5$ cm has been experimentally measured for the 1200 km path and $75°$ zenith angle, with the input telescope aperture varied from 10 cm to 100 cm [76]. For terrestrial atmospheric turbulence C_n^2 is equal to 10^{-12} m$^{-\frac{2}{3}}$ for the 1 km path, where the wind velocity was 10 m/s and the telescope aperture was 20 cm [77]. For desert atmospheric turbulence, C_n^2 is equal to $10^{-13.2}$ m$^{-\frac{2}{3}}$ (for an average wind velocity of 6 m/s) for 1.2 km path at *Edward Air Force Base*, Mojave Desert, CA, USA [78]. For the atmospheric turbulence measured by our team in collaboration with our Austrian colleagues at the 1.2 km intra-city link in Vienna, Austria, the coherence radius was equal to $r_0 = 1.6$ cm for the laser wavelength $\lambda = 0.532$ μm and $r_0 = 2.65$ cm for $\lambda = 0.81$ μm; the wind velocity varied from 5 to 10 m/s for the receiving aperture D = 140 mm.

However, C_n^2 can be calculated from the known receiving aperture diameter D and Fried radius r_0. In this research we set $\frac{D}{r_0} = 10$, the wavelength $\lambda = 1$ μm, and the wind velocity $v = 6$ m/s. C_n^2 varies roughly from 9×10^{-15} m$^{-\frac{2}{3}}$ to 8×10^{-14} m$^{-\frac{2}{3}}$ for the path length from 500 m to 3 km.

Figure 4 presents the results of the Zernike approximation of the phase screen simulated with the Kolmogorov spectrum and the results of the reconstruction of this phase screen with the stacked-actuator deformable mirror.

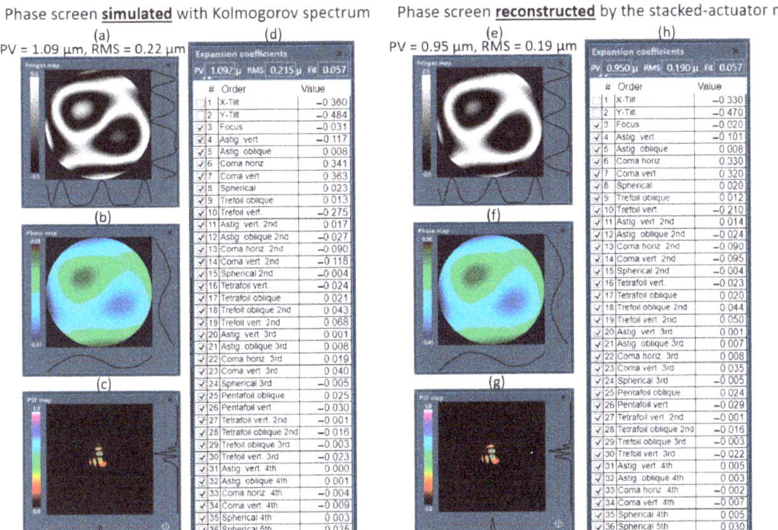

(d) Expansion coefficients — PV 1.092 µ RMS 0.215 µ Fit 0.057

#	Order	Value
1	X-Tilt	−0 360
2	Y-Tilt	−0 484
3	Focus	−0 031
4	Astig. vert.	−0 117
5	Astig. oblique	0 008
6	Coma horiz	0 341
7	Coma vert	0 363
8	Spherical	0 023
9	Trefoil oblique	0 013
10	Trefoil vert.	−0 275
11	Astig. vert. 2nd	0 017
12	Astig. oblique 2nd	−0 027
13	Coma horiz. 2nd	−0 090
14	Coma vert. 2nd	−0 118
15	Spherical 2nd	−0 004
16	Tetrafoil vert.	−0 024
17	Tetrafoil oblique	0 021
18	Trefoil oblique 2nd	0 043
19	Trefoil vert. 2nd	0 068
20	Astig. vert. 3rd	0 001
21	Astig. oblique 3rd	0 008
22	Coma horiz. 3rd	0 019
23	Coma vert. 3rd	0 040
24	Spherical 3rd	−0 005
25	Pentafoil oblique	0 025
26	Pentafoil vert	−0.030
27	Tetrafoil vert. 2nd	−0 001
28	Tetrafoil oblique 2nd	−0 016
29	Trefoil oblique 3rd	−0 003
30	Trefoil vert. 3rd	−0 023
31	Astig. vert. 4th	0 000
32	Astig. oblique 4th	0 001
33	Coma horiz. 4th	−0 004
34	Coma vert. 4th	−0 009
35	Spherical 4th	0 003
36	Spherical 5th	0 035

(h) Expansion coefficients — PV 0.950 µ RMS 0.190 µ Fit 0.057

#	Order	Value
1	X-Tilt	−0 330
2	Y-Tilt	−0 470
3	Focus	−0 020
4	Astig. vert	−0 101
5	Astig. oblique	0 008
6	Coma horiz	0.330
7	Coma vert	0 320
8	Spherical	0 020
9	Trefoil oblique	0 012
10	Trefoil vert	−0 210
11	Astig. vert. 2nd	0 014
12	Astig. oblique 2nd	−0 024
13	Coma horiz. 2nd	−0 090
14	Coma vert. 2nd	−0 095
15	Spherical 2nd	−0 004
16	Tetrafoil vert.	−0 023
17	Tetrafoil oblique	0 020
18	Trefoil oblique 2nd	0 044
19	Trefoil vert. 2nd	0 050
20	Astig. vert. 3rd	0 001
21	Astig. oblique 3rd	0 007
22	Coma horiz. 3rd	0 008
23	Coma vert. 3rd	0 035
24	Spherical 3rd	−0 005
25	Pentafoil oblique	0 024
26	Pentafoil vert	−0 029
27	Tetrafoil vert. 2nd	−0 001
28	Tetrafoil oblique 2nd	−0 016
29	Trefoil oblique 3rd	−0 003
30	Trefoil vert. 3rd	−0 022
31	Astig. vert. 4th	0 005
32	Astig. oblique 4th	0 003
33	Coma horiz. 4th	−0 002
34	Coma vert. 4th	−0 007
35	Spherical 4th	0 005
36	Spherical 5th	0 030

Figure 4. Results of the approximation of the phase screen using Zernike polynomials. (**a**) Fringes map, (**b**) phase map, (**c**) calculated far field intensity distribution—for the approximated phase screen, (**d**) Zernike decomposition and results of the reconstruction of the phase screen by means of stacked-actuator mirror, (**e**) fringes map, (**f**) phase map, (**g**) calculated far field intensity distribution—for the approximated phase screen, (**h**) Zernike decomposition.

The input parameters for the phase screens simulation algorithm were as follows:

It can be seen from Figure 4 that the stacked-actuator deformable mirror reconstructs the simulated phase surface with an error of about 13%. The amplitude of the wavefront distortions of the approximated phase screen was equal to 1.09 µm (root mean square error RMS = 0.22 µm). The amplitude of the reconstructed surface was equal to 0.95 µm (RMS = 0.19 µm). The voltages on the actuators of the stacked-actuator mirror required to reconstruct such a wavefront are presented in Figure 5.

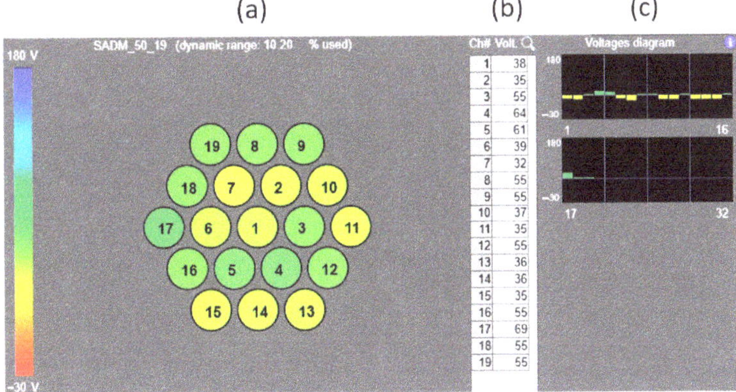

Figure 5. Actuator's voltages calculated to reconstruct the phase screen. (**a**) Schematical representation of the actuators of the stacked-actuator mirror where each color from the palette corresponds to the voltage value, (**b**) table of absolute voltages values on each actuator, (**c**) bar diagram of the voltages.

It should be noted that the voltage range for the stacked-actuator mirror is -30 V$-+180$ V. This means that in order to use the whole dynamic range, we have to set the offset voltages for each actuator. In our case, the offset voltage was 50 V. It can be seen from Figure 5 that the voltage values required to reconstruct the desired wavefront shape were not very high—the voltage variation was about ± 20 V. It took only 10% of the dynamic range of the mirror, where 100% of the dynamic range means that either -30 V or $+180$ V is applied for each actuator. This is due to high-sensitivity of the piezoceramic actuators used in the mirror. The 13% error of the phase screen reconstruction is mainly because of the limited number of the actuators used in the mirror.

In order to estimate the accuracy of the reconstruction, we subtracted the raw phase data of the reconstructed screen from the raw phase data of the simulated phase screen. The resultant phase difference surface is presented in Figure 6.

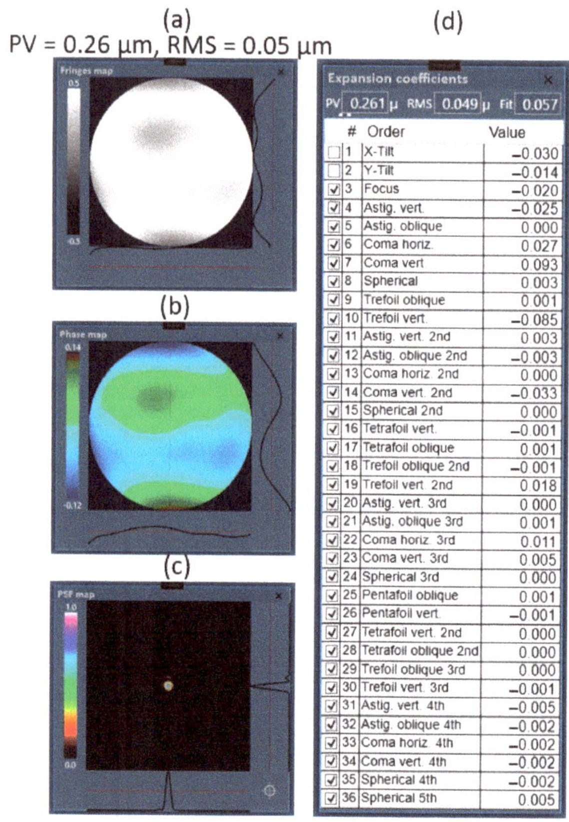

#	Order	Value
1	X-Tilt	−0.030
2	Y-Tilt	−0.014
3	Focus	−0.020
4	Astig. vert.	−0.025
5	Astig. oblique	0.000
6	Coma horiz.	0.027
7	Coma vert	0.093
8	Spherical	0.003
9	Trefoil oblique	0.001
10	Trefoil vert.	−0.085
11	Astig. vert. 2nd	0.003
12	Astig. oblique 2nd	−0.003
13	Coma horiz. 2nd	0.000
14	Coma vert. 2nd	−0.033
15	Spherical 2nd	0.000
16	Tetrafoil vert.	−0.001
17	Tetrafoil oblique	0.001
18	Trefoil oblique 2nd	−0.001
19	Trefoil vert. 2nd	0.018
20	Astig. vert. 3rd	0.000
21	Astig. oblique 3rd	0.001
22	Coma horiz. 3rd	0.011
23	Coma vert. 3rd	0.005
24	Spherical 3rd	0.000
25	Pentafoil oblique	0.001
26	Pentafoil vert.	−0.001
27	Tetrafoil vert. 2nd	0.000
28	Tetrafoil oblique 2nd	0.000
29	Trefoil oblique 3rd	0.000
30	Trefoil vert. 3rd	−0.001
31	Astig. vert. 4th	−0.005
32	Astig. oblique 4th	−0.002
33	Coma horiz. 4th	−0.002
34	Coma vert. 4th	−0.002
35	Spherical 4th	−0.002
36	Spherical 5th	0.005

Figure 6. Residual error of phase screen reconstruction by means of the stacked-actuator mirror: (**a**) fringes map, (**b**) phase map, (**c**) calculated far field intensity distribution, (**d**) Zernike decomposition.

Such a difference (PV = 0.26 µm) between the simulated and reconstructed phase surfaces may be due, first of all, to a hysteresis effect of piezoceramics used in the actuators of the mirror [79]. In essence, the response of the actuators of the stacked-actuator mirror is on average 15% smaller/bigger than the expected response for each act of voltage application. Thus, it takes 2–3 iterations to compensate for the hysteresis and reduce the phase delay between the expected and real phase screen.

The second step was to compensate for the introduced wavefront distortions with the bimorph deformable mirror. Figure 7 shows the result of the compensation of the phase screen with the bimorph mirror.

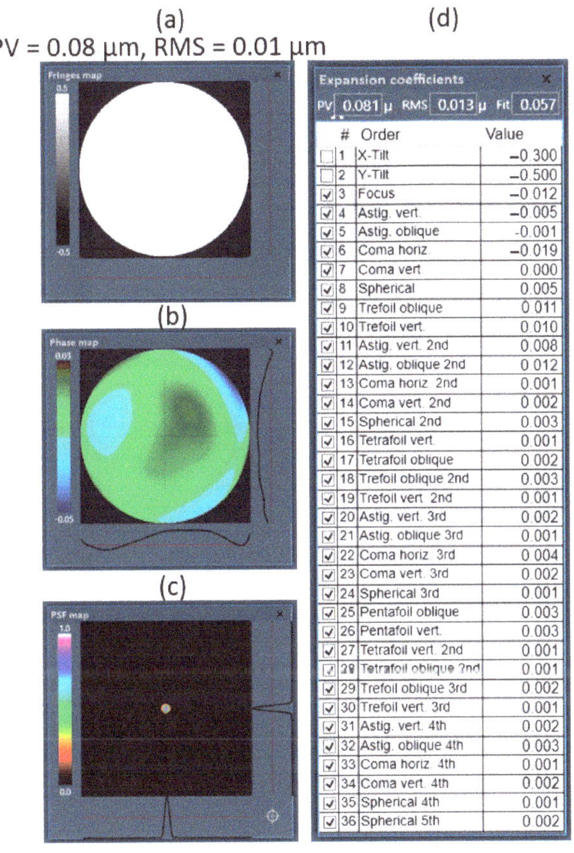

Expansion coefficients

PV 0.081 μ   RMS 0.013 μ   Fit 0.057

#	Order	Value
1	X-Tilt	−0.300
2	Y-Tilt	−0.500
3	Focus	−0.012
4	Astig. vert.	−0.005
5	Astig. oblique	-0.001
6	Coma horiz.	−0.019
7	Coma vert.	0.000
8	Spherical	0.005
9	Trefoil oblique	0.011
10	Trefoil vert.	0.010
11	Astig. vert. 2nd	0.008
12	Astig. oblique 2nd	0.012
13	Coma horiz. 2nd	0.001
14	Coma vert. 2nd	0.002
15	Spherical 2nd	0.003
16	Tetrafoil vert.	0.001
17	Tetrafoil oblique	0.002
18	Trefoil oblique 2nd	0.003
19	Trefoil vert. 2nd	0.001
20	Astig. vert. 3rd	0.002
21	Astig. oblique 3rd	0.001
22	Coma horiz. 3rd	0.004
23	Coma vert. 3rd	0.002
24	Spherical 3rd	0.001
25	Pentafoil oblique	0.003
26	Pentafoil vert.	0.003
27	Tetrafoil vert. 2nd	0.001
28	Tetrafoil oblique 2nd	0.001
29	Trefoil oblique 3rd	0.002
30	Trefoil vert. 3rd	0.001
31	Astig. vert. 4th	0.002
32	Astig. oblique 4th	0.003
33	Coma horiz. 4th	0.001
34	Coma vert. 4th	0.002
35	Spherical 4th	0.001
36	Spherical 5th	0.002

Figure 7. Residual error of phase screen correction by means of the bimorph mirror: (**a**) fringes map, (**b**) phase map, (**c**) calculated far field intensity distribution, (**d**) Zernike decomposition.

The 35 mm diameter bimorph deformable mirror with 32 control electrodes compensates for the introduced distortions with high efficiency. The residual wavefront amplitude was equal to 0.08 μm, while the RMS was only 0.01 μm ($\lambda/100$).

The voltages on the electrodes of the bimorph mirror were required to compensate for the wavefront distortions introduced by the stacked-actuator mirror are presented in Figure 8.

It can be seen from Figure 8 that the voltage values required to compensate for the induced wavefront distortions were higher than for the stacked-actuator mirror—the voltage change was about ±100 V. It took about 9% of the dynamic range of the mirror, where 100% of the dynamic range means that either −300 V or +300 V is applied to each electrode.

Since the efficiency of the currently used 19-channel stacked-actuator mirror as a phase screen reconstruction device was not very high (the reconstruction error was about 13%), the first step for future research will be to increase the number of actuators of the stacked-actuator mirror. In addition, the amplitude of the phase fluctuations of the simulated phase screens can be increased in order to test both the efficiency and stability of the compensation loop.

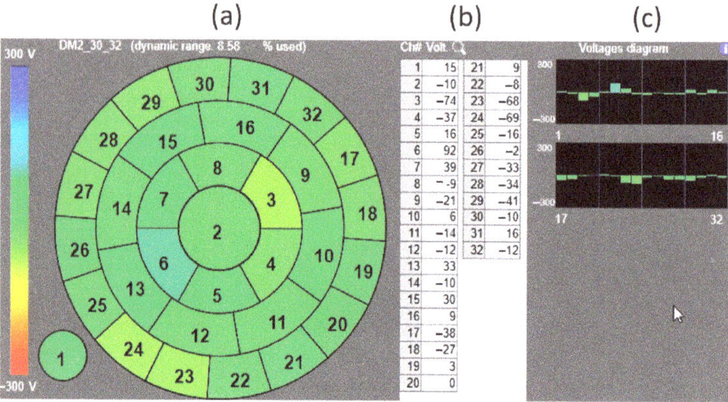

Figure 8. Electrode voltages calculated to compensate for the induced wavefront distortions: (**a**) schematical representation of the bimorph mirror electrodes where each color from the palette corresponds to the voltage value, (**b**) table of absolute voltages values on each electrode, (**c**) bar diagram of the voltages.

4. Conclusions

In this research, turbulence simulator software was developed and tested to generate phase screens with Kolmogorov spectra. The set of phase screens was generated for the ratio of receiving telescope aperture to Fried radius equal to 10, a wavelength equal to 1 μm, and a wind velocity equal to 6 m/s. The refractive index structure parameter C_n^2 varied roughly from 9×10^{-15} m$^{-\frac{2}{3}}$ to 8×10^{-14} m$^{-\frac{2}{3}}$ for the path length from 500 m up to 3000 m. The generated set of phase screens was reconstructed using the 19-channel stacked-actuator deformable mirror and then compensated for using the 32-channel bimorph deformable mirror. It was shown that the residual amplitude of the wavefront reconstructed by the 19-channel stacked-actuator mirror was 0.26 λ, while the residual amplitude of the wavefront compensated by the 32-channel bimorph mirror was 0.08 λ.

Author Contributions: Conceptualization, A.K. and J.S.; methodology, I.G.; software, I.G.; validation, J.S., I.G. and A.K.; formal analysis, A.K.; investigation, I.G.; resources, V.S. and V.T.; data curation, J.S.; writing—original draft preparation, I.G.; writing—review and editing, I.G., V.T. and V.S.; visualization, I.G.; supervision, A.K.; project administration, A.K.; funding acquisition, A.K. All authors have read and agreed to the published version of the manuscript.

Funding: This research was carried out (1) within the state assignment of Ministry of Science and Higher Education of the Russian Federation No. 122032900183-1—research and development of the phase screen generation algorithm; (2) within the Russian Science Foundation project # 19-19-00706P—development of the stacked-actuator deformable mirror; and (3) within the scientific program of the National Center of Physics and Mathematics (project "High Energy Density Physics. Stage 2023-2025")—development of the turbulence generator.

Institutional Review Board Statement: Not applicable.

Informed Consent Statement: Not applicable.

Data Availability Statement: Data sharing is not applicable to this article.

Acknowledgments: Special thanks to Ann Lylova for technical, theoretical and software support.

Conflicts of Interest: The authors declare no conflict of interest.

References

1. Tatarsky, V.I. *Waves Propagation in Turbulent Atmosphere*; Nauka: Moscow, Russia, 1967.
2. Andrews, L.C.; Phillips, R.L. *Laser Beam Propagation Through Random Media*, 2nd ed.; SPIE Press: Bellingham, WA, USA, 2005.
3. Huang, Q.; Liu, D.; Chen, Y.; Wang, Y.; Tan, J.; Chen, W.; Liu, J.; Zhu, N. Secure free-space optical communication system based on data fragmentation multipath transmission technology. *Opt. Express* **2018**, *26*, 13536–13542. [CrossRef] [PubMed]
4. Nafria, V.; Han, X.; Djordjevic, I. Improving free-space optical communication with adaptive optics for higher order modulation. In Proceedings of the Optics and Photonics for Information Processing XIV, Online, CA, USA, 24 August–4 September 2020.
5. Vorontsov, M.; Weyrauch, T.; Carhart, G.; Beresnev, L. Adaptive Optics for Free Space Laser Communications. In *Lasers, Sources and Related Photonic Devices, OSA Technical Digest Series (CD)*; Optica Publishing Group: Washington, DC, USA, 2010; p. LSMA1.
6. Weyrauch, T.; Vorontsov, M. Free-space laser communications with adaptive optics: Atmospheric compensation experiments. *J. Optic. Comm. Rep.* **2004**, *1*, 355–379. [CrossRef]
7. Lema, G. Free space optics communication system design using iterative optimization. *J. Opt. Commun.* **2020**, 000010151520200007. [CrossRef]
8. Zhang, Y.; Wang, Y.; Deng, Y.; Du, A.; Liu, J. Design of a Free Space Optical Communication System for an Unmanned Aerial Vehicle Command and Control Link. *Photonics* **2021**, *8*, 163. [CrossRef]
9. Majumdar, A. Fundamentals of Free-Space Optical (FSO) Communication System. In *Advanced Free Space Optics (FSO)*; Series in Optical Sciences; Springer: New York, NY, USA, 2015; Volume 186.
10. Lu, M.; Bagheri, M.; James, A.P.; Phung, T. Wireless Charging Techniques for UAVs: A Review, Reconceptualization, and Extension. *IEEE Access* **2018**, *6*, 29865–29884. [CrossRef]
11. Geoffrey, A.; Landis, H. Laser beamed power—Satellite demonstration applications. In Proceedings of the 43rd International Astronautical Congress, Washington, DC, USA, 28 August–5 September 1992.
12. Vorontsov, M.; Weyrauch, T.; Carhart, G. *Lasers, Sources and Related Photonic Devices*; OSA Technical Digest Series (CD); Optica Publishing Group: San Diego, USA, 2010; p. LSMA1.
13. Galaktionov, I.; Kudryashov, A.; Sheldakova, J.; Nikitin, A. Laser beam focusing through the scattering medium by means of adaptive optics. In Proceedings of the Adaptive Optics and Wavefront Control for Biological Systems III, San Francisco, CA, USA, 28 January–2 February 2017.
14. Wang, R.; Wang, Y.; Jin, C.; Yin, X.; Wang, S.; Yang, C.; Cao, Z.; Mu, Q.; Gao, S.; Xuan, L. Demonstration of horizontal free-space laser communication with the effect of the bandwidth of adaptive optics system. *Opt. Commun.* **2018**, *431*, 167–173. [CrossRef]
15. Chittoor, P.K.; Chokkalingam, B.; Mihet-Popa, L. A Review on UAV Wireless Charging: Fundamentals, Applications, Charging Techniques and Standards. *IEEE Access* **2021**, *9*, 69235–69266. [CrossRef]
16. Wang, C.; Ma, Z. Design of wireless power transfer device for UAV. In Proceedings of the 2016 IEEE International Conference on Mechatronics and Automation, Harbin, China, 7–10 August 2016; pp. 2449–2454.
17. Bennet, F.; Conan, R.; D'Orgeville, C.; Dawson, M.; Paulin, N.; Price, I.; Rigaut, F.; Ritchie, I.; Smith, C.; Uhlendorf, K. Adaptive optics for laser space debris removal. In Proceedings of the Adaptive Optics Systems III, Amsterdam, The Netherlands, 1–6 July 2012.
18. Phipps, C.R.; Baker, K.L.; Libby, S.B.; Liedahl, D.A.; Olivier, S.S.; Pleasance, L.D.; Rubenchik, A.; Trebes, J.E.; George, E.V.; Marcovici, B.; et al. Removing orbital debris with lasers. *Adv. Space Res.* **2012**, *49*, 1283–1300. [CrossRef]
19. Shen, S.; Jin, X.; Hao, C. Cleaning space debris with a space-based laser system. *Chin. J. Aeronaut.* **2014**, *27*, 805–811. [CrossRef]
20. Sheldakova, J.; Galaktionov, I.; Nikitin, A.; Rukosuev, A.; Kudryashov, A. LC phase modulator vs deformable mirror for laser beam shaping: What is better? In Proceedings of the Laser Beam Shaping XVIII, San Diego, CA, USA, 19–23 August 2018; Volume 10744, pp. 156–162.
21. Galaktionov, I.; Nikitin, A.; Sheldakova, J.; Toporovsky, V.; Kudryashov, A. Focusing of a Laser Beam Passed through a Moderately Scattering Medium Using Phase-Only Spatial Light Modulator. *Photonics* **2022**, *9*, 296. [CrossRef]
22. Galaktionov, I.; Kudryashov, A.; Sheldakova, J.; Nikitin, A. The use of modified hill-climbing algorithm for laser beam focusing through the turbid medium. In Proceedings of the Laser Resonators, Microresonators, and Beam Control XIX, San Francisco, CA, USA, 28 January–2 February 2017.
23. Barros, R.; Keary, S. Experimental setup for investigation of laser beam propagation along horizontal urban path. *Proc. SPIE* **2014**, *9242*, 396–404.
24. Mata-Calvo, R. Transmitter diversity verification on Artemis geostationary satellite. In Proceedings of the Free-Space Laser Communication and Atmospheric Propagation XXVI, San Francisco, CA, USA, 1–6 February 2014.
25. Mosavi, N.; Marks, B.; Boone, B.; Menyuk, C. Optical beam spreading in the presence of both atmospheric turbulence and quartic aberration. In Proceedings of the Free-Space Laser Communication and Atmospheric Propagation XXVI, San Francisco, CA, USA, 1–6 February 2014.
26. Murty, S.S.R. Laser beam propagation in atmospheric turbulence. *Proc. Indian Acad. Sci.* **1979**, *2*, 179–195. [CrossRef]
27. Kwiecień, J. The effects of atmospheric turbulence on laser beam propagation in a closed space—An analytic and experimental approach. *Opt. Commun.* **2019**, *433*, 200–208. [CrossRef]
28. Searles, S.; Hart, G.; Dowling, J.; Hanley, S. Laser beam propagation in turbulent conditions. *Appl. Opt.* **1991**, *30*, 401–406. [CrossRef]
29. Gareth, D.; Naven, C. Experimental analysis of a laser beam propagating in angular turbulence. *Open Phys.* **2022**, *20*, 402–415.

30. Summerer, L.; Purcell, O. Concepts for wireless energy transmission via laser. *J. Br. Interplanet. Soc.* **2005**, *58*, 1–10.
31. Fahey, T.; Islam, M.; Gardi, A.; Sabatini, R. Laser Beam Atmospheric Propagation Modelling for Aerospace LIDAR Applications. *Atmosphere* **2021**, *12*, 918. [CrossRef]
32. Zohuri, B. Atmospheric Propagation of High-Energy Laser Beams. In *Directed Energy Weapons*; Springer: Cham, Switzerland, 2016.
33. Shuto, Y. Effect of Water and Aerosols Absorption on Laser Beam Propagation in Moist Atmosphere at Eye-Safe Wavelength of 1.57 μm. *J. Electr. Electron. Eng.* **2023**, *11*, 15–22.
34. Rosen, L.; Ipser, J. High Energy Laser Beam Scattering by Atmospheric Aerosol Aureoles. In Proceedings of the Nonlinear Optical Beam Manipulation and High Energy Beam Propagation through the Atmosphere, Los Angeles, CA, USA, 5–20 January 1989.
35. Oosterwijk, A.; Heikamp, S.; Manders-Groot, A.; Lex, A.; Eijk, J. Comparison of modelled atmospheric aerosol content and its influence on high-energy laser propagation. In Proceedings of the Laser Communication and Propagation through the Atmosphere and Oceans VIII, San Diego, CA, USA, 11–15 August 2019.
36. Kudryashov, A.; Rukosuev, A.; Samarkin, V.; Galaktionov, I.; Kopylov, E. Fast adaptive optical system for 1.5 km horizontal beam propagation. In Proceedings of the Unconventional and Indirect Imaging, Image Reconstruction, and Wavefront Sensing 2018, San Diego, CA, USA, 19–23 August 2018.
37. Soloviev, A.; Kotov, A.; Perevalov, S.; Esyunin, M.; Starodubtsev, M.; Alexandrov, A.; Galaktionov, I.; Samarkin, V.; Kudryashov, A.; Ginzburg, V.; et al. Adaptive system for wavefront correction of the PEARL laser facility. *Quantum Electron.* **2022**, *50*, 1115–1122. [CrossRef]
38. Samarkin, V.; Alexandrov, A.; Galaktionov, I.; Kudryashov, A.; Nikitin, A.; Rukosuev, A.; Toporovsky, V.; Sheldakova, Y. Large-aperture adaptive optical system for correcting wavefront distortions of a petawatt Ti:sapphire laser beam. *Quantum Electron.* **2022**, *52*, 187–194. [CrossRef]
39. Buck, A.L. Effects of the Atmosphere on Laser Beam Propagation. *Appl. Opt.* **1967**, *6*, 703–708. [CrossRef] [PubMed]
40. Galaktionov, I.; Kudryashov, A.; Sheldakova, J.; Nikitin, A.; Samarkin, V. Laser beam focusing through the atmosphere aerosol. In Proceedings of the Unconventional and Indirect Imaging, Image Reconstruction, and Wavefront Sensing 2017, San Diego, CA, USA, 6–10 August 2017; Volume 10410, pp. 152–165.
41. Lylova, A.; Kudryashov, A.; Sheldakova, J.; Borsoni, G. The real-time atmospheric turbulence modeling and compensation with the use of adaptive optics. In Proceedings of the Optics in Atmospheric Propagation and Adaptive Systems XVIII, Toulouse, France, 21–24 September 2015.
42. Gonglewski, J.; Highland, R.; Dayton, D.; Sandven, S.; Rogers, S.; Browne, S. ADONIS: Daylight Imaging through Atmospheric Turbulence. In Proceedings of the Digital Image Recovery and Synthesis III, Denver, CO, USA, 4–9 August 1996.
43. Marchi, G.; Weiß, R. Evaluations and Progress in the Development of an Adaptive Optics System for Ground Object Observation. In Proceedings of the Optics in Atmospheric Propagation and Adaptive Systems X, Florence, Italy, 17–20 September 2007.
44. Marchi, G.; Scheifling, C. Adaptive optics concepts and systems for multipurpose applications at near horizontal line of sight: Developments and results. In Proceedings of the Optics in Atmospheric Propagation and Adaptive Systems XII, Berlin, Germany, 31 August–3 September 2009.
45. Segel, M.; Gladysz, S.; Stein, K. Optimal modal compensation in gradient-based wavefront sensorless adaptive optics. In Proceedings of the Laser Communication and Propagation through the Atmosphere and Oceans VIII, San Diego, CA, USA, 11–15 August 2019.
46. Segel, M.; Zepp, A.; Anzuola, E.; Gladysz, S.; Stein, K. Optimization of wavefront-sensorless adaptive optics for horizontal laser beam propagation in a realistic turbulence environment. In Proceedings of the Laser Communication and Propagation through the Atmosphere and Oceans VI, San Diego, CA, USA, 6–10 August 2017.
47. Gonglewski, J.; Dayton, D. Least squares blind deconvolution of air to ground imaging. In Proceedings of the Optics in Atmospheric Propagation and Adaptive Systems VIII, Bruges, Belgium, 19–22 September 2005.
48. Mackey, R.; Dainty, C. Adaptive optics correction over a 3km near horizontal path. In Proceedings of the Optics in Atmospheric Propagation and Adaptive Systems XI, Cardiff, UK, 15–18 September 2008.
49. Sheldakova, J.; Toporovsky, V.; Galaktionov, I.; Nikitin, A.; Rukosuev, A.; Samarkin, V.; Kudryashov, A. Flat-top beam formation with miniature bimorph deformable mirror. In Proceedings of the Laser Beam Shaping XX, Online, CA, USA, 24 August–4 September 2020.
50. Samarkin, V.; Alexandrov, A.; Galaktionov, I.; Kudryashov, A.; Nikitin, A.; Rukosuev, A.; Toporovsky, V.; Sheldakova, J. Wide-Aperture Bimorph Deformable Mirror for Beam Focusing in 4.2 PW Ti:Sa Laser. *Appl. Sci.* **2022**, *12*, 1144. [CrossRef]
51. Galaktionov, I.; Nikitin, A.; Samarkin, V.; Sheldakova, J.; Kudryashov, A. Laser beam focusing through the scattering medium-low order aberration correction approach. In Proceedings of the Unconventional and Indirect Imaging, Image Reconstruction, and Wavefront Sensing 2018, San Diego, CA, USA, 19–23 August 2018.
52. Galaktionov, I.; Kudryashov, A.; Sheldakova, J.; Nikitin, A. Laser beam focusing through the dense multiple scattering suspension using bimorph mirror. In Proceedings of the Adaptive Optics and Wavefront Control for Biological Systems V, San Francisco, CA, USA, 20 February 2019; Volume 10886, p. 1088619.
53. Platt, B.; Shack, R.J. History and principles of Shack-Hartmann wavefront sensing. *Refr. Surg.* **2001**, *17*, 15. [CrossRef]
54. Liang, J.; Grimm, B.; Goelz, S.; Bille, J.F. Objective measurement of the wave aberrations of the human eye with the use of a Hartmann-Shack wave-front sensor. *J. Opt. Soc. Am. A* **1994**, *11*, 1949. [CrossRef]
55. Lane, R.G. Wave-front reconstruction using a Shack–Hartmann sensor. *Appl. Opt.* **1992**, *31*, 6902. [CrossRef]
56. Primot, J. Theoretical description of Shack–Hartmann wave-front sensor. *Opt. Commun.* **2003**, *222*, 81–92. [CrossRef]

57. Konnik, M.; Doná, J. Waffle mode mitigation in adaptive optics systems: A constrained Receding Horizon Control approach. In Proceedings of the American Control Conference, Washington, DC, USA, 17–19 June 2013; pp. 3390–3396.
58. Mauch, S.; Reger, J. Real-Time Spot Detection and Ordering for a Shack–Hartmann Wavefront Sensor with a Low-Cost FPGA. *Proc. IEEE Trans. Instrum. Meas.* **2014**, *63*, 2379–2386. [CrossRef]
59. Nikitin, A.; Galaktionov, I.; Sheldakova, J.; Kudryashov, A.; Baryshnikov, N.; Denisov, D.; Karasik, V.; Sakharov, A. Absolute calibration of a Shack-Hartmann wavefront sensor for measurements of wavefronts. In Proceedings of the Photonic Instrumentation Engineering VI, San Francisco, CA, USA, 2–7 February 2019.
60. Galaktionov, I.; Sheldakova, J.; Nikitin, A.; Toporovsky, V.; Kudryashov, A. A Hybrid Model for Analysis of Laser Beam Distortions Using Monte Carlo and Shack–Hartmann Techniques: Numerical Study and Experimental Results. *Algorithms* **2023**, *16*, 337. [CrossRef]
61. Galaktionov, I.; Kudryashov, A.; Sheldakova, J.; Byalko, A.; Borsoni, G. Laser beam propagation and wavefront correction in turbid media. In Proceedings of the Unconventional Imaging and Wavefront Sensing 2015, San Diego, CA, USA, 9–13 August 2015.
62. Galaktionov, I.; Nikitin, A.; Sheldakova, J.; Kudryashov, A. B-spline approximation of a wavefront measured by Shack-Hartmann sensor. In Proceedings of the Laser Beam Shaping XXI, San Diego, CA, USA, 1–5 August 2021.
63. Malacara-Hernandez, D. Wavefront fitting with discrete orthogonal polynomials in a unit radius circle. *Opt. Eng.* **1990**, *29*, 672–675. [CrossRef]
64. Wyant, J.C.; Creath, K. Basic wavefront aberration theory for optical metrology. In *Applied Optics and Optical Engineering*; Academic Press: Cambridge, MA, USA, 1992; pp. 28–39.
65. Genberg, V.; Michels, G.; Doyle, K. Orthogonality of Zernike polynomials. *Proc. SPIE* **2002**, *4771*, 276–287.
66. Lakshminarayanan, V.; Fleck, A. Zernike polynomials: A guide. *J. Mod. Opt.* **2011**, *58*, 545–561. [CrossRef]
67. Linnik, J.V. *Least Square Error Method and the Basics of Mathematical-Statistical Theory*, 2nd ed.; Mir: Moscow, Russia, 1962.
68. Kandidov, V.; Shlenov, A. *Atmospheric Turbulence Simulation. Grid Representation. Technical Summary*; International Laser Center: Moscow, Russia, 1995.
69. Vorontsov, M.; Shmalhausen, V. *Principles of Adaptive Optics*; Nauka: Moscow, Russia, 1985.
70. Dudorov, V.; Kolosov, V.; Filimonov, G. Algorithm for formation of an infinite random turbulent screen. In Proceedings of the Twelfth Joint International Symposium on Atmospheric and Ocean Optics/Atmospheric Physics, Tomsk, Russia, 27–30 June 2005.
71. Lane, R.G.; Glindemann, A.; Dainty, J.C. Simulation of Kolmogorov phase screen. *Waves Random Media* **1992**, *2*, 209–224. [CrossRef]
72. Burger, L.; Litvin, I.; Forbes, A. Simulating atmospheric turbulence using a phase-only spatial light modulator. *South Afr. J. Sci.* **2008**, *104*, 129–137.
73. Laslandes, M.; Patterson, K.; Pellegrino, S. Optimized actuators for ultrathin deformable primary mirrors. *Appl. Opt.* **2015**, *54*, 1559. [CrossRef] [PubMed]
74. Liu, D.; Wang, Z.; Liu, J.; Tan, J.; Yu, L.; Mei, H.; Zhou, Y.; Zhu, N. Performance analysis of 1-km free-space optical communication system over real atmospheric turbulence channels. *Opt. Eng.* **2017**, *56*, 106111. [CrossRef]
75. Li, M.; Cvijetic, M. Coherent free space optics communications over the maritime atmosphere with use of adaptive optics for beam wavefront correction. *Appl. Opt.* **2015**, *54*, 1453–1462. [CrossRef]
76. Wright, M.W.; Morris, J.F.; Kovalik, J.M.; Andrews, K.S.; Abrahamson, M.J.; Biswas, A. Adaptive optics correction into single mode fiber for a low Earth orbiting space to ground optical communication link using the OPALS downlink. *Opt. Express* **2015**, *23*, 33705–33712. [CrossRef] [PubMed]
77. Thompson, C.A.; Kartz, M.W.; Flath, L.M.; Wilks, S.C.; Young, R.A.; Johnson, G.W.; Ruggiero, A.J. Free space optical communications utilizing MEMS adaptive optics correction. In Proceedings of the Free-Space Laser Communication and Laser Imaging II, Seattle, WA, USA, 7–11 July 2002.
78. Majumdar, A.K.; Eaton, F.D.; Jensen, M.L.; Kyrazis, D.T.; Schumm, B.; Dierking, M.P.; Shoemake, M.A.; Dexheimer, D.; Ricklin, J.C. Atmospheric Turbulence Measurements over Desert site using groundbased instruments, kite/tethered-blimp platform and aircraft relevant to optical communications and imaging systems: Preliminary Results. In Proceedings of the Free-Space Laser Communications VI, San Diego, CA, USA, 13–17 August 2006.
79. Cornelissen, S.; Hartzell, A.; Stewart, J.; Bifano, T.; Bierden, T. MEMS Deformable Mirrors for Astronomical Adaptive Optics. In Proceedings of the Adaptive Optics Systems II, San Diego, CA, USA, 27 June–2 July 2010.

photonics

Article

Spektr–UF Mission Spectrograph Space Qualified CCD Detector Subsystem

Andrey Shugarov *,† and Mikhail Sachkov †

Institute of Astronomy of the Russian Academy of Sciences, 119017 Moscow, Russia; msachkov@inasan.ru
* Correspondence: shugarov@inasan.ru
† These authors contributed equally to this work.

Abstract: Spektr–UF (World Space Observatory Ultraviolet, WSO-UV) is a Russian-led international collaboration aiming to develop a large space-borne 1.7 m Ritchey–Chretién telescope with science instruments to study the Universe in ultraviolet wavelengths. The WSO-UV spectrograph (WUVS) consists of three channels: two high-resolution channels (R = 50,000) with spectral ranges of 115–176 nm and 174–310 nm, and a low-resolution (R = 1000) channel with a spectral range of 115–305 nm. Each of the three channels has an almost identical custom detector consisting of a CCD inside a vacuum enclosure, and drive electronics. The main challenges of the WUVS detectors are to achieve high quantum efficiency in the FUV-NUV range, to provide low readout noise (3 e⁻ at 50 kHz) and low dark current (<12 e⁻/pixel/hour), to operate with integral exposures of up to 10 h and to provide good photometric accuracy. A custom vacuum enclosure and three variants of a custom CCD272-64 sensor with different UV AR coatings optimised for each WUVS channel were designed. The enclosure prevents contamination and maintains the CCD at the operating temperature of −100 °C, while the temperature of the WUVS optical bench is +20 °C. A camera electronics box (CEB) that houses the CCD drive electronics was developed. Digital correlated double sampling technology allows for extremely low readout noise and flexible frequency for normal and binned pixel readout modes. This paper presents the WUVS detector design drivers, methods for extending the service life of the CCD sensors working with low signals in a space radiation environment and the key calculated parameters and results of the engineering qualification model qualification campaign.

Keywords: ultraviolet; WSO–UV; CCD; quantum efficiency; gradient anti-reflection coating

Citation: Shugarov, A.; Sachkov, M. Spektr–UF Mission Spectrograph Space Qualified CCD Detector Subsystem. *Photonics* **2023**, *10*, 1032. https://doi.org/10.3390/photonics10091032

Received: 30 July 2023
Revised: 31 August 2023
Accepted: 5 September 2023
Published: 8 September 2023

1. Introduction

The Spektr–UF (Spectrum–UV, World Space Observatory–Ultraviolet, WSO–UV) observatory is designed to study the Universe in 115 nm to 310 nm ultraviolet (UV) wavelengths. A 170 cm aperture space telescope has two main science instruments—spectrographs and field cameras [1–3].

The main scientific objectives of WSO-UV are: evolution of the early Universe, star formation, galaxy evolution, extrasolar planets and their atmospheres, stellar physics, compact bodies, astrophysical accretion processes, etc. [4].

WUVS (WSO-UV Spectrograph) consists of two high-resolution spectrographs (R = 50,000) covering the far-UV range of 115–176 nm (Vacuum UltraViolet Echellé Spectrograph, VUVES) and the near-UV range of 174–310 nm (UltraViolet Echellé Spectrograph, UVES). The third WUVS channel is a long-slit spectrograph (LSS) that has two sub-channels operated with a single detector, the total coverage wavelength range being 115–305 nm (Figure 1). WUVS high-resolution FUV and NUV channels have a classical Echellé optical design (Figure 2), and the low-resolution channel has a Rowland optical design (Figure 3) with two subchannels for FUV and NUV, both operated with one detector. The VUVES and UVES channel optical schemes are optimised to utilize the whole photosensitive surface of the CCD detectors [5]. The LSS channel has two long narrow spectra located along the bottom long side of the detector while the remaining part of the CCD remains unusable [6].

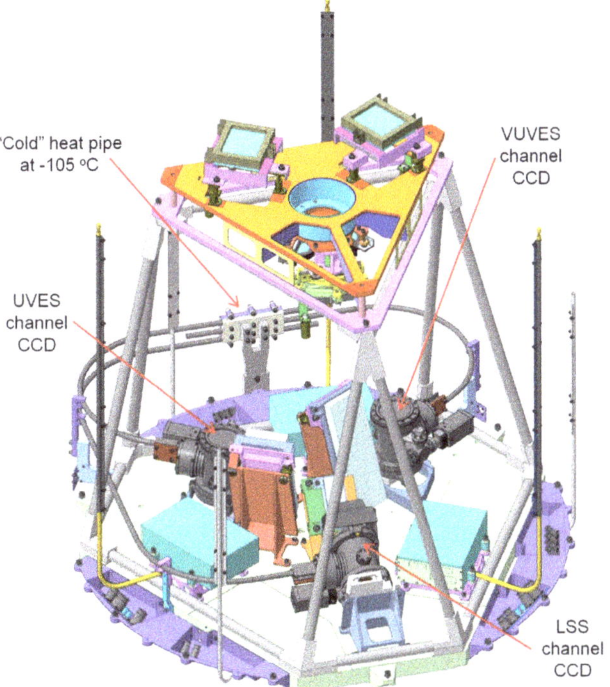

Figure 1. WUVS spectrograph with three CCD detectors.

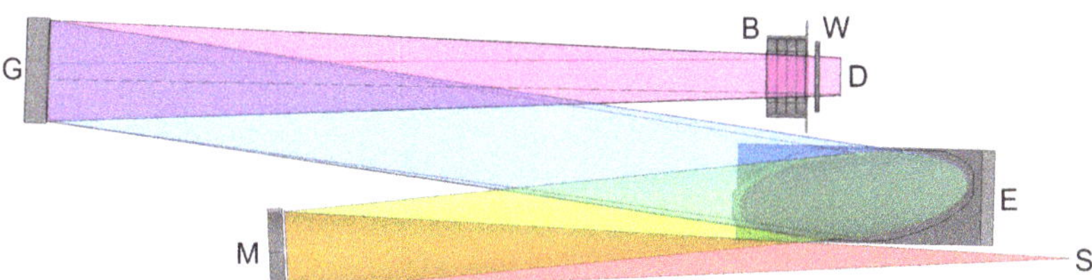

Figure 2. UVES optical layout: S—entrance slit, M—collimator mirror, E—echellé grating, G—cross-dispersion grating, W—entrance window of CCD, D—CCD surface, B—detector baffle.

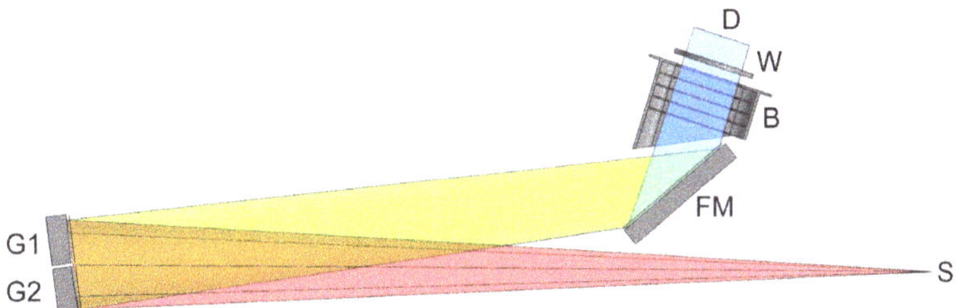

Figure 3. LSS Layout: S—entrance slit, G1—toroidal grating of NUV branch, G2—toroidal grating of FUV branch, FM—flat mirror, W—entrance window of CCD, D—CCD surface, B—detector baffle.

The spectrographs' slits have a size of 1 arcsec, while the telescope PSF is three times smaller; therefore, all spectrographs operate in a slit-less mode.

WSO-UV will operate on geosynchronous orbit. One of the advantages of the WUVS spectrograph over similar HST spectrographs operating on the low Earth orbit is the possibility of having long uninterrupted observations of up to 10 h or even longer.

Each WUVS channel has a mechanical shutter close to the slit to block the light from the telescope to allow for CCD readout, which could be as long as 2 min at the lowest CCD readout speed. Another goal of the shutter is to obtain accurate CCD dark current calibration images, which may take up to several hours.

The WUVS does not have an internal lamp for wavelength calibration; therefore, well-known stars are the only source used to calibrate the spectrograph.

Each long WUVS exposure should be split into a series of short sub-exposures to implement cosmic ray rejection algorithms and dithering (if needed) and to compensate for possible minor spectra displacement on the CCD due to the spectrograph thermal stability and spacecraft pointing system drift. A sub-exposure should be no longer than 10–20 min while the total observation time may reach 10 h.

T-170M telescope fine guidance sensors (FGSs) were mounted close to the spectrographs' entrance slits to ensure the proper pointing and stabilisation of the telescope in order to keep the star at the center of the slit. For each 10 min subexposure, the FGS provides information about the actual position of the star within the slit. In case of unexpected problems with the telescope pointing stability during WUVS long exposure (up to 10 h), this information will be used to prevent degradation of the spectral resolution.

To cool down detectors, the WSO–UV spacecraft provides an isolated cold heat pipe connected to a large external radiator. This system maintains the temperature of the CCD enclosure's cold finger at around -105 °C. Low-power heaters are installed on the cold finger and near the CCD for precise temperature stabilisation.

The WUVS data processing unit uses lossless compression algorithms while co-adding images will be a part of the science data processing pipeline at the ground scientific center.

With this article, we want to give visibility to the current progress of developing and qualifying world-class space-qualified custom UV detectors for a large WSO-UV space mission.

The main aim is to present general scientific and technological drivers and key design solutions. Special attention was paid to the problem of CCD operation with very low signals in a space radiation environment and methods for extending the CCD service life. At the end of the paper, key measured parameters of the engineering qualification model are presented.

We hope that our experience in UV technologies will help astronomers to develop other new UV instruments.

2. WUVS Detector System General Design

The WUVS detectors should provide high sensitivity in a UV, high dynamic range and low dark current.

The previous design of the WSO-UV spectrograph was based on micro channel plate (MCP) detectors. At that time, it was the only way to achieve high sensitivity in the UV range [7]. The disadvantages of the MCP detectors are limited local and global count rates, lifespan and resolution. MCP detectors limit the maximum signal-to-noise ratio (SNR) of the spectra by 30 approximately. To eliminate the lifespan problem, it was suggested to install a spare detector in each channel as well as a corresponding remotely controlled folding mirror to activate the spare detector.

Thanks to the improvements in CCD technology, especially in UV coatings and readout noise, replacing an MCP detector by a CCD is made possible.

Before choosing the CCD technology to be used in the WUVS instead of MCP, the use of a custom design scientific complementary metal–oxide–semiconductor (sCMOS) was briefly considered. The main possible advantages of the sCMOS are a potentially lower readout noise (down to 1 e$^-$ RMS) and the absence of charge transfer degradation due to the radiation. On the other hand, there are several disadvantages, such as an uncertainty of the lowest achievable dark current even with deep cooling, some problems with photometric accuracy, a lack of sCMOS heritage in space and the overall developing risks related to a new custom UV-optimised sCMOS. Because of these reasons, a more conservative approach to using mature CCD technology was selected.

The main advantages of the CCD detectors compared to MCP are high sensitivity in the NUV range, a large format and geometrical stability, high dynamical range and the possibility to obtain UV spectra with a high SNR from relatively bright targets without the risk of damaging the detector.

In the FUV range (120–200 nm), the sensitivity of the CCD is decreased due to the absence of suitable space-qualified CCD AR coatings operated in this range. Therefore, for weak sources, the CsI MCP operated in photon-counting mode still has better sensitivity than CCD in FUV.

The WUVS detector subsystem was designed based on the Institute of Astronomy RAS (INASAN) statement of work and produced by Teledyne e2v and RAL space companies, and the main components are shown in Figure 4. It consists of three channels (Figure 5), each optimised for a specific range of wavelengths. For design reasons, all three detectors should be identical, except for minor changes such as anti-reflection coating on the CCD, the selection of the readout area of the CCD and active output amplifiers. Three variants of a custom CCD272-64 sensor with different UV AR coatings, each optimised for a specific WUVS channel, were designed by the team.

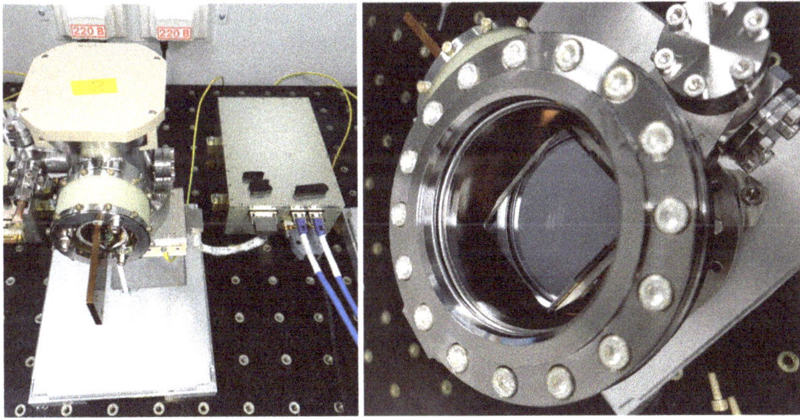

Figure 4. The engineering qualification model (EQM) of the enclosure with interconnection module (ICM) and camera electronics box (CEB).

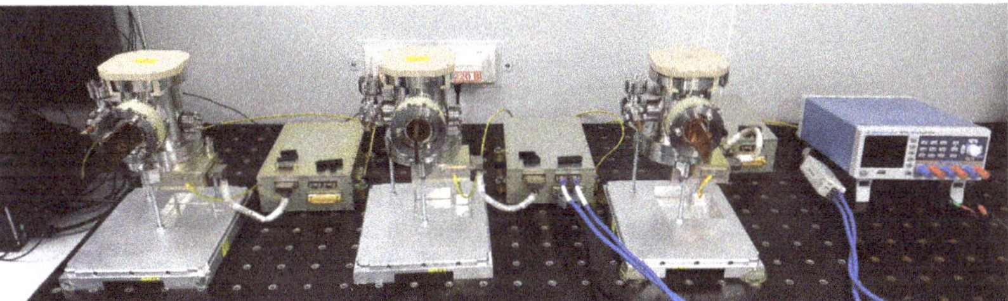

Figure 5. Three WUVS detector assemblies ready for installation.

The CCD quantum efficiency (QE) in NUV and especially in FUV is very sensitive to the CCD surface molecular contamination. The WUVS optical assembly and T-170M telescope scientific instrument compartment are not clear enough to allow for the operation of an open-face cooled CCD in the UV range. To prevent contamination of the CCD and maintain the CCD operating temperature at $-100\ °C$, a custom vacuum enclosure with an input window made of MgF_2, identical for all WUVS channels, was designed. The drawbacks of this design approach are the light losses on the window and additional thermal load on the CCD cooling system. The MgF_2 sample demonstrated a transmittance of 67% at 120 nm and 94% at 300 nm wavelengths.

The CCD enclosures and CCD drive electronics are located inside the WUVS optical assembly because the distance between the CCD and electronics should be minimised.

To prevent the WUVS optical bench being heated by CCD drive electronics, electronic boxes are mounted on the thermal inserts and each electronic box has a heat pipe to evacuate its heat (about 10 W) to the WUVS wall that acts as a large common radiator. The WUVS optical scheme will be aligned at a $+20\ °C$ temperature and the WUVS optical assembly should be stabilised at this temperature after the launch; therefore, the CCD enclosure mounting feet will also be at $+20\ °C$.

Mechanical shutters are located on the top of the WUVS optical assembly slightly above the spectrographs' entrance slits. Their main goals are to block the light during the CCD readout procedure and to obtain CCD bias and dark calibration frames [8].

The WUVS spectral resolution element (resel) has a size of about 70 μm. It was a tradeoff in the WUVS custom CCD pixel size between 12 μm and 24 μm. For a small CCD pixel, the WUVS resel has a size of six pixels and, for a large pixel, three pixels. The choice is not so obvious, especially if the CCD operates in a space environment. The main differences with ground-based spectrographs are the presence of cosmic rays on each frame, quick degradation of CCD charge transfer efficiency (CTE) with time, especially for low-level signals, and quick general degradation of the CCD cosmetics.

In the case of having a large pixel with a notch channel, the low-level signal CTE for a combined signal within the resel will potentially be higher and the CCD readout noise contribution will potentially be lower.

With a small pixel, these two factors act in opposite (negative) directions. But small pixels also have advantages in operating the CCD in a space environment. Firstly, it is possible to reject cosmic rays more gently, which is a large problem for WUVS long-exposure images with a close to zero background. Therefore, increasing the subexposure time slightly might be considered, which leads to a decrease in the amount of CCD readouts within the long exposure, and it will reduce the CCD readout noise contribution. The other benefit of a small pixel is to have a higher tolerance for the appearance of bad columns and hot pixels after 5–10 years of operation in space. Indeed, if the resel has a size of 6×6 pixels, the presence of several remarkably defected pixels may be tolerated. And, finally, a small pixel always provides the best spectra sampling.

WUVS science and technical teams have decided to choose the small pixel CCD option. It will maximize the scientific performance during the first few years of the WSO-UV mission when the CCD will not be seriously degraded by radiation. During the second half of the WUVS life in space with a degraded CCD, it provides the best WUVS performance for observing medium and bright sources.

There is a tradeoff between using a classical single-end CCD video signal output and pseudo-differential video output. When using both real and dummy CCD outputs, there is better common-mode noise rejection and reduced susceptibility to electromagnetic interference and interference from the CCD clocking. To minimise the WSO-UV mission risk, pseudo-differential video output was selected.

Each WUVS detector has a small tungsten lamp to illuminate the CCD in optical wavelength. Before each sub-exposure, this system provides a preflash of the CCD at a level of about $10\ e^-/$pixel to fill the traps to improve charge transfer efficiency. The second purpose of the lamps is to obtain more or less uniform flat field images at a level of a few thousand electrons per pixel. This helps for the quick evaluation of the CCD pixel health as it identifies new bad or hot pixels because of the quick CCD degradation in space. Due to the illumination system operating in an optical wavelength, these flat fields cannot be used to calibrate scientific UV spectra.

CCD detectors are very sensitive to optical wavelength radiation; therefore, all internal elements of the WUVS have a black coating optimised for visible light suppression. Some residual light from the Sun may enter the telescope's scientific instrument compartment through its side walls via small holes and gaskets. The WUVS optical assembly was designed to be light-tight, so, together with the telescope, there is double-layer protection for the WUVS detectors from Sun radiation.

Because of the compact and crowd optical design, the WUVS has a limited amount of space to install internal baffles and diaphragms. Each detector has a baffle in front of the CCD to reject out-of-band spectral orders; nevertheless, VUVES and UVES detectors remain partially open for scattered light inside the WUVS. The LSS channel has an additional baffle to avoid direct illumination of the detector from the grating.

3. The WUVS CCD

The WUVS CCD272-64 is a derivation of CCD273 used in the ESA's EUCLID mission; therefore, we can classify the WUVS CCD as a semi-custom device [9,10]. The main modifications are the higher gain of the output amplifiers to achieve better noise performance and back surface optimisation to operate in the UV range.

WUVS semi-custom CCD272-64 is a back-thinned back-illuminated two-phase device pixel array, the format is 4096 columns by 3112 rows and the pixel size is 12 μm square. Depending on the WUVS scientific observation program, pixels can be combined in 2×2 groups to give an effective 24 μm pixel size for minimisation of the readout noise within the resolution element of the spectrum and/or to decrease the readout time.

CCD272-64 has two serial registers with two output amplifiers each, but the drive electronics only have two analog circuits to read the CCD. Because of this restriction, the CCD readout schemes are different for different channels in order to achieve the best performance.

The LSS version has been designed to read only a half of the chip using one serial register with its two (left and right) output amplifiers. The VUVES and UVES CCDs have been designed to read the whole chip using two serial registers, but each register reading uses only one amplifier.

One of the key scientific design drivers for the WUVS CCD is to optimise its performance in order to operate at very-low-level signals down to a few electrons per pixel with a very long integration time of up to 10 h. For this purpose, the CCD operation temperature has been chosen at $-100\ °C$. At this temperature, an increase of $1\ °C$ in CCD temperature will result in about a 1.5 e$^-$/pixel/hour increase in dark current. The CCD uses a low-voltage process to minimise power consumption both on the device and in the drive electronics.

To implement cosmic ray rejection algorithms, the WUVS very long exposure will be split into a series of short sub-exposures of about 10 min each. This naturally leads to an increase in the readout noise.

One of the challenges of the WUVS CCD is to achieve a detector lifespan of up to 10 years while operating at the geosynchronous orbit. The radiation degradation of the CCD leads to the increasing of a dark current, number of cosmetic defects and, the most critical parameter for the WUVS spectrograph, the degradation of charge transfer efficiency (CTE) for low-level signals on a zero background. To prolong the long-term stability of the WUVS CCD detector performance in a radiation environment, several techniques and design features are used:

- Shielding of the CCD by enclosure and other mechanical elements of the WUVS and spacecraft;
- Using the Teledyen e2v "radiation hard" process;
- Putting the CCD parallel transfer direction perpendicular to the dispersion of each WUVS channel. This design choice helps to minimise degradation of the spectral

resolution due to CTE, as parallel transfer CTE is higher than serial transfer CTE, but the consequence of this choice is a higher degradation of spectral line intensity because of parallel transfer CTE, e.g., smearing and charge losses;

- Putting the LSS channel FUV spectrum directly next to the serial register to minimise the number of parallel transfers, and the NUV spectrum next to the FUV, because, typically, the FUV spectrum is weaker than the NUV spectrum;
- Using a split frame transfer CCD readout mode to minimise the number of parallel transfers for VUVES and UVES channels;
- Using an optical preflash system to illuminate the CCD at a level of about $10\,\mathrm{e}^-$ for partially filling charge traps;
- Heating the CCD up to about $+20\,^\circ\mathrm{C}$ periodically for annealing.

For the WUVS science program, the best possible readout noise is more important than the full well capacity, so the conversion factor of the CCD output amplifiers increased to $7\,\mu\mathrm{V}/\mathrm{e}^-$, and, accordingly, the maximum signal in a pixel decreased to $30{,}000\,\mathrm{e}^-$. Such a full well capacity would not limit the ability to obtain high SNR spectra, because one resolution element has a size range from 3×3 to 6×6 pixels, and all exposures will be split into sub-exposures. The CCD gain is $0.8\,\mathrm{e}^-/\mathrm{ADU}$ in a 16 bits format. To improve common mode noise suppression, a differential output architecture of the CCD output circuits is used.

One of the problems of the CCD operation in space is general degradation of the CTE. Additionally, the low-level signal CTE with zero background degradation needs to be considered. For the WUVS, the second aspect is the most important because typical spectra have a very low intensity while the spaces between echelle spectral lines are completely dark.

Low-level signal CTE with zero background degradation occurs because of the radiation degradation of the silicon, which causes many small traps to appear. If the traps are empty, they can temporarily catch several electrons during integration or CCD readout; therefore, some low-level signal details on the image may smear or even completely disappear. After a few years of operation in space, the low-level signal CTE with zero background may be more than an order of magnitude lower than CTE for a high-level signal with non-zero background.

The CCD272-64 pixel structure forms a potential well within the pixel that acts as a "notch" channel. While the amount of collecting electrons in the pixel is small, they are concentrated in the center of the pixel and, therefore, the probability to be captured by traps is reduced. During an on-ground validation campaign, WUVS CCD272-64 was tested before and after irradiation to measure both standard CTE and low-level signal CTE with zero background.

One of the main challenges for the WUVS CCD is to provide the best possible quantum efficiency over a very challenging spectral range from 120 nm to 320 nm. UV radiation is the most difficult to detect with CCD detectors because the absorption depth in silicon is the smallest at these wavelengths. The CCD must be thinned to minimise any dead layer at the back surface and a different kind of treatment of the back surface may be implemented to further improve the quantum efficiency. Teledyne e2v has chosen to use a laser annealing treatment of the WUVS CCD. The drawback of this technique is the presence of a fixed pattern variation of QE in the UV range.

A novel process is used whereby the CCD has an anti-reflection HfO coating of different thicknesses over a part of the image area to match the operation wavelength at each point of the detector with the spectrograph's dispersion. The coating should be removed for the shortest wavelengths (115–180 nm), where the presence of the coating would degrade the quantum efficiency. Therefore, the VUVS channel's variant of the CCD has no coating at all, the UVES channel's CCD is fully coated with gradient AR coating and the LSS channel's CCD has two sections: an uncoated area for the FUV spectrum and a gradient AR coating area for the NUV spectrum.

4. The Hermetic CCD Enclosure

Due to the CCD cold surface being very sensitive to contamination, a custom cryostat (enclosure) was designed and manufactured to maintain a very clean environment for the CCD. The enclosure was designed according to ultra-high vacuum standards. The purposes of the enclosure are to:

- Protect the CCD surface from any kind of contamination;
- Provide an additional radiation shielding for the CCD;
- Provide a UV transparent window for illuminating the CCD;
- Provide a thermal path for cooling the CCD.

One of the challenges of the enclosure is to guarantee the absence of any condensation on the cold-sensitive CCD surface, which may impact the QE, for 9 years of operation. According to our calculation, the partial pressure of water within the enclosure should be maintained below 1.6×10^{-5} mbar at $-100\ °C$. Before the final assembling of the WUVS flight model, all enclosures will pass through a pump and bake procedure again and be filled up with ultra clean Ar.

The enclosure input window is maintained at $+22\ °C$ by utilising an additional dedicated heater in order to minimise contamination on the window side facing towards the WUVS optical assembly. A spacecraft battery directly powers the CCD window heater to ensure the protection of the window against contamination. This is essential when the WUVS remains unpowered during the launch and commissioning phases and if the spacecraft enters a safe mode operation.

To prolong the detectors' lifespan in a radiation environment, the annealing procedure will be carried out monthly. For this purpose, a 500 W heater is installed on the external radiator for temporal deactivation of the CCD passive cooling system.

5. CCD Drive Electronics

The camera electronics box (CEB) was designed according to the INASAN statement of work and built by RAL Space, STFC (UK) based on its heritage of the production of CCD camera electronics for several projects, such as STEREO SCIP/HI, SDO AIA/HMI, GOES-R SUVI. The CEB is connected to the WUVS by a redundant SpaceWire link that is used for CCD image transmission, telemetry and the commanding of the CEB. The WUVS also provides redundant 27 V power for the CEB.

Three identical CEBs are located inside the optical-mechanical unit of the WUVS close to the dedicated CCD enclosures to minimise the analog cable lengths. The CEB power consumption is 11 W per channel and the heat produced by each CEB is evacuated by a dedicated heat pipe. The CEB box dimensions are $150 \times 220 \times 65$ mm, the weight is 3.2 kg and the wall thinness is 6 mm to improve protection against radiation (Figure 6).

The CEB houses three camera electronic cards: a power supply card, bridge card and CCD camera card. The CCD camera card provides the majority of the functionality within the CEB system, including the SpaceWire interface, video digitisation, CCD bias voltage generation and CCD clock driving.

A key technology implemented in the CEB is a digital correlated double sampling. It enables several readout speeds of 50 kHz, 100 kHz and 500 kHz, each in normal and binned pixel readout modes, with each mode optimised to have the lowest possible readout noise. The CEB 14-bit ADCs continuously sample the entire CCD video signal at a rate of 25 MHz, including settling and clamp periods. At a pixel frequency of 50 kHz, the CEB obtains about 160 valid samples to measure each pixel's signal. The average pixel value is calculated with high precision and then rounded to 16 bits to avoid an increase in system noise due to the quantisation error of 14 bits format data.

An additional interconnect module was also designed, which is mounted directly on the CCD video signal connector on the bottom of the enclosure. It provides the CCD video signal amplification to reduce the system noise and additional filtration of the CCD voltages. The interconnect module is connected to the CEB by a short electrical harness.

Two types of harnesses (VUVES/UVES and LSS) were developed in order to keep them as short as possible within the WUVS mechanical constraints.

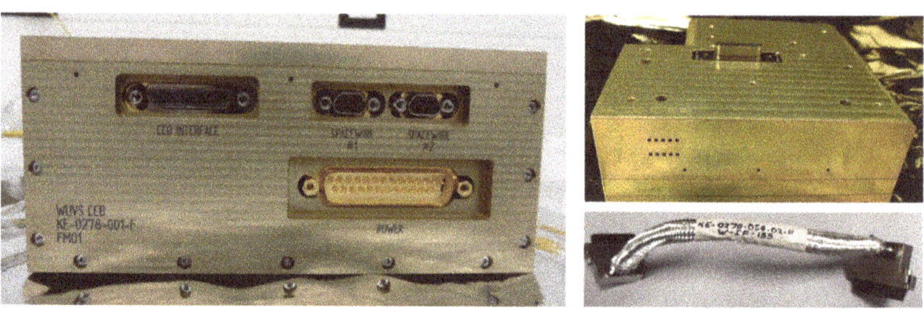

Figure 6. Flight models of the camera electronic box (CEB), interconnection module (ICM) and analog video cable.

6. Qualification Campaign Main Results

The assessment of quantum efficiency in FUV and NUV ranges for modern CCD and CMOS is still limited due to the complexity of the necessary equipment, especially the source of monochromatic, well-regulated and uniform UV radiation.

WUVS CCD reflectivity was measured to predict QE in the UV range using Teledyne e2v's own technique. In 2019, a direct QE measurement of the LSS channel engineering qualification model CCD was carried out on the metrological station "Kosmos" at the Budker Institute of Nuclear Physics. As a source of UV light, a synchrotron radiation from the VEPP-4M storage ring was used with a MgF_2 filter to cut off high-energy radiation and with the UV monochromator to select the working wavelength. All the measurements were performed at the CCD nominal working temperature of $-100\ ^\circ$C. In addition to the QE measurement, the efficiency of the CCD enclosure cooling system was checked.

The LSS device has both an uncoated region and a gradient AR-coated region (Figure 7). In accordance with the WUVS spectrograph dispersion, the AR coating thickness is optimised for 180 nm on the left side of the CCD and for 310 nm on the right side.

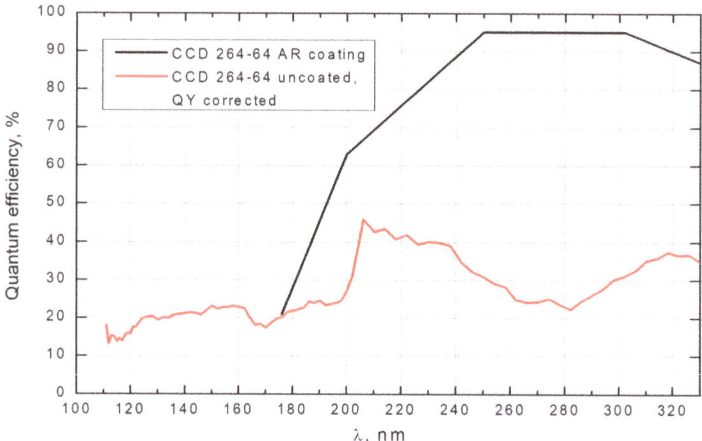

Figure 7. EQM LSS enclosure quantum efficiency measurement results for the uncoated and gradient AR-coated regions of the CCD.

The QE was measured in the 111–320 nm range for the uncoated region and at 10 points of different thicknesses for the AR-coated region. The measurement results show a good compliance with our prediction based on a reflectivity method. Our measurements show up to a four times QE increase due to the AR coating, which is a great improvement in the CCD responsivity in UV. For future FUV missions, it is a great opportunity to develop new kinds of space-qualified anti-reflection coatings to operate below 180 nm.

Enclosure cooling system verification results confirm that the temperature difference between the CCD and the "cold finger" is 5 °C.

During the test of the FM CEB, a readout noise of about 2.6 e⁻ RMS at 50 kHz and about 3 e⁻ RMS at 100 kHz was measured. The linearity is better than 0.2% and the cross-talk between the channels is less than 29 ppm. The CEB linearity is about 0.1% within the full CCD input voltage range. The CEB electronics contribution to the crosstalk between two CCD readout channels is about 40 ppm, and was measured using a CCD signal emulator.

7. Summary and the Current Status

The WUVS detector has a custom design to maximise the scientific efficiency of the WSO-UV mission. Several methods for extending the service life of the CCD detectors in a space radiation environment were implemented.

The WUVS spectrograph CCD detector system has been developed based on the Teledyne e2v and RAL space expertise, considering specific requirements of the WSO-UV mission to operate with very low signals for a long exposure time.

The critical design review (CDR) of the camera electronics box and enclosure were passed in 2016 and 2019, respectively, while, between 2020 and 2021, the WUVS detector subsystem successfully passed the qualification campaign and the key parameters were verified. In addition to the Teledyne e2v factory validation and verification campaign, the quantum efficiency of the LSS EQM CCD and the cooling system efficiency were measured. In 2019, four CEB FM (camera electronics box flight model) units were successfully delivered to Russia and passed incoming inspection.

According to the qualification campaign results, the current WUVS detector design is mature enough to assemble the WUVS flight model.

Author Contributions: Investigation, writing—original draft preparation, A.S.; writing—review and editing, supervision, M.S. All authors have read and agreed to the published version of the manuscript.

Funding: This research received no external funding.

Institutional Review Board Statement: INASAN expertise on the possibility of publication N98/23 from 9 August 2023.

Data Availability Statement: Data available on request due to restrictions.

Acknowledgments: We express our gratitude to the Teledyne e2v and RAL Space companies for the production of the custom CCD272-64, custom enclosures and low readout noise electronics for the WSO-UV space project, and to the Lebedev Physical Institute for their continued project support and for organising the verification campaign. Without hard work and significant contributions from the whole WUVS project team the development of the WUVS detectors would have not been possible, in addition special thanks goes toward the BINP team for the provision of the facilities for characterisation of the CCD in UV.

Conflicts of Interest: The authors declare no conflict of interest.

References

1. Sachkov, M.; Gómez de Castro, A.I.; Shustov, B.; Sichevsky, S.; Shugarov, A. World Space Observatory: Ultraviolet mission: Status 2022. In Proceedings of the SPIE Astronomical Telescopes + Instrumentation, Montreal, QC, Canada, 17–23 July 2022; Volume 12181. [CrossRef]
2. Shustov, B.M.; Sachkov, M.E.; Sichevsky, S.G.; Arkhangelsky, R.N.; Beitia-Antero, L.; Bisikalo, D.V.; Bogachev, S.A.; Buslaeva, A.I.; Vallejo, J.C.; Gomez de Castro, A.I.; et al. WSO-UV Project: New Touches. *Sol. Syst. Res.* **2021**, *55*, 677–687. [CrossRef]
3. Shustov, B.; Gómez de Castro, A.I.; Sachkov, M.; Vallejo, J.C.; Marcos-Arenal, P.; Kanev, E.; Savanov, I.; Shugarov, A.; Sichevskii, S. The World Space Observatory Ultraviolet (WSO-UV), as a bridge to future UV astronomy. *Astrophys. Space Sci.* **2018**, *363*, 62.

4. Boyarchuk, A.A.; Shustov, B.M.; Savanov, I.S.; Sachkov, M.E.; Bisikalo, D.V.; Mashonkina, L.I.; Wiebe, D.Z.; Shematovich, V.I.; Shchekinov, Y.A.; Ryabchikova, T.A.; et al. Scientific problems addressed by the Spektr-UV space project (world space Observatory—Ultraviolet). *Astron. Rep.* **2016**, *60*, 1–42. [CrossRef]

5. Sachkov, M.; Panchuk, V.; Yushkin, M.; Fatkhullin, T. Optical design of WUVS instrument: WSO-UV spectrographs. *Space Telesc. Instrum. 2016 Ultrav. Gamma Ray* **2016**, *9905*, 1030–1035.

6. Panchuk, V.; Yushkin, M.; Fatkhullin, T.; Sachkov, M. Optical layouts of the WSO-UV spectrographs. *Astrophys. Space Sci.* **2014**, *354*, 163–168. [CrossRef]

7. Reutlinger, A.; Sachkov, M.; Gál, C.; Brandt, C.; Haberler, P.; Zuknik, K.H.; Sedlmaier, T.; Shustov, B.; Moisheev, A.; Kappelmann, N.; et al. Using the CeSiC material for the WSO-UV spectrographs. *Astrophys. Space Sci.* **2011**, *335*, 311–316. [CrossRef]

8. Shugarov, A.; Savanov, I.; Sachkov, M.; Jerram, P.; Moody, I.; Pool, P.; Turner, P.; Pittock, R.; Kuzin, S.; Waltham, N. UV detectors for spectrographs of WSO-UV project. *Astrophys. Space Sci.* **2014**, *354*, 169–175. [CrossRef]

9. Shugarov, A.S.; Sachkov, M.; Bruce, G.; Robbins, M.; Walker, A.; Waltham, N.; Clapp, M.; Salter, M.; Patel, G.; Kuzin, S.; et al. WSO-UV mission WUVS instrument FUV-UV CCD detectors qualification campaign main results. *Int. Conf. Space Opt. 2020* **2021**, *11852*, 2223–2234.

10. Hayes-Thakore, C.; Spark, S.; Pool, P.; Walker, A.; Clapp, M.; Waltham, N.; Shugarov, A. Sensor system development for the WSO-UV (World Space Observatory-Ultraviolet) space-based astronomical telescope. *Sensors Syst. Next-Gen. Satell. XIX* **2015**, *9639*, 212–221.

Article

Benefits of Intelligent Fuzzy Controllers in Comparison to Classical Methods for Adaptive Optics

Victor Costa [†] and Wesley Beccaro *[,†]

Department of Electronic Systems Engineering, Polytechnic School, University of São Paulo, São Paulo 05508-010, Brazil; costa.victor@usp.br
* Correspondence: wesley@lme.usp.br
[†] These authors contributed equally to this work.

Abstract: Adaptive Optics (AO) systems have been developed throughout recent decades as a strategy to compensate for the effects of atmospheric turbulence, primarily caused by poor astronomical seeing. These systems reduce the wavefront distortions using deformable mirrors. Several AO simulation tools have been developed, such as the Object-Oriented, MATLAB, and Adaptive Optics Toolbox (OOMAO), to assist in the project of AO. However, the main AO simulators focus on AO models, not prioritizing the different control techniques. Moreover, the commonly applied control strategies in ground-based telescopes are based on Integral (I) or Proportional-Integral (PI) controllers. This work proposes the integration of OOMAO models to Simulink to support the development of advanced controllers and compares traditional controllers with intelligent systems based on fuzzy logic. The controllers were compared in three scenarios of different turbulence and atmosphere conditions. The simulations were performed using the characteristics/parameters of the Southern Astrophysical Research (SOAR) telescope and assessed with the Full Width at Half Maximum (FWHM), Half Light Radius (HLR), and Strehl ratio metrics to compare the performance of the controllers. The results demonstrate that adaptive optics can be satisfactorily simulated in OOMAO adapted to Simulink and thus further increase the number of control strategies available to OOMAO. The comparative results between the MATLAB script and the Simulink blocks designed showed a maximum relative error of 3% in the Strehl ratio and 1.59% in the FWHM measurement. In the assessment of the control algorithms, the fuzzy PI controller reported a 25% increase in the FWHM metrics in the critical scenario when compared with open-loop metrics. Furthermore, the fuzzy PI controller outperformed the results when compared with the I and PI controllers. The findings underscore the constraints of conventional control methods, whereas the implementation of fuzzy-based controllers showcases the promise of intelligent approaches in enhancing control performance under challenging atmospheric conditions.

Keywords: adaptive optics; intelligent controllers; instrumentation; simulation; wavefront sensors; point-spread functions

Citation: Costa, V.; Beccaro, W. Benefits of Intelligent Fuzzy Controllers in Comparison to Classical Methods for Adaptive Optics. *Photonics* **2023**, *10*, 988. https://doi.org/10.3390/photonics10090988

Received: 15 June 2023
Revised: 13 August 2023
Accepted: 28 August 2023
Published: 30 August 2023

1. Introduction

Telescopes are instruments designed to observe distant celestial bodies. Normally, a few photons from these bodies reach the instrument, which requires the development of an infrastructure for capturing and correcting noise and distortions [1–5].

Atmospheric turbulence is a key factor in the degradation of the image obtained by ground-based optical telescopes. This phenomenon, called seeing, limits the resolution of these instruments. Since the 1950s, AO techniques have been used to mitigate this effect [6]. Atmospheric turbulence can be compensated by geometric modifications of deformable mirrors employing real-time systems and wavefront sensors [7–14].

Typically, the final technology employed to implement AO systems in telescopes demands the use of numerical simulations before the direct integration of the control systems [15]. To support the design of AO systems, several models of atmospheric turbulence and astronomical instruments have been developed over the years. YAO, developed in Yorick, is a classic implementation of AO simulation [16]. Currently, other simulators have shown excellent results, such as Simulation 'Optique Adaptative' with Python (SOAPY) [17], AOtools [18], both developed in Python, and the Object-Oriented, MATLAB and Adaptive Optics Toolbox (OOMAO), which is developed in MATLAB® [19]. OOMAO is a toolbox dedicated to AO systems that uses the object-oriented paradigm and features an atmospheric turbulence generation model proposed by [20]. Since it is numerically more efficient than traditional Monte Carlo models, it allows the simulation of atmospheric turbulence effects on larger telescopes. The OOMAO framework also implements models of source, atmosphere, telescope, wavefront sensor (Shack–Hartmann), and deformable mirror.

OOMAO mainly focuses on adaptive optics models, not prioritizing the different control strategies. Although it is possible to implement alternative strategies, there is a difficulty in adapting control techniques other than those already implemented in the toolbox (i.e., integral controller using the reconstruction obtained from Singular Value Decomposition (SVD), open-loop controller, and modal controller). One way to overcome this limitation is integrating OOMAO with Simulink—MathWorks software focusing on modeling, simulation, and analysis of dynamic systems—which enables the designer to focus on the controller project at a high level of abstraction and maintaining AO models encapsulated in blocks. By employing this approach, it becomes possible to assess the performance enhancements of AO systems by evaluating various control strategies, such as intelligent controllers [21]. It is worth emphasizing that this integration is not a straightforward task. First, it requires the reconfiguration of functions and operations carried out during simulations. Second, despite Simulink being a specialized tool within MATLAB, it can employ a different set of solvers and methods for dynamical simulation. To ensure proper functionality, it is necessary to validate the integration of MATLAB functions in Simulink. This verification confirms the seamless interaction between MATLAB functions and Simulink, ensuring that the desired behavior of the control system is achieved as intended.

This work proposes the integration of the models developed in OOMAO into Simulink and compares classical and intelligent control techniques for adaptive optics. A comparative analysis between the open-loop system and closed-loop systems implemented with Integral (I), Proportional-Integral (PI), and Fuzzy-based controllers is presented. The main contributions of the paper can be summarized as follows:

- The integration and evaluation of OOMAO models into Simulink.
- As proof of concept, intelligent controllers have been developed and compared with traditional ones, illustrating how to assess controllers in scenarios with different turbulence and atmospheric conditions using performance metrics commonly employed in AO projects.
- We have demonstrated the potential of fuzzy logic controllers over conventional controllers in the AO design, reporting some advantages of the intelligent approach such as the wider range of operating conditions, the capacity to include the experience in the controller rules, and the better performance when compared with conventional controllers.

The remainder of the article is organized as follows: Section 2 briefly presents an overview of classical and hybrid fuzzy logic controllers. Section 3 describes the simulation parameters, the adaptive optics quality metrics, and the scenarios considered for the simulations. Section 4 presents the results and discussions of the integration of OOMAO with Simulink and the comparative evaluation of the controllers. Finally, Section 5 describes the conclusion.

2. Overview of Classical and Intelligent Controllers

2.1. PID Controllers

PID controllers are applied to most modern process control systems today [22]. The parallel PID algorithm is described as

$$u(t) = K_p \left[e(t) + \frac{1}{T_i} \int_0^t e(\tau)d\tau + T_d \frac{de(t)}{dt} \right],$$ (1)

where $e(t)$ and $u(t)$ are, respectively, the control error ($e(t) = r(t) - y(t)$, reference set point minus the measured process variable) and the control signal. K_p is the proportional gain, T_i is the integral time, and T_d is the derivative time. Another way to express the PID is in terms of K_p, $K_i = K_p/T_i$, and $K_d = K_p T_d$, respectively, proportional, integral, and derivative gains.

The digital implementation of the PID requires discretizing the integral and derivative actions using approximations (e.g., forward, backward, or trapezoidal approximations). It is possible to discretize the integral action separately from the derivative action. To avoid sharp changes in integral action when sudden changes occur in setpoint or process variable signals, the trapezoidal integration (also known as Tustin or Bilinear transformation) can be applied to the integral action, obtaining

$$u_I(k+1) = u_I(k) + \frac{K_p T_s}{T_i} \left[\frac{e(k+1) + e(k)}{2} \right],$$ (2)

where T_s is the sampling time and k is the discrete time [23]. The I-part can be precalculated right after the PID calculation. The derivative action is usually discretized by the backward approximation technique [23], which results in the expression

$$u_D(k) = \frac{T_d}{T_d + NT_s} u_D(k-1) - \frac{K_p T_d N}{T_d + NT_s} [y(k) - y(k-1)],$$ (3)

considering the ideal derivative action filtered by a first order system with the time constant T_d/N (often N is between 8 and 20), which acts as a low-pass filter, limiting the high-frequency measurement noise. Alternatively, $u_D(k) = \frac{K_p T_d}{T_s}[e(k) - e(k-1)]$ in its most basic form.

The proportional term is implemented by replacing the continuous variables with their sampled versions as $u_P(k) = K_p e(k)$ and the control signal is given by $u(k) = u_P(k) + u_I(k) + u_D(k)$.

Normally, when designing a PID controller, the dynamics of the sensors and actuators are taken into account to determine the response and accommodation times of the closed-loop system. However, in some cases, such as the deformable mirror described by OOMAO, considering the sampling rates normally used in AO systems, the dynamics are fast enough for its accommodation to be considered instantaneous.

2.2. Fuzzy Based Controllers

The PID controller, due to its robustness and simplicity, is widely employed in the industry; however, the algorithm has low performance when applied to non-linear systems. An alternative to PID that introduces nonlinearities to the controller is the incorporation of fuzzy logic strategies. Fuzzy controllers formalize human reasoning in control systems. In general, they are composed of three blocks: fuzzification, fuzzy inference engine, and defuzzification.

The fuzzification changes the input variables into a fuzzy value using different membership functions typically built with sine, trapezoid, and triangle functions. The expert-defined fuzzy rules, commonly expressed as "if-then" statements, are integrated into the fuzzy inference engine. The inference procedure utilizes this knowledge base to determine a reasonable output based on linguistic terms. Finally, the defuzzification converts the output into a crisp value, in the case of the hybrid fuzzy PID controller, the K_p, T_i, and T_d

parameters, for example. Various defuzzification methods, including the center of mass, the center of area, minimum and maximum, and others, can be employed for this purpose [24].

We have proposed a hybrid fuzzy PI architecture, as presented in Figure 1, since it incorporates the fuzzy inference system as a supervisor of the PI controller. The determination of the parameters of the controller (i.e., proportional gain and integral time) relies on the fuzzy logic, employing linguistic parameters and the Mamdani fuzzy inference system.

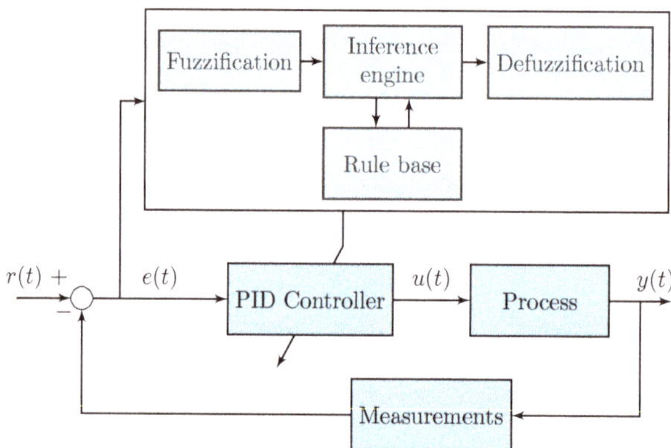

Figure 1. Hybrid Fuzzy PID control block diagram. In detail, the steps of fuzzy reasoning comprise the fuzzification, inference engine, and defuzzification blocks.

2.3. Intelligent Adaptive Optics

There exist some innovative proposals that involve the integration of intelligent techniques, including fuzzy logic, machine learning, and other intelligent algorithms in the context of AO. These strategies aim to enhance the capabilities of AO systems by leveraging intelligent approaches to address the challenges posed by atmospheric turbulence and wavefront distortions.

Within the framework of fuzzy controllers, Ke and Zhang [25] employed a fuzzy control algorithm to adjust the parameters of the PID controller within the deflection mirror loop. Florez-Meza et al. [26] presented a study that focused on the application of fuzzy control in the AO tip-tilt system. The authors demonstrated the effectiveness of the fuzzy approach, showcasing high stability and robustness indices. Regarding machine learning methods, Nousiainen et al. [21] put forth the concept of modeling a closed-loop AO system as a Markov decision process (MDP). They conducted an evaluation of the performance of conventional deep reinforcement learning algorithms on the AO system, considering a telescope diameter equal to 8 m. A similar control strategy was implemented in [27]. Pou et al. [28] introduce a control algorithm based on multi-agent reinforcement learning (MARL). This approach allows the controller to learn a non-linear policy without requiring prior knowledge of atmospheric dynamics. The results obtained, conducted on an 8 m telescope with a 40×40 WFS, demonstrate a performance improvement compared to the integrator baseline. Furthermore, the performance achieved by the MARL-based approach is comparable to a model-based predictive strategy that utilizes a linear quadratic Gaussian controller and has prior knowledge of atmospheric conditions. Guo et al. [29] conducted a comprehensive review of AO focusing on machine learning algorithms. The authors examined the blocks that constitute the AO systems and explored the potential integration of machine learning algorithms into each of these parts.

3. Materials and Methods

Figure 2 illustrates the flow diagram adopted for analyzing the controllers. The AO control loop simulation requires configuring the various parameters of the models, such as the simulation scenarios, which describe general atmospheric conditions, the turbulence intensity, the light source, the telescope characteristics, and the control metrics adopted.

Three distinct simulation scenarios were adopted: typical, worst, and critical cases. They are based on nominal atmospheric conditions observed at the SOAR telescope site, situated at an elevation of 2738 m in Cerro-Pachón, Chile. The SOAR telescope has been designed to operate across a broad spectral range, from the atmospheric cut-off in the blue wavelength (320 nm) to the near-infrared. Simulation scenarios and adaptive optics metrics are described in Sections 3.2 and 3.3.

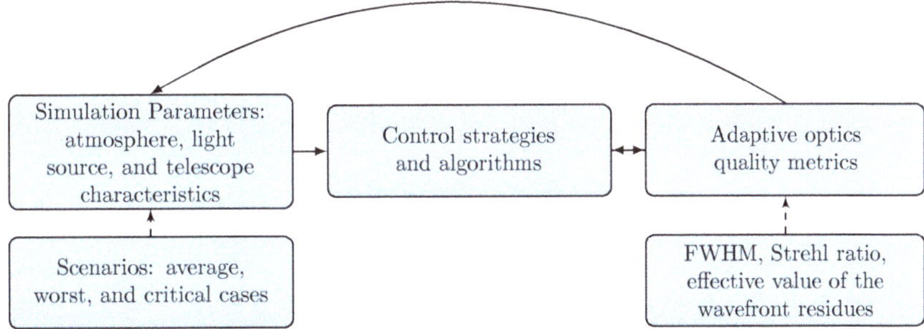

Figure 2. Flow diagram for testing and validating AO controllers simulations.

As mentioned previously, we have applied I, PI, and fuzzy PI controllers instead of PID and fuzzy PID controllers (i.e., disregarding the derivative action). In general, PI and fuzzy PI controllers are more adequate to reduce the steady-state error than their PID versions.

3.1. Simulation Parameters: Atmosphere, Light Source, and Telescope Characteristics

Prior to any simulation in OOMAO, it is essential to create and define certain simulation classes such as light source, atmosphere, telescope, Shack–Hartmann wavefront sensor (WFS), and deformable mirror. Those objects are instantiated by calling their constructors and aggregated through overloaded operators [30].

Considering the atmospheric and instrumentation information extracted from the SOAR telescope, an object from the telescope class was created with the diameter of the primary mirror, D, equal to 4.1 m and field of view of 3×3 arcmin, according to the SOAR adaptive optics module (SAM) and its upgraded counterpart, SAMplus. SAM was developed to compensate for the effects of ground-layer atmospheric turbulence across a wide range of wavelengths, including near-infrared (IR) and visible wavelengths [8]. It was initially developed utilizing the bimorph DM technology, featuring 60 actuators. SAM incorporated a Shack–Hartmann WFS with a 10×10 configuration, utilizing an 80×80 pixel CCD-39. SAMplus represents an upgraded iteration of the SAM that features an increased number of DM and WFS.

The laser guide star (LGS) was projected at an altitude of 7062 m in the U band. The second source simulates, in the I band, the star to be corrected, positioned at infinity. The Shack–Hartmann WFS was instantiated with 16×16 lenslets, with each lenslet consisting of 8×8 pixels. A deformable mirror was created with 17×17 actuators. Both the WFS and the deformable mirror actuators are arranged in square grids. Additionally, a camera with dimensions of 128×128 pixels was utilized. Moreover, in the description of the telescope class, the central obscuration ratio was established at 0.228 (22.8%). The influence of spider structures has been disregarded.

The atmospheric data consists of different scenarios simulating atmospheric turbulence ranging from a typical case to a critical case. The Fried parameter, r_0, ranges from 10 to 14 cm while the wind speed goes from 9 to 40 m/s. The simulations were performed at the I band (approximately 790 nm). The stellar magnitude was set to 3.3.

3.2. Adaptive Optics Quality Metrics

In general, the goal of an AO control loop is to minimize the residual phase of the measured wavefront. Several techniques measure the quality of an adaptive optics control loop, depending on the type of astronomical observation one intends to perform. Among the metrics, this report employs the following methods to analyze and compare the outcomes of various AO control strategies put forth: FWHM of the point-spread function (PSF), which is the diffraction pattern formed by the wavefront; the Strehl ratio, measuring the effective value of the wavefront residues; and the HLR [31]. In the conducted simulations, the Strehl ratio is represented in percentage form. Those metrics were chosen once they represent a mix between metrics commonly used in AO (i.e., FHWM and Strehl ratio) and control metrics that are more readily obtainable in real time (e.g., the effective value from wavefront residues). Specifically, the FWHM was estimated by taking the average of the FWHM values computed in the xz- and yz-plane, corresponding to the x- and y-axis, respectively.

3.3. Simulation Scenarios

As shown in Figure 2, three simulation scenarios were established. In each scenario, five atmospheric layers were created at altitudes ranging from 0 to 8 km. Each atmospheric layer represents different partial contributions to the atmospheric turbulence and wind velocity, following the same criterion established in simulations for the SAMplus [8]. The way in which these parameters were grouped is described in Table 1. The wind speed in each atmospheric turbulence layer was also grouped into typical, worst, and critical cases. In the typical case, winds of at most 25 m/s were considered while the winds reach 40 m/s in the critical case. Table 2 presents a description of the wind speeds simulated in each scenario.

Table 1. Indication of r_0 and total contribution of each atmospheric layer in the seeing phenomenon. Considering the typical case (I), the worst case (II), and the critical case (III) [8].

Layer	1	2	3	4	5	R_0 [cm]
Altitude [km]	0	1	2	4	8	-
Fractional r_0—I	0.74	0.02	0.02	0.10	0.12	14
Fractional r_0—II	0.70	0.03	0.07	0.10	0.10	11
Fractional r_0—III	0.65	0.05	0.09	0.11	0.10	10

Table 2. Wind speed in each atmospheric layer, considering the typical case (I), the worst case (II), and the critical case (III) [8].

Layer	1	2	3	4	5
Altitude [km]	0	1	2	4	8
Wind speed—I	9 m/s	10 m/s	15 m/s	25 m/s	25 m/s
Wind speed—II	15 m/s	25 m/s	25 m/s	30 m/s	35 m/s
Wind speed—III	25 m/s	25 m/s	30 m/s	35 m/s	40 m/s

4. Results and Discussion

4.1. Integration of OOMAO with Simulink

As above mentioned, OOMAO is focused on AO modeling strategies, while Simulink has a standard library with many available controllers and is fully integrated with MATLAB. To speed up the process of simulating non-native controllers (e.g., fuzzy-based controllers) in OOMAO, an adaptation layer between OOMAO and Simulink was developed. Figure 3 presents an example of a block diagram implemented using the adapted OOMAO blocks in Simulink. To showcase the capabilities of the integration, Figure 4a depicts the total incident wavefront, while Figure 4b,c illustrate the residual wavefront and the accumulated wavefront after the I controller action, respectively. Figure 4a exhibits a significant variation in phase ranging from −20 to 40 µm, which deviates from the ideal scenario as we desire values closer to zero. This is demonstrated in Figure 4b, which represents the controlled phase of the system. In addition to providing the capability to implement different control strategies, the integration also enhances the visualization of the results.

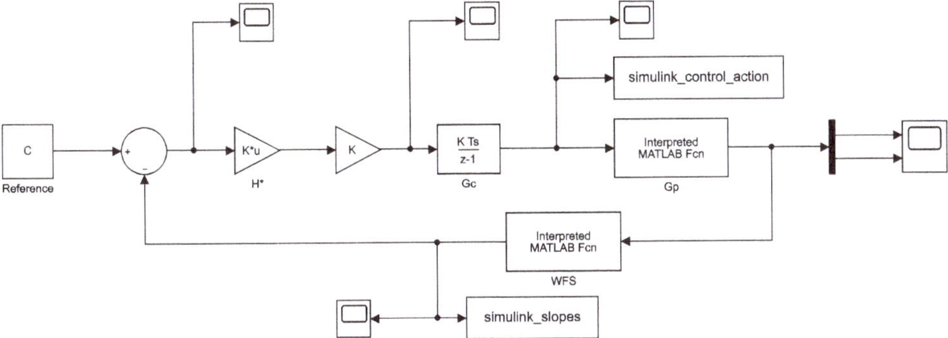

Figure 3. Block diagram of OOMAO simulation integrated with Simulink. G_c indicates the controller (in the example, a discrete-time Integral controller), G_p is the atmospheric turbulence process, $\mathbf{H}^*$ indicates the reconstruction matrix, and WFS the model of the Shack-Hartman wavefront sensor.

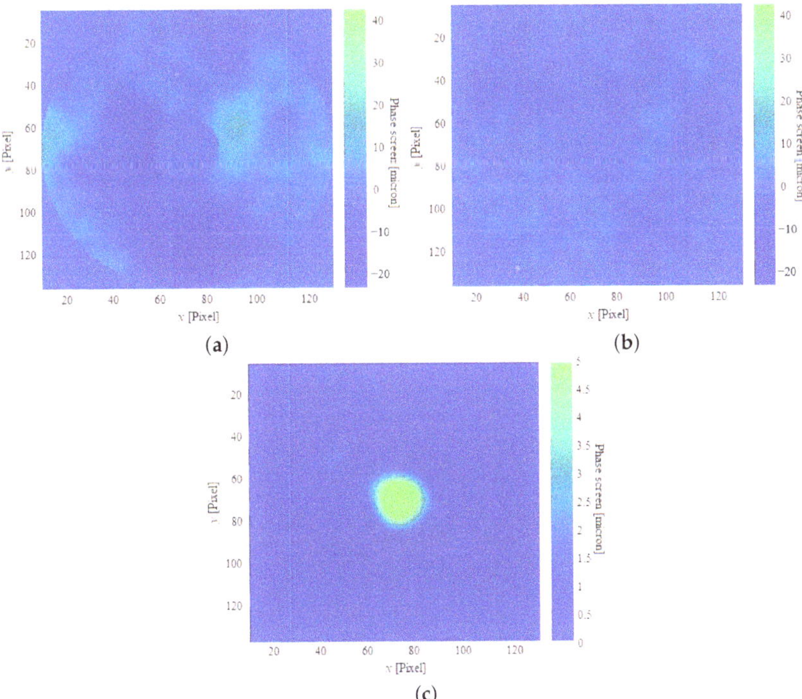

Figure 4. The total incident wavefront is represented in (**a**). The residual wavefront and the accumulated wavefront after the Integral controller action are indicated, respectively, in (**b**,**c**). The axes x and y represent the camera with 128×128 pixels and the z axis is the phase [µm].

Additionally, to validate this integration, the Integral controller was implemented, and the tests were replicated in both OOMAO and OOMAO integrated with Simulink. The versions of the same controllers were compared under the conditions previously described in Section 3. The stochastic method used to simulate the atmospheric turbulence was fixed to the Mersenne twister using the seed identified as *mt19937ar* under the MathWorks environment. Figure 5 presents comparative graphs of wavefront RMS values versus time under the typical, worst, and critical cases, respectively. The results demonstrated a higher concordance between the curves obtained with the controller implemented in Simulink and with the native implementation of OOMAO in MATLAB, except for a short and constant 4-sample delay between the signal acquired from MATLAB and the one from Simulink. Simulink initiates from a starting point that is four steps ahead of the MATLAB code, as it executes the functions enclosed within the "Interpreted MATLAB Function" blocks multiple times. There is no need to include a delay block in Simulink, as the phase turbulence simulation yields valid results.

Table 3 compares the MSE of some variable computed between the implementation without (i.e., in the native code of OOMAO) and with the Simulink approach, here proposed.

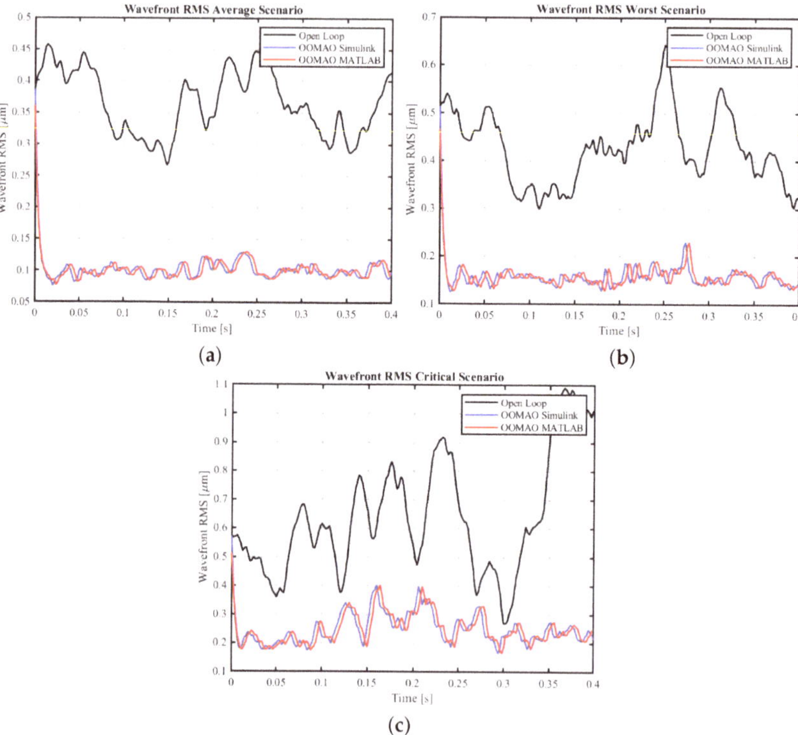

Figure 5. Wavefront RMS values over time for an open loop (black line), with the controller implemented in Simulink using the adaptation layer (blue line), and with the controller directly implemented in MATLAB without the adaptation layer (red line). Simulation scenarios: (**a**) typical case, (**b**) worst case, and (**c**) critical case. The delay between the curves of "OOMAO Simulink" and "OOMAO MATLAB" occurs due to internal processing aspects of the Simulink, not affecting the calculation.

Table 3. MSE of control variables calculated between implementation with and without the Simulink adaptation layer.

	Typical Scenario	Worst Scenario	Critical Scenario
MSE—Wavefront RMS values [μm^2]	7.49×10^{-5}	2.60×10^{-4}	9.82×10^{-4}
MSE—Slopes on first WFS lenslet [μm^2]	1.44×10^{-1}	7.17×10^{-1}	1.26×10^{0}
MSE—Slopes on last WFS lenslet [μm^2]	9.37×10^{-2}	2.93×10^{-1}	7.29×10^{-1}
MSE—DM first actuator control action amplitudes	3.11×10^{-16}	8.05×10^{-16}	1.09×10^{-15}
MSE—DM last actuator control action amplitudes	3.93×10^{-16}	6.95×10^{-16}	1.30×10^{-15}
MSE—DM control action RMS value amplitudes	1.18×10^{-17}	2.07×10^{-17}	5.41×10^{-17}

4.2. Numerical Simulation of Classical and Intelligent Controllers

4.2.1. Open-Loop Simulation

The open-loop simulations were carried out in each of the respective proposed scenarios. Figure 6a–c are the PSFs of the first open-loop turbulence profile in the typical, worst, and critical cases, respectively. The metrics of FWHM, HLR, and Strehl ratio in three turbulence profiles (T_1, T_2, and T_3) are shown in Table 4.

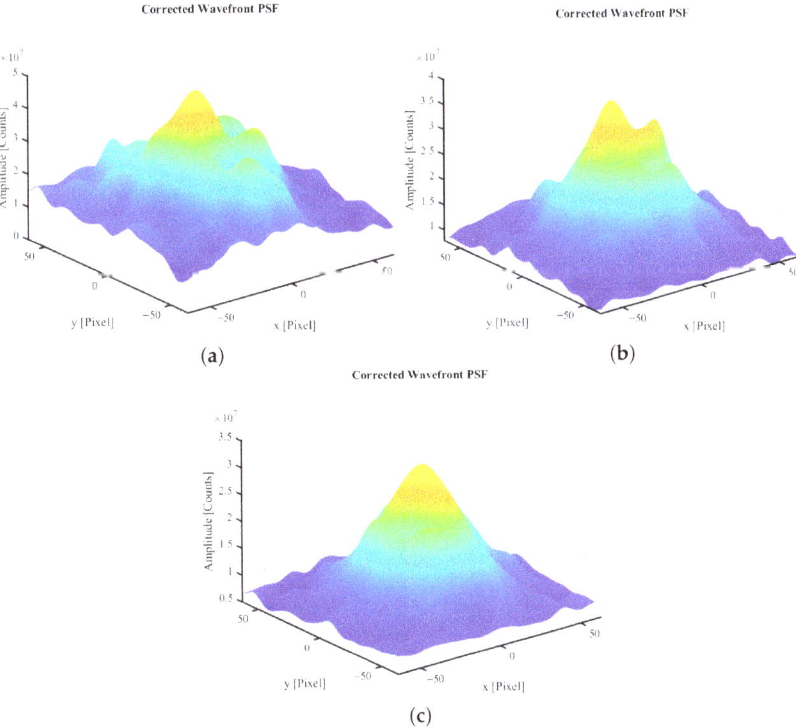

Figure 6. PSF of the first turbulence profile in an open loop for: (**a**) typical case, (**b**) worst case, and (**c**) critical case.

Comparing the results of the PSFs presented in Figure 6 with the value of an ideal PSF, it is possible to verify that the open-loop system is significantly noisier and less able to concentrate light when compared to the ideal one. One of the consequences of this fact is that the open-loop system does not have a PSF that is easily parameterized on a Gaussian or Moffat surface, which are more conventional surfaces for the characterization of PSF profiles [31]. Due to this difficulty, the proposed metrics may not be described effectively.

Therefore, not all metrics necessarily worsen due to atmospheric conditions; this implies that it is possible in some metrics of the open-loop system, the worst or the critical cases may have better indexes than the same metrics in the typical case, as is the case of FWHM and HLR. The Strehl ratio, which briefly indicates the ratio between the maximum points of the experimental PSF with the maximum point of the ideal PSF, can capture that the ability to concentrate energy worsens according to the atmospheric conditions.

Table 4. System performance in an open-loop for the cases analyzed in three turbulence profiles (T_1, T_2, and T_3) according to the metrics of FWHM [Pixel], HLR [Pixel], and Strehl ratio [%].

Metrics	Typical Scenario			Worst Scenario			Critical Scenario		
	T_1	T_2	T_3	T_1	T_2	T_3	T_1	T_2	T_3
FWHM	65.875	85	53.25	72.375	64.000	54.375	66.375	63.250	66.125
HLR	33.00	34.00	32.00	30.00	29.25	28.75	29.25	28.75	29.00
Strehl ratio	0.175%	0.184%	0.218%	0.124%	0.121%	0.157%	0.104%	0.110%	0.111%

4.2.2. Integral Controller

Figure 7a–c shows the PSFs of the first turbulence profile of the simulations of the I controller for the typical, worst, and critical scenarios, respectively. Table 5 presents a summary of the astronomical metrics evaluated. In the typical scenario, a reduction of the FWHM of around 40 pixels can be observed while no meaningful reduction was perceived either in the worst-case scenario or in the critical scenario. This conclusion can also be drawn by checking the PSF for each scenario, in which only the PSF of the typical case has a Gaussian-shaped surface with a high peak at the center and a symmetric bell-shaped surface.

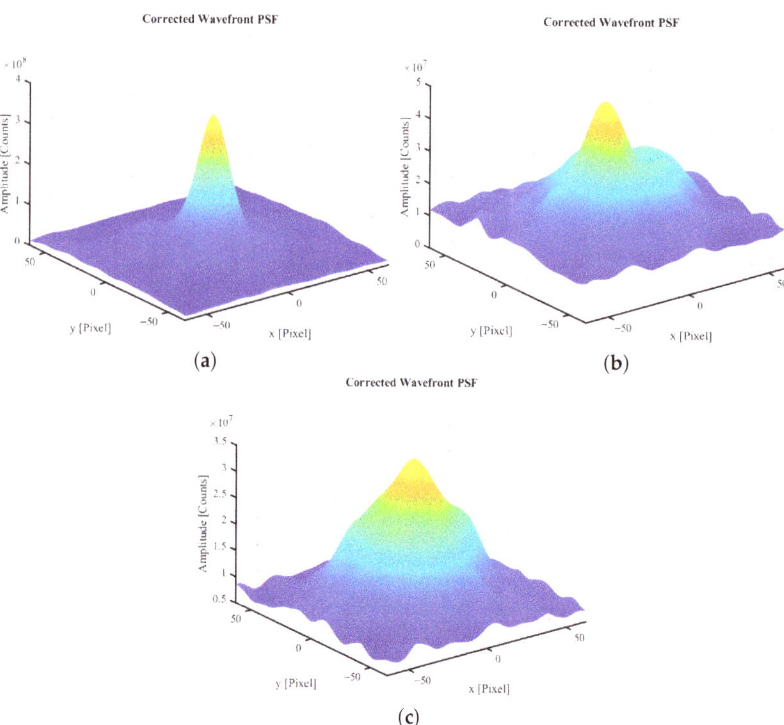

Figure 7. PSF of the first turbulence profile using an Integral action controller for: (**a**) typical case, (**b**) worst case, and (**c**) critical case.

In comparison with the metrics presented in Table 4, the I controller, in the typical case, performs significantly better than the open-loop system under the same conditions. This, however, is not necessarily true in the worst and most critical cases. Therefore, the I controller is sufficient to control the system only in good atmospheric conditions, since low-quality correction metrics were obtained in the worst and critical scenarios. Therefore, based on the results, conducting astronomical observations using AO (specifically, the I controller) on days with unfavorable atmospheric conditions would yield comparable results to systems without AO (i.e., open-loop).

Table 5. System performance with I controller for the cases analyzed in three turbulence profiles (T_1, T_2, and T_3) according to the metrics of FWHM [Pixel], HLR [Pixel], and Strehl ratio [%].

Metrics	Typical Scenario			Worst Scenario			Critical Scenario		
	T_1	T_2	T_3	T_1	T_2	T_3	T_1	T_2	T_3
FWHM	20.875	17.875	21.500	65.750	57.750	61.625	66.625	67.625	59.875
HLR	12.50	11.75	14.75	30.25	31.25	30.00	28.75	29.00	29.75
Strehl ratio	1.077%	1.303%	1.039%	0.162%	0.153%	0.129%	0.1097%	0.101%	0.113%

4.2.3. Proportional-Integral Controller

Figure 8a–c are the PSFs of the first turbulence profile controlled by the PI controller in the typical, worst, and critical cases, respectively. When comparing these PSFs with the ones obtained with the I controller, a greater concentration of energy is observed in all scenarios; moreover, the performance of the controller deteriorates significantly with the deterioration of the simulated atmospheric conditions.

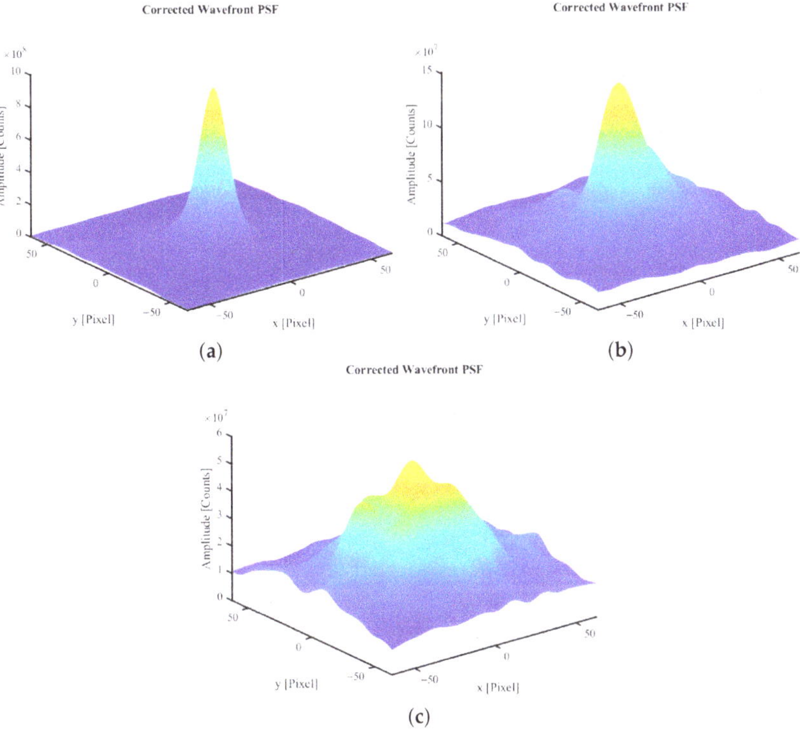

Figure 8. PSF of the first turbulence profile using PI action controller for: (**a**) typical case, (**b**) worst case, and (**c**) critical case.

The same conclusion can be reached by verifying the metrics in Table 6 with those provided in Tables 4 and 5, in which the metrics of FHWM and Strehl ratio obtained with the PI controller consistently performed better than the metrics of I controller and open-loop. The HLR metric obtained with the PI controller also performed generally better (in the typical and worst scenarios), however, had similar results to the HLR metric obtained with the I controller in the critical scenario.

Table 6. System performance with PI controller for the cases analyzed in three turbulence profiles (T_1, T_2, and T_3) according to the metrics of FWHM [Pixel], HLR [Pixel], and Strehl ratio [%].

Metrics	Typical Scenario			Worst Scenario			Critical Scenario		
	T_1	T_2	T_3	T_1	T_2	T_3	T_1	T_2	T_3
FWHM	15.625	15.375	15.500	36.625	29.500	30.875	66.750	72.750	72.625
HLR	7.75	7.75	7.50	22.00	22.75	24.00	30.75	31.75	31.75
Strehl ratio	3.091%	3.552%	3.428%	0.494%	0.528%	0.396%	0.193%	0.182%	0.184%

4.2.4. Hybrid Fuzzy Proportional Integral Controller

The hybrid fuzzy PI controller was designed using the structure of Figure 1, a PI controller with a gain scheduler based on fuzzy logic. As an input for the gain scheduler, the RMS value of the residual and the total wavefront error were adopted. The primary justification for selecting these metrics as inputs in the fuzzy logic system lies in their facility to be computed in real-time systems.

For each scheduler input, three triangular membership functions were adopted, categorizing the inputs into the linguistic terms high, medium, and low, as pointed out in Figure 9a. The residual wavefront membership functions assume values within the range of [0, +0.5], and for the total wavefront membership functions the interval was designed within the range of [0.4, 1.2]. The gain scheduler has two outputs, referring to the proportional gain and integral time, in which membership functions were also fitted in triangular functions with the same linguistic terms.

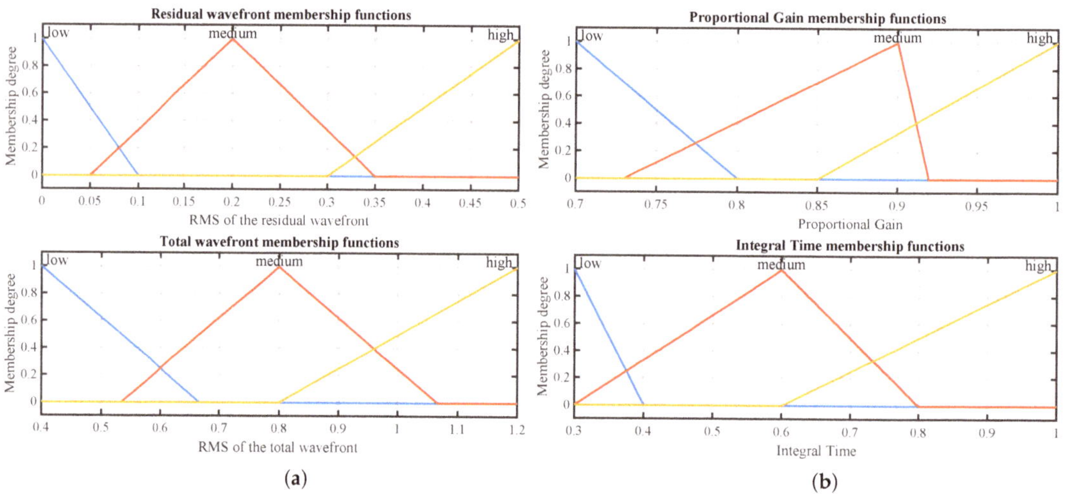

Figure 9. Triangular membership functions with the linguistic terms low, medium, and high for: (**a**) input variables (RMS value of the residual and total wavefront error), and (**b**) output variables (proportional gain and integral time).

Figure 9b depicts the membership functions for the proportional gain and integral time, respectively. Considering that traditional PID control tuning methods are not readily applicable to the proposed multi-input, multiple-output (MIMO), and non-linear system, the PI parameters were empirically tested, adjusted, and fine-tuned to enhance the performance of the AO system, with a focus on improving the metrics of FWHM, HLR, and Strehl ratio. During the tests, the integral time was systematically adjusted within the range of 0.1 to 1, while the proportional action was varied between 0.5 and 5. The optimal parameters for the proportional-integral action controller are determined to be $K_p = 1$ and $T_i = 0.4$. Moreover, it has been observed that when the proportional gain values are set within the range of 0.7 to 1 and the integral time falls between 0.3 and 1, a stable condition is attained, resulting in effective AO control performance.

Figure 10a,b shows the fuzzy control surface that maps the given inputs (RMS values of the residual and total wavefront) to the outputs (i.e., the proportional and integral time). Lookup tables can be utilized to approximate the nonlinear control surfaces, resulting in simplified control code, and enhancing the execution speed. Furthermore, this approach facilitates the embedding of the designed fuzzy controller in real-time systems. In Figure 10a, there is a noticeable proportional gain scheduler, which is divided into three plateaus whereby the residual and total values remain in the middle ranges. The integral time schedule, conversely, presents a greater variation compared with the proportional gain schedule, with a minimum value being reached when the total effective values (i.e., RMS of the total wavefront) are around 0.7 µm. Both outputs have their maximum values if the dominant membership function reaches the linguistic variable "high".

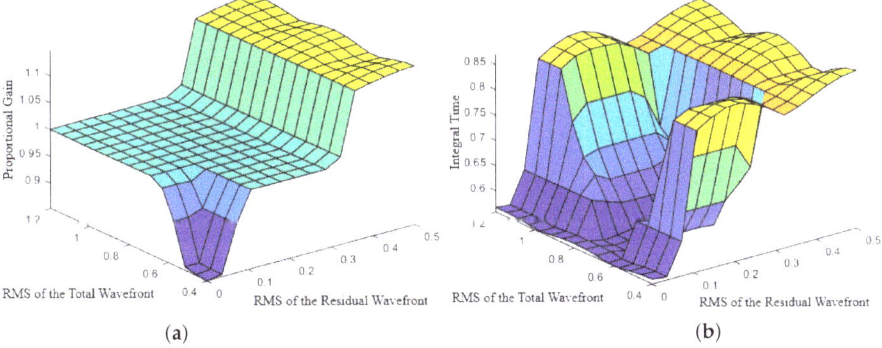

(a)

(b)

Figure 10. Fuzzy output surface (control surface plot) for: (**a**) proportional gain and (**b**) integral time.

Figure 11a–c corresponds to the PSFs of the first turbulence profile controlled by the fuzzy PI controller in the typical, worst, and critical scenarios, respectively. By comparing the surface plots of the typical case depicted in Figures 6–8 and 11, one can observe peak amplitudes (in counts) of about 5×10^7 (open loop), 4×10^8 (I controller), 10×10^8 (PI controller), and 15×10^8 (fuzzy PI controller), respectively. That is, in the fuzzy PI controller, the peak amplitude is found to be 30 times larger than that of the open-loop case, 3 times larger than the I controller, and 1.5 times larger than the PI controller. Comparing the PSFs of this control strategy with those presented by the PI controller, Figure 8, the fuzzy PI has a better overall performance. Additionally, it is noteworthy that the performance of the controllers, in all cases assessed, is significantly limited by the deterioration of the simulated atmospheric conditions. As can be seen in the performance of the critical case, that does not replicate the improvement observed in the typical case, even with the fuzzy PI controller.

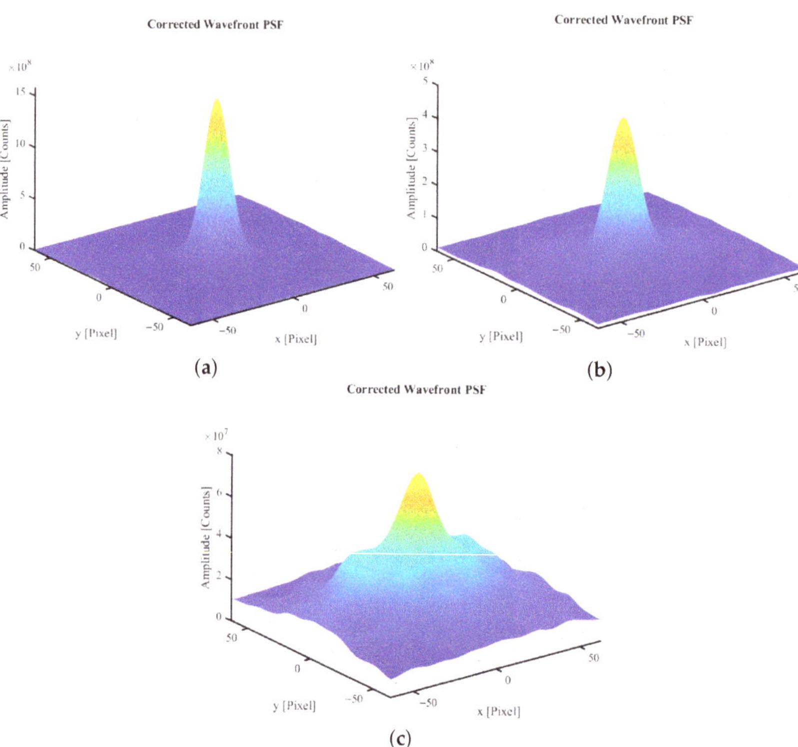

Figure 11. PSF of the first turbulence profile using fuzzy Proportional-Integral action controller for: (**a**) typical case, (**b**) worst case, and (**c**) critical case.

The metrics of FWHM, HLR, and Strehl ratio of this controller in all turbulence profiles are presented in Table 7.

Table 7. System performance with fuzzy PI controller for the cases analyzed in three turbulence profiles (T_1, T_2, and T_3) according to the metrics of FWHM [Pixel], HLR [Pixel], and Strehl ratio [%].

Metrics	Typical Scenario			Worst Scenario			Critical Scenario		
	T_1	T_2	T_3	T_1	T_2	T_3	T_1	T_2	T_3
FWHM	14.500	14.375	14.625	18.250	19.375	19.750	50.000	52.750	45.125
HLR	6.50	6.50	6.50	9.50	12.50	13.50	29.25	29.50	29.25
Strehl ratio	4.869%	5.112%	5.090%	1.377%	1.028%	0.844%	0.251%	0.240%	0.249%

4.2.5. Discussions

Figure 12a–c represents the wavefront RMS versus time of the open-loop system and closed-loop system under I, PI, and fuzzy PI controllers for all the turbulence profiles adopted. The fuzzy-based controller has a smaller RMS value in every single simulation and thus outperforms its peers. The open-loop simulations serve as a performance baseline for comparison, presenting greater values of wavefront RMS in the simulations. Between the classical controllers (I and PI algorithms), the results indicate that the PI version presents better results than the I controller.

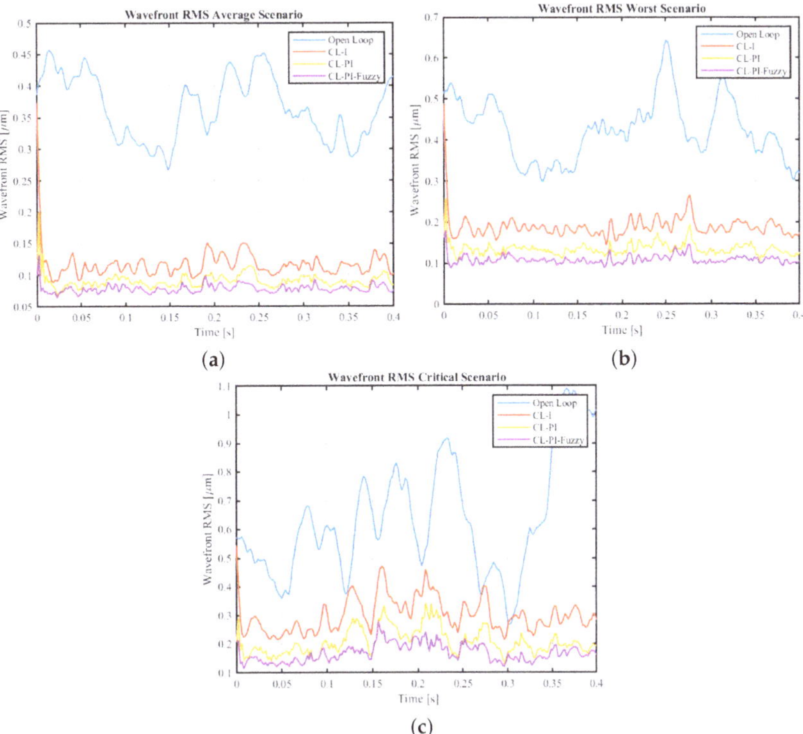

Figure 12. Wavefront RMS values over time for all the control strategies implemented: open loop, in a closed loop with I (CL-I), PI (CL-PI), and fuzzy PI (CL-PI-Fuzzy) controllers. Simulation scenarios: (**a**) typical case, (**b**) worst case, and (**c**) critical case.

Table 8 presents a summary of the mean and standard deviation values of the residual wavefront for the open-loop and closed-loop systems. The PI controller presents values consistently smaller than the I controller and open-loop system values. There is an improvement of about 75% to 80% in the mean value of the residual wavefront in the mean case while an improvement by about 70% occurs in the critical case. By analyzing the standard deviations of effective values of the wavefront, it is concluded that the PI controller has smaller deviations than the I action controller. However, it is observed that the standard deviations significantly grow with the worsening of atmospheric conditions, going from 0.008 μm in the first turbulence profile in the typical case to 0.045 μm in the first turbulence profile in the critical case. Based on the analysis conducted, it can be inferred that the I and PI controllers are insufficient for effectively controlling the system during critical scenarios.

The hybrid fuzzy PI controller, in turn, presents better results than the ones reported by the PI controller, with an improvement of about 80% when compared to the mean effective values of the wavefronts in open-loop in the typical case and about 75% compared to the critical case. In addition, the fuzzy PI strategy has a performance of about 10% better than the PI algorithm in the typical case and about 15% better in the critical scenario. This again reveals and reaffirms the potential of using intelligent controllers with dynamic gain scaling in AO systems.

Table 8. Performance of open loop systems (OL), in a closed loop with I controller (CL-I), with PI controller (CL-PI), with fuzzy PI controller (CL-PI-Fuzzy), for the scenarios analyzed in three turbulence profiles (T_1, T_2, and T_3) as a function of the metrics of the mean (RMS Mean) [μm] and the standard deviation of RMS values (RMS Std. Dev.) [μm].

Metrics	Controller	Typical Scenario			Worst Scenario			Critical Scenario		
		T_1	T_2	T_3	T_1	T_2	T_3	T_1	T_2	T_3
RMS Mean	OL	0.358	0.461	0.330	0.418	0.737	0.637	0.662	0.601	0.613
	CL-I	0.115	0.115	0.115	0.187	0.203	0.203	0.305	0.276	0.282
	CL-PI	0.090	0.088	0.087	0.134	0.151	0.148	0.212	0.190	0.195
	CL-PI-Fuzzy	0.078	0.079	0.077	0.108	0.125	0.125	0.172	0.161	0.163
RMS Std. Dev.	OL	0.049	0.104	0.075	0.076	0.232	0.197	0.213	0.123	0.101
	CL-I	0.013	0.017	0.016	0.018	0.030	0.026	0.057	0.031	0.043
	CL-PI	0.008	0.012	0.010	0.013	0.027	0.018	0.045	0.024	0.031
	CL-PI-Fuzzy	0.006	0.009	0.007	0.009	0.019	0.018	0.030	0.018	0.024

5. Conclusions

This work demonstrates the results of the integration of the OOMAO toolbox with Simulink. As proof of concept, we have evaluated AO control strategies based on I, PI, and hybrid PI fuzzy controllers, analyzing them using realistic atmospheric data in the context of the SOAR telescope. The hybrid fuzzy PI controller reported a 25% increase in the FWHM metrics in the critical scenario if compared with open-loop metrics in the critical case. The results highlight the limitations of classical control techniques when applied to AO systems. Moreover, the results of the hybrid fuzzy PI controller demonstrate the potential of intelligent control techniques to improve control performance in hostile atmospheric conditions, thus expanding the viability of AO applications in regions of high atmospheric turbulence. In that sense, fixed-gain controllers are limited in conditions of fast-changing scenarios, as presented by atmospheric turbulence. In these cases, adaptive strategies using gain scheduling and intelligent controllers might be more suitable as a control technique. Moreover, it is expected that the fuzzy PI controller retains their superior performance compared to classical approaches (I and PI controllers) in different scenarios (e.g., modification of the telescope and site) since the fuzzy strategy itself allows greater flexibility for tuning and control. Further tests to evaluate different scenarios will be conducted in future works.

Author Contributions: Conceptualization, V.C. and W.B.; methodology, V.C. and W.B.; software, V.C.; validation, V.C. and W.B.; formal analysis, V.C. and W.B.; investigation, V.C. and W.B.; resources, V.C. and W.B.; data curation, V.C. and W.B.; writing—original draft preparation, V.C. and W.B.; writing—review and editing, V.C. and W.B.; supervision, V.C. and W.B.; project administration, V.C. and W.B.; funding acquisition, W.B. All authors have read and agreed to the published version of the manuscript.

Funding: This study was financed in part by the Coordenação de Aperfeiçoamento de Pessoal de Nível Superior-Brasil (CAPES)-Finance Code 001. This research also received partial support from the Fundação de Amparo à Pesquisa do Estado de São Paulo (FAPESP) under Grants 2022/10909-5 and 2023/04428-7.

Institutional Review Board Statement: Not applicable.

Informed Consent Statement: Not applicable.

Data Availability Statement: The data presented in this study are available on request from the corresponding author.

Acknowledgments: The authors acknowledge the support of the Astronomy Instrumentation Group of the Institute of Astronomy, Geophysics and Atmospheric Sciences (IAG) of the University of São Paulo (USP).

Conflicts of Interest: The authors declare no conflict of interest.

References

1. Pullen, W.C.; Wang, T.; Choi, H.; Ke, X.; Negi, V.S.; Huang, L.; Idir, M.; Kim, D. Statistical Tool Size Study for Computer-Controlled Optical Surfacing. *Photonics* **2023**, *10*, 286. [CrossRef]
2. Hippler, S. Adaptive Optics for Extremely Large Telescopes. *J. Astron. Instrum.* **2019**, *8*, 1950001. [CrossRef]
3. Hardy, J.W. *Adaptive Optics for Astronomical Telescopes*; Oxford University Press: New York, NY, USA , 1998.
4. Tyson, R.K. *Topics in Adaptive Optics*; IntechOpen: Rijeka, Croatia, 2012. [CrossRef]
5. Davies, R.; Kasper, M. Adaptive Optics for Astronomy. *Annu. Rev. Astron. Astrophys.* **2012**, *50*, 305–351. [CrossRef]
6. Madec, P.Y. Overview of Deformable Mirror Technologies for Adaptive Optics and Astronomy. In Proceedings of the SPIE—The International Society for Optical Engineering, Amsterdam, The Netherlands, 2 June 2012 ; Volume 844705. [CrossRef]
7. Ellerbroek, B.L.; Vogel, C.R. Simulations of Closed-loop Wavefront Reconstruction for Multiconjugate Adaptive Optics on Giant Telescopes. In Proceedings of the Astronomical Adaptive Optics Systems and Applications, International Society for Optics and Photonics, SPIE, San Diego, CA, USA, 24 December 2003; Tyson, R.K., Lloyd-Hart, M., Eds.; Volume 5169, pp. 206–217. [CrossRef]
8. Faes, D.M.; Tokovinin, A.; Vieira, T.; Mello, A.; Domingues, M.; Andrade, D.; Quint, B.C.; dos Santos, J.B., Sr. SAMplus: Adaptive Optics at Optical Wavelengths for SOAR. In Proceedings of the Adaptive Optics Systems VI, International Society for Optics and Photonics, SPIE, Austin, TX, USA, 11 July 2018; Close, L.M., Schreiber, L., Schmidt, D., Eds.; Volume 10703, pp. 983–992. [CrossRef]
9. Basden, A.G.; Jenkins, D.; Morris, T.J.; Osborn, J.; Townson, M.J. Efficient Implementation of Pseudo Open-Loop Control for Adaptive Optics on Extremely Large Telescopes. *Mon. Not. R. Astron. Soc.* **2019**, *486*, 1774–1780. [CrossRef]
10. Gu, H.; Zhao, Z.; Zhang, Z.; Cao, S.; Wu, J.; Hu, L. High-Precision Wavefront Reconstruction from Shack-Hartmann Wavefront Sensor Data by a Deep Convolutional Neural Network. *Meas. Sci. Technol.* **2021**, *32*, 085101. [CrossRef]
11. Shatokhina, I.; Hutterer, V.; Ramlau, R. Review on Methods for Wavefront Reconstruction from Pyramid Wavefront Sensor Data. *J. Astron. Telesc. Instrum. Syst.* **2020**, *6*, 1–39. [CrossRef]
12. Bond, C.Z.; Hadi, K.E.; Sauvage, J.F.; Correia, C.; Fauvarque, O.; Rabaud, D.; Lamb, M.; Neichel, B.; Fusco, T. Experimental Study of an Optimised Pyramid Wave-front Sensor for Extremely Large Telescopes. In Proceedings of the Adaptive Optics Systems V, International Society for Optics and Photonics, SPIE, Edinburgh, UK, 27 July 2016; Marchetti, E., Close, L.M., Véran, J.P., Eds.; Volume 9909, pp. 1783–1795. [CrossRef]
13. Schatz, L.; Males, J.R.; Correia, C.M.; Neichel, B.; Chambouleyron, V.; Codona, J.L.; Fauvarque, O.; Sauvage, J.F.; Fusco, T.; Hart, M.; et al. Three-Sided Pyramid Wavefront Sensor, Part 1: Simulations and Analysis for Astronomical Adaptive Optics. *J. Astron. Telesc. Instrum. Syst.* **2021**, *7*, 1–19. [CrossRef]
14. Yang, H.; Jin, Y.; Hu, Y.; Zhang, D.; Yu, Y.; Liu, J.; Li, J.; Jiang, X.; Yu, X. Image Degradation Model for Dynamic Star Maps in Multiple Scenarios. *Photonics* **2022**, *9*, 673. [CrossRef]
15. Folcher, J.P.; Carbillet, M.; Ferrari, A.; Abelli, A. Adaptive Optics Feedback Control. *Eur. Astron. Soc. Publ. Ser.* **2013**, *59*, 93–130. [CrossRef]
16. Rigaut, F.; Van Dam, M. Simulating Astronomical Adaptive Optics Systems Using Yao. In Proceedings of the Third AO4ELT Conference, Florence, Italy, 26–31 May 2013; Esposito, S., Fini, L., Eds.; p. 18. [CrossRef]
17. Reeves, A. Soapy: An Adaptive Optics Simulation Written Purely in Python for Rapid Concept Development. In Proceedings of the Adaptive Optics Systems V, International Society for Optics and Photonics, SPIE, Edinburgh, UK, 27 July 2016; Marchetti, E., Close, L.M., Véran, J.P., Eds.; Volume 9909, pp. 2173–2183. [CrossRef]
18. Townson, M.J.; Farley, O.J.D.; de Xivry, G.O.; Osborn, J.; Reeves, A.P. AOtools: A Python Package for Adaptive Optics Modelling and Analysis. *Opt. Express* **2019**, *27*, 31316–31329. [CrossRef] [PubMed]
19. Conan, R.; Correia, C. Object-oriented Matlab Adaptive Optics Toolbox. In Proceedings of the Adaptive Optics Systems IV, International Society for Optics and Photonics, SPIE, Montréal, QC, Canada, 7 August 2014; Marchetti, E., Close, L.M., Véran, J.P., Eds.; Volume 9148, pp. 2066–2082. [CrossRef]
20. Assémat, F.; Wilson, R.W.; Gendron, E. Method for Simulating Infinitely Long and Non Stationary Phase Screens with Optimized Memory Storage. *Opt. Express* **2006**, *14*, 988–999. [CrossRef]
21. Nousiainen, J.; Rajani, C.; Kasper, M.; Helin, T. Adaptive Optics Control Using Model-Based Reinforcement Learning. *Opt. Express* **2021**, *29*, 15327–15344. [CrossRef] [PubMed]
22. Ogata, K. *Modern Control Engineering*, 5th ed.; Prentice Hall: Upper Saddle River, NJ, USA, 2009.
23. Åström, K.J.; Wittenmark, B. *Computer-Controlled Systems—Theory and Design*; Prentice-Hall, Inc.: Upper Saddle River, NJ, USA, 1997.
24. AlSabbah, S.; AlDhaifallah, M.; Al-Jarrah, M. Design of Multiregional Supervisory Fuzzy PID Control of pH Reactors. *J. Control Sci. Eng.* **2015**, *2015*, 396879. [CrossRef]
25. Ke, X.; Zhang, D. Fuzzy Control Algorithm for Adaptive Optical Systems. *Appl. Opt.* **2019**, *58*, 9967–9975. [CrossRef] [PubMed]
26. Flores-Meza, R.; Sotelo, P.D.; Garfias, F.; Cuevas, S.; Sanchez, L.J. Adaptive Optics Tip-Tilt System with Fuzzy Control. In Proceedings of the Adaptive Optical Systems Technology, International Society for Optics and Photonics, SPIE, Munich, Germany, 7 July 2000; Wizinowich, P.L., Ed.; Volume 4007, pp. 671–681. [CrossRef]
27. Nousiainen, J.; Rajani, C.; Kasper, M.; Helin, T.; Haffert, S.Y.; Vérinaud, C.; Males, J.R.; Van Gorkom, K.; Close, L.M.; Long, J.D.; et al. Toward On-Sky Adaptive Optics Control Using Reinforcement Learning—Model-based Policy Optimization for Adaptive Optics. *A&A* **2022**, *664*, A71. [CrossRef]

28. Pou, B.; Ferreira, F.; Quinones, E.; Gratadour, D.; Martin, M. Adaptive Optics Control with Multi-Agent Model-Free Reinforcement Learning. *Opt. Express* **2022**, *30*, 2991–3015. [CrossRef] [PubMed]
29. Guo, Y.; Zhong, L.; Min, L.; Wang, J.; Wu, Y.; Chen, K.; Wei, K.; Rao, C. Adaptive Optics Based on Machine Learning: A Review. *Opto-Electron. Adv.* **2022**, *5*, 200082-1–200082-20. [CrossRef]
30. Conan, R.; Bennet, F.; Bouchez, A.H.; van Dam, M.A.; Espeland, B.; Gardhouse, W.; d'Orgeville, C.; Parcell, S.; Piatrou, P.; Price, I.; et al. The Giant Magellan Telescope Laser Tomography Adaptive Optics System. In Proceedings of the Adaptive Optics Systems III, International Society for Optics and Photonics, SPIE, Amsterdam, The Netherlands, 13 September 2012; Ellerbroek, B.L., Marchetti, E., Véran, J.P., Eds.; Volume 8447, pp. 1271–1287. [CrossRef]
31. Andersen, D.; Stoesz, J.; Morris, S.; Lloyd-Hart, M.; Crampton, D.; Butterley, T.; Ellerbroek, B.; Jolissaint, L.; Milton, N.M.; Myers, R.; et al. Performance Modeling of a Wide-Field Ground-Layer Adaptive Optics System. *Publ. Astron. Soc. Pac.* **2006**, *118*, 1574–1590. [CrossRef]

Article

Astronomical Camera Based on a CCD261-84 Detector with Increased Sensitivity in the Near-Infrared

Irina Afanasieva [1,*], Valery Murzin [1], Valery Ardilanov [1], Nikolai Ivaschenko [1], Maksim Pritychenko [1], Alexei Moiseev [1,2], Elena Shablovinskaya [1] and Eugene Malygin [1]

[1] Special Astrophysical Observatory, Russian Academy of Sciences, Nizhniy Arkhyz 369167, Russia; vamur@sao.ru (V.M.); valery@sao.ru (V.A.); moisav@sao.ru (A.M.); shabli@sao.ru (E.S.); male@sao.ru (E.M.)

[2] Sternberg Astronomical Institute, Moscow M.V. Lomonosov State University, Universitetskij Pr. 13, Moscow 119992, Russia

* Correspondence: riv@sao.ru

Abstract: Herein, we describe the design, implementation and operation principles of an astronomical camera system, based on a large-format CCD261-84 detector with an extremely thick 200 μm substrate. The DINACON-V controller was used with the CCD to achieve high performance and low noise. The CCD system photometric characteristics are presented. A spatial autocorrelation analysis of flat-field images was performed to reveal the dependence of substrate voltage on the lateral charge spreading. The investigation of the dispersion index for the optimal choice of exposure time is discussed. Studies of the patterns of fringes were carried out in comparison with previous detectors. The amplitude of fringes with CCD261-84 was significantly lower, compared to previous-generation detectors. The results of using a new camera for imaging and spectral observations at the Russian 6 m telescope with the SCORPIO-2 multimode focal reducer are considered. The developed CCD camera system makes it possible to significantly increase the sensitivity in the 800–1000 spectral range.

Keywords: instrumentation; fully depleted scientific detectors; instrument optimisation; statistics; autocorrelation; imaging spectroscopy

Citation: Afanasieva, I.; Murzin, V.; Ardilanov, V.; Ivaschenko, N.; Pritychenko, M.; Moiseev, A.; Shablovinskaya, E.; Malygin, E. Astronomical Camera Based on a CCD261-84 Detector with Increased Sensitivity in the Near-Infrared. *Photonics* **2023**, *10*, 774. https://doi.org/10.3390/photonics10070774

Received: 25 May 2023
Revised: 25 June 2023
Accepted: 1 July 2023
Published: 4 July 2023

1. Introduction

The near-infrared (NIR) band (from 750 nm to 2500 nm) is of considerable interest to astronomers [1]. To achieve maximum sensitivity in this range, astronomical camera systems mainly use matrix multiplexed hybrid IR (infrared) detectors. Such detectors are much more sensitive than silicon devices, but they are vastly more expensive and not widely available [2].

For wavelengths up to 1100 nm, silicon CCD detectors have recently become an alternative to hybrid devices. Conventional CCD detectors have very high sensitivity in the visible range, but, at wavelengths greater than 700–800 nm, they show a significant decrease in sensitivity [3]. This is due to the small thickness of the silicon device substrate, which leads to a decrease in the absorption of long-wave photons in silicon.

The progress in the development of silicon CCD detectors in the field of increasing sensitivity in the red and near-infrared bands can be observed by the example of improving the line of scientific detectors manufactured by Teledyne E2V [4]. Conventional backside illuminated (BSI) CCD detectors manufactured using epitaxial silicon technology with a substrate thickness of no more than 20 μm have a sharp decrease in sensitivity starting from 700 nm [5]. The use of a thicker substrate up to 40 μm with deep depletion technology improves red sensitivity. The next class of devices with additionally improved IR sensitivity are CCD detectors with medium substrate thickness. These devices, based on bulk silicon with substrates up to 70 μm, demonstrate high red sensitivity and good photometric characteristics [6]. The last level of development is devices with a thick substrate (more than 100 μm) of high-resistance full depletion silicon, called High-Rho CCD detectors [7,8]. They have the highest sensitivity in the near-infrared range for their class [9].

In 2018, the Special Astrophysical Observatory of the Russian Academy of Sciences (SAO RAS) acquired the High-Rho CCD261-84, a full-frame scientific BSI CCD detector with a frame format of 2048 × 4104 and 200 μm substrate thickness, manufactured by Teledyne E2V [10]. The detector has very high sensitivity in the red and near-infrared bands. In this paper, we present the implementation of an acquisition system with this photodetector (in this paper, referred to as CCD System) and demonstrate the results of studying its photometric characteristics and the features associated with its operation on the 6 m Big Telescope Alt-azimuthal (BTA) of the SAO RAS as a part of the SCORPIO-2 multimode focal reducer [11].

2. CCD System Construction

Over the past 20 years, SAO RAS has created 5 generations of CCD controllers able to manage all kinds of CCD detectors, including various mosaic configurations [12,13]. On the one hand, the creation of new generations of CCD controllers is caused by the appearance of new photosensitive devices and tasks to be solved, and by the emergence of new electronic components with qualitatively increased characteristics on the other hand.

DINACON-V is the latest generation of CCD controllers built on the basis of modern electronic components [13]. The controller architecture allows the handling of both single and mosaic photodetectors with a total number of video channels up to 256 and a maximum data throughput between the controller and the host computer up to 10 Gbit/s. One of the main features of the controller is the ability to implement camera systems using High-Rho CCD detectors, which require high control voltage to create a sufficient electric field on a thick substrate [14].

The camera system with the CCD261-84 detector (Figure 1) consists of two parts: a nitrogen-cooled CCD camera and a power supply unit (PSU).

Figure 1. General view of the CCD261-84 camera system: the camera (**left**) and the PSU (**right**).

Structurally, the camera and the power supply unit can be separated from each other by up to 1.5 m, which is limited by the length of the connecting cables. The distance between the CCD camera and the host computer can be up to 200 m and is determined by the required length of the fiber optic communication line. The system is designed to operate in an ambient temperature range from −40 to +40 ° C.

3. CCD Camera Design

The left part of Figure 2 shows the design of the CCD camera, including an optical cryostat and a camera electronics unit (CEU).

The optical cryostat consists of the liquid nitrogen (LN2) cryostat itself and the optical head (Figure 2, right). The nitrogen cryostat includes a nonspillable type of nitrogen tank, an adsorption cryopump, and a cold radiator for connecting the heat load. The filler neck for filling liquid nitrogen is located at the top of the cryostat. The optical head includes a detector support with a CCD detector, a preamplifier (PA) board connected to the CCD detector, and a flexible heat conductor connected to a cold heatsink.

The CEU includes a hermetic case, which also provides the removal of heat generated by the electronics placed inside the case, a video processor–generator board (VP), and a level shaper board (LS). The electrical connection between the camera electronics unit and the PA board is carried out through a sealed connector on the optical head.

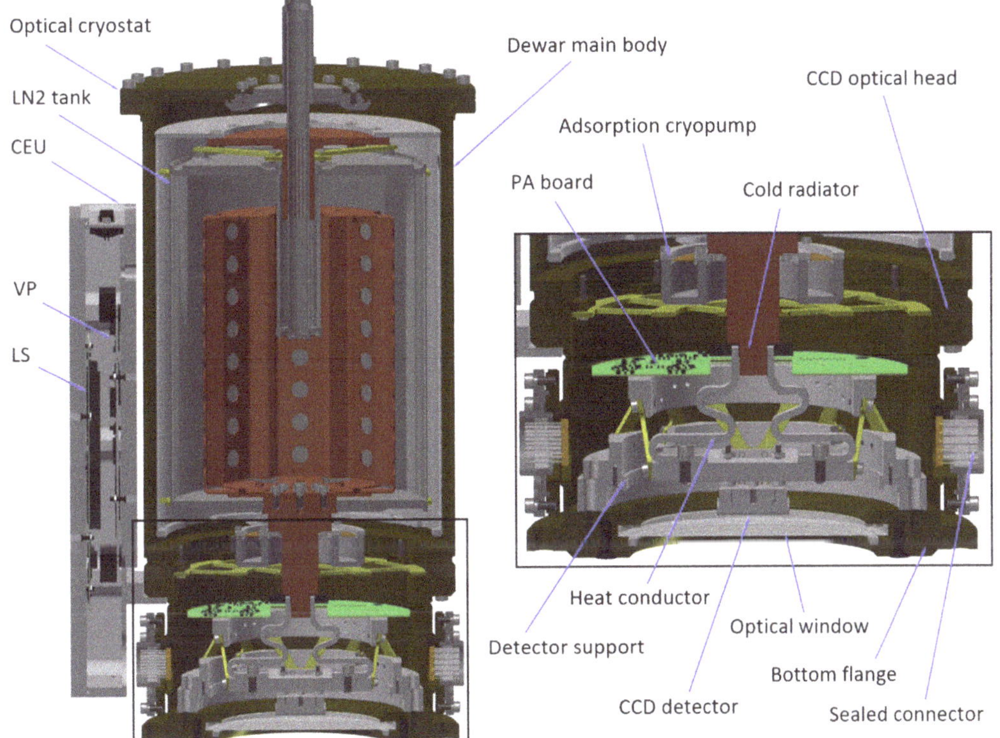

Figure 2. A 3D cut view of the CCD camera: full view of the camera (**left**) and zoomed view of the optical head (**right**).

The CCD camera has an input optical window and can be mounted at a telescope or spectrograph focus in an arbitrary spatial position using mounting holes in the bottom flange.

4. CCD Controller Electronics

The controller includes the PA board, the VP, the LS, and the communication module (CM) with the power source. The PA board is located in the CCD optical head, and the VP and LS are in the camera electronics unit, which is mounted directly on the optical cryostat. The CM with a power source is structurally placed in a separate block (Figure 1, right) and connected to the CEU with two connecting cables.

Placing a preliminary amplifier in the camera is necessary to amplify the video signal as close as possible to the outputs of the CCD detector. This approach enables one to achieve the maximum signal-to-noise ratio as well as to perform signal conversion to a differential form, which is necessary to minimize interference during video signal transmission and for subsequent analog-to-digital conversion (ADC). The PA board also includes electrostatic protection circuits and protection against increased operating voltages on the CCD detector. Inside the camera optical head, there is also a temperature sensor and photodetector base heater. The temperature sensor and heater are the elements of the photodetector temperature stabilization circuitry.

The operation of the DINACON-V CCD controller can be explained using the block diagram shown in Figure 3.

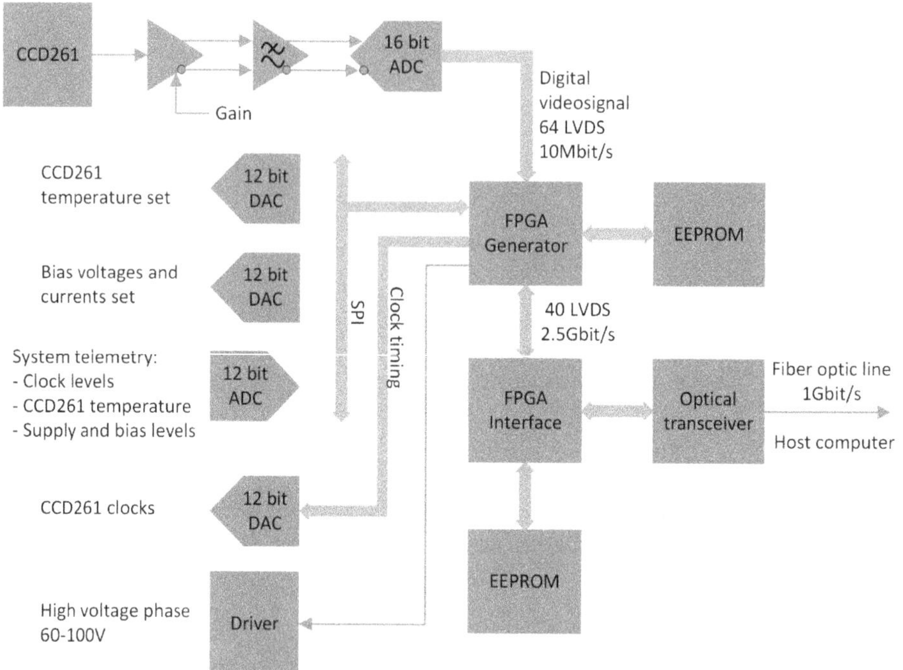

Figure 3. Block diagram of the DINACON-V CCD controller.

In the CEU, digital circuitry is implemented in a Field Programmable Gate Array (FPGA), and the analog part of the circuits is controlled by DAC (digital-to-analog converter) and ADC. The CEU is housed in a hermetically sealed enclosure and performs the following tasks:

- Formation of control voltages and signals of the CCD detector and their telemetry;
- Stabilization of the operating temperature of the CCD detector;
- Processing of analog video signal with subsequent ADC and noise-reducing digital filtering;
- Organization of an interface with the communication module for digital data exchange.

The generated timebase logic signals are then further converted to the required levels to control the CCD detector. In addition to standard voltage levels, a high-voltage level (up to -100 V) is generated and applied to the back side of the CCD detector substrate. This voltage can be adjusted to minimize the effect of charge spreading.

CCD261-84 has two video outputs, and, accordingly, two video channels are implemented. The video signal processing channel uses components with low leakage currents, low noise and high speed. Due to the low leakage currents, signal clamping occurs once per line, which significantly improves performance compared to the need to clamp the signal in each pixel.

The processing channel also has stages with gain switching and low-pass filtering. For further video signal conversion into digital form, a high-precision 10 MHz 16-bit ADC is used. The ADC–FPGA interface is organized by serial differential lines, which minimizes interference from digital circuits.

At the stage of digital processing and filtering of the video signal performed in the FPGA, three video signal readout rates—fast, normal or slow with 4, 16 or 64 signal samples, respectively, in one pixel—are implemented. These are processed by an antinoise optimal digital filter with a finite impulse response. The filter coefficients are determined in advance based on the noise spectrum measurement of the CCD output stage. The CCD controller provides not only very low read noise but also high stability of the video channel transfer characteristic under conditions of a large ambient temperature variation around the telescope, which is essential for photometry tasks.

The communication module synchronizes processes in the CCD controller and performs the following tasks:

- Receiving and executing commands from the host computer;
- Receiving video data coming from camera electronics and sending video and telemetry data to the control computer;
- Exposure synchronization with external precise time service;
- External shutter control.

Data packet exchange between the CCD controller and the host computer occurs via a fiber optic interface using the Ethernet 1 Gbit standard [15] and BPF (Berkeley Packet Filter) technology [16]. To exchange the control program with the CCD controller, a subset of protocols of the GigE (gigabit ethernet) Vision 2.0 standard is used [17]. Two types of logical channels are implemented—a control channel and a video channel. The control channel is based on the GigE Vision Control Protocol (GVCP). The video channel is a stream of video data received from the system to the computer's internal memory and is based on the GigE Vision Streaming Protocol (GVSP).

The power source generates seven highly stable supply voltages, performs telemetry of these voltages and additionally implements pressure telemetry by connecting a cryostat pressure sensor.

5. Camera Control Software

The software for data acquisition, reading and processing [18] is programmed using VC++ and QT, and operates under Windows 7/10 x64. The software provides the following capabilities:

- Control of the CCD System and exposure parameters' setup;
- Visualization and analysis of video data;
- Image storage in the Flexible Image Transport System (FITS) standard [19];
- Interactive and automatic observation modes;
- Telemetry and diagnostics of the CCD System;
- Software development kit (SDK).

A standalone program runs multiple threads in parallel: an interface and service thread, a control thread, a visualization thread and a sorting thread. The main thread provides the user interface and execution of commands. The service thread provides the observation process. The control thread receives packets and retrieves data. The sorting thread forms a frame with an image.

The implemented structure allows you to instantly respond to commands and execute them, while the process of visualization and control is simultaneously carried out independently according to its own timeline.

6. Study of the Effect of Charge Spreading

CCD261-84 has a particularly thick (200 μm) high-resistance silicon substrate. Such photodetectors are characterized by the so-called lateral diffusion effect, or charge spreading into adjacent image elements (Figure 4), leading to deterioration in image quality. To reduce this effect, it is necessary to increase the depth of the substrate depletion region.

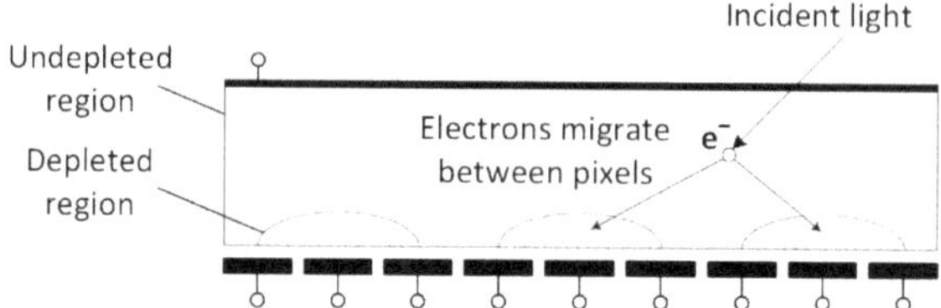

Figure 4. Cross-section of the CCD, showing the effect of photoelectron charge collection.

The use of High-Rho technology makes it possible to achieve full depletion of a thick high-resistance substrate. This is achieved by applying high negative potential (up to -100 V) to the back side of the device. At the same time, the bias on the front side of the substrate remains the same, maintaining the usual voltage level for the clock phases and the video signal output circuits.

Previously, the effect of charge spreading was investigated in [20]. The essence of the estimation method is to calculate the two-dimensional spatial autocorrelation function $R_{m,n}$ of the studied image element (m, n) using the difference D between two uniformly illuminated image arrays of size $M \times N$. The value of the function is calculated by the autocorrelation expression:

$$R_{m,n} = \frac{\sum_{j=1}^{N-n}\sum_{i=1}^{M-m} D_{i,j}D_{i+m,j+n}}{\sum_{j=1}^{N}\sum_{i=1}^{M} D_{i,j}^2}. \tag{1}$$

The left plot of Figure 5 shows the results of the 2D spatial autocorrelation analysis of two flat-field images without binning in low-gain mode with an average flux of $\sim$90 ke$^-$. The central pixel is off-scale and is of value 100%. The correlation is clearly visible and amounts to 0.64% in the horizontal and 1.49% in the vertical direction. The effect of charge spreading in CCD261-84 is lower than in E2V CCD44-82 with the same flux: 1.4% in the horizontal and 2.2% in the vertical direction (for details, see [20]). The right plot of Figure 5 is the result of an analysis of two $\sim$17 ke$^-$ flat-field images without binning in high-gain mode. The vertical correlation is slightly visible and amounts to 0.35%.

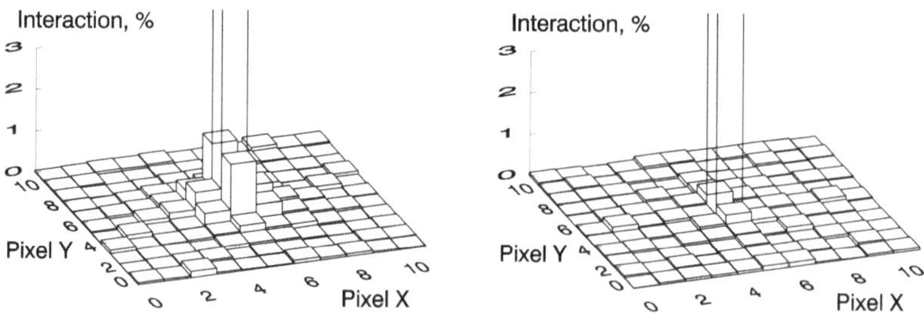

Figure 5. Results of the 2D spatial autocorrelation analysis of data taken with CCD261-84 with a median signal value $\sim$28,000 ADU: in low-gain mode (**left**) and high-gain mode (**right**).

By controlling the substrate voltage V_{bs}, one can find the optimal value for reducing the lateral charge diffusion. Figure 6 shows the relationship between adjacent image elements, depending on the signal strength at different photodetector back substrate voltages at incident radiation wavelengths of 400 and 700 nm. Testing images were obtained using a flat-field stand, including the CCD System and an integrating sphere calibration standard (Gooch & Housego). The operating temperature of the photodetector was -130 ± 0.1 °C.

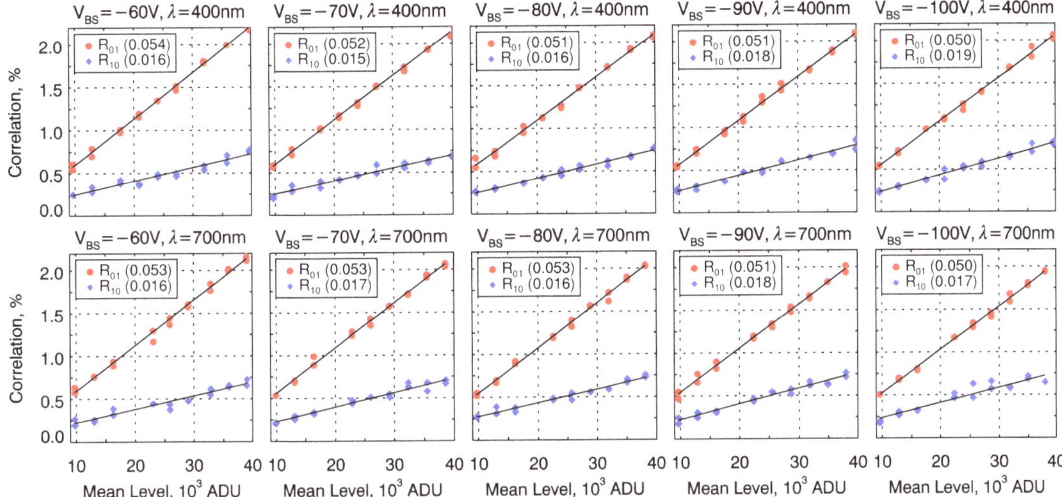

Figure 6. Measured flat-field correlation coefficients R_{01} and R_{10} for the CCD261-84 in low-gain mode and their best-fit line slope factors (given in parentheses).

We see a linear increase in correlation values while increasing the light output. The vertical nearest neighbor coefficient R_{01} is systematically much higher than the other coefficients. This behavior is observed for all values of V_{bs}. The dependence of charge diffusion on voltage is weakly expressed. By lowering the voltage on the substrate to -100 V, we reduce this dependence but within small limits (0.04–0.15%).

As mentioned above, the CCD controller provides high stability of the video channel transfer characteristic. The transfer characteristic is represented by the gain in the video channel, the so-called charge-to-voltage conversion factor of the CCD system. The conversion factor characterizes the relationship between the charge collected in each pixel and the analog-to-digital unit (ADU) value in the output image [21] and is used, in almost all calculations, to obtain real luminous flux values.

Taking into account the summed correlation between pixels, it is possible to more accurately calculate the gain in the system.

7. The Index of Dispersion

Contrary to popular misconceptions, the signal registered by a CCD does not strictly follow Poisson statistics. Regarding astronomical research, the registration of weak signals whose statistics are distorted by the readout noise introduced by electronics is of particular interest.

As a criterion for checking the deviation from the Poisson distribution, the dispersion index, the so-called Fano factor, is effective [22]. By definition, the dispersion index is the ratio of the variance of counts to the average value of the registered signal (concerning the CCD study, the method is described in [23]). For Poisson distribution, this ratio is equal to one and corresponds only to a certain range of registered values.

Figure 7 shows the dependence of the dispersion index on the magnitude of the registered signal in various modes for CCD261-84 in comparison with E2V CCD42-90 (2 K × 4.5 K with a pixel size of 13.5 μm), the detector previously used with the SCORPIO-2 device. The left and right panels correspond to two gain modes—low (×1) and high (×4), respectively. We see that, compared to the previous detector (CCD42-90), with the new CCD, the Poisson statistics can be accepted for relatively low fluxes—100–200 ADU at the 'normal' and 'slow' readout rates.

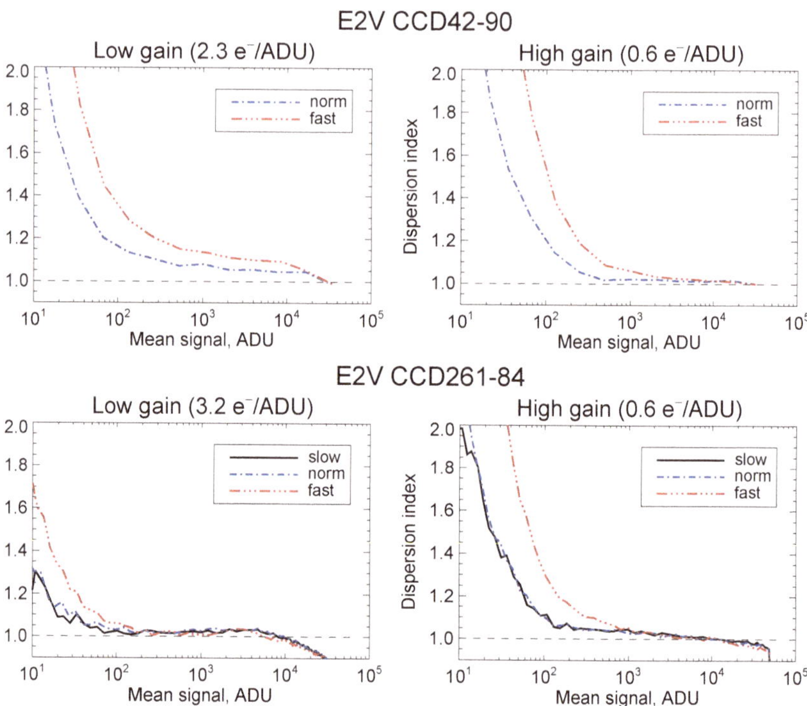

Figure 7. Graphs of the dependence of the dispersion index on the flux of the recorded signal in various readout modes for the CCD261-84 (**bottom**) in comparison with the CCD42-90 (**top**), according to [23].

The plots presented allow one to choose the optimal choice of exposure time and readout mode to prevent distortion of count statistics during observations of astrophysical objects with different brightness levels.

8. Defringing

One of the disadvantages of BSI CCDs is the effect of the interference of incident and reflected waves in the NIR range, the so-called fringes. The reverse side silicon bulk entering wave may be reflected from the front surface of the substrate and, returning back, interfere with the incident wave. As a result, depending on how many times the wavelength fits on the thickness of the pixel, the interference can either increase or decrease the amplitude of the resulting wave. The dependence of the pixel sensitivity on the wavelength becomes a periodic function (Figure 8).

Teledyne E2V uses thicker substrates in its devices to remove this effect. This makes it possible to shift interference effects to wavelengths by about 1 μm, where the sensitivity of the silicon substrate is negligible. Additionally, to reduce interference, multilayer coatings of the substrates with various materials are used to prevent wave reflections. Figure 8 clearly demonstrates the very low amplitude of the fringes detected by the new device: it reaches 1–2% at a wavelength 900–950 nm, while, in detectors of the previous generation, we have significantly higher values (5–10% for CCD42-90 and even more than 50% for CCD42-40). The CCD42-90 and CCD42-40 have a basic midband anti-reflection (AR) coating, and the CCD261-84 has a Multi-2 AR coating.

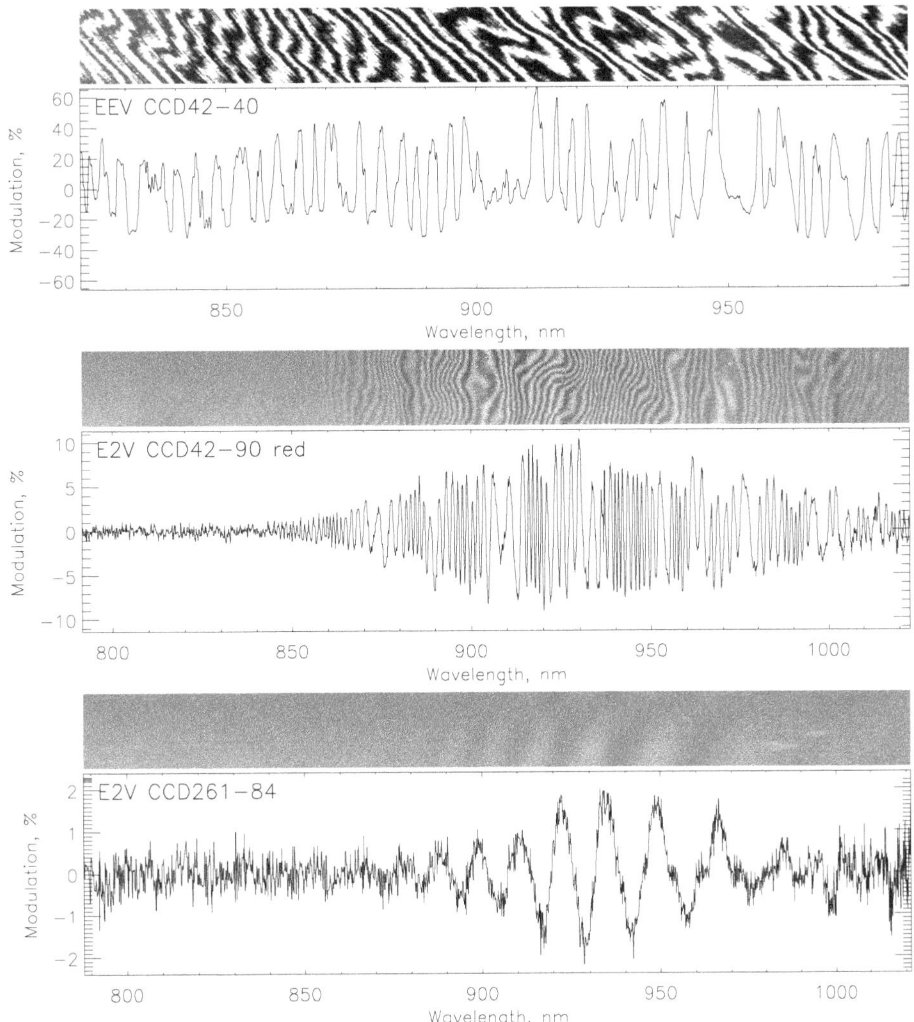

Figure 8. Pattern of fringes in the red spectral range taken from flat-field observations with CCD42-20 at the SCORPIO [24] (**top**), CCD42-90 (**middle**), and CCD261-84 (**bottom**). The last two datasets were obtained with the same grism on SCORPIO-2 illuminated by a system of LEDs in the standard integration sphere of the standard SCORPIO-2 calibration system [25]. The flat field in the archival CCD42-40 was illuminated by a halogen lamp in the previous version of a similar calibration system. Each panel contains a grayscale representation of a fragment of normalized flat field (vertical width is 200 pixels) at a scale of $\pm25\%$ of the average intensity. The plot below shows a horizontal cut along this fragment.

9. Observations and Processing

Since 2020, the camera with CCD261-84 has been used as a detector on the multimode focal reducer SCORPIO-2 [11] at the 6 m telescope of the SAO RAS. This system operates in different readout modes, depending on the type of observation mode of the focal reducer. The full-format and original binning 1×1 is accepted only in the case of spectral observations with the integral-field unit (SCORPIO-2/IFU, [26]). In the most commonly used long-slit spectroscopic mode, the full-format CCD is read in 1×2 binning mode, which provides an optimal spatial sampling 0.''4/pix along the spectrograph slit. For the same reason, reading with 2 × 2 binning is accepted in the SCORPIO-2 direct image mode, whereas only the central square fragment (2048 × 2048 pix in the original binning) is used. In the last case, the detector provides the 6.'8 field of view with the same spatial scale of 0.''4/pix.

Compared to the previous detector CCD42-90, the newest one has significantly higher sensitivity with low contrast of fringes at a wavelength of more than 800 nm (Figure 9). Both of these advantages allow us to use SCORPIO-2 to solve new observational tasks in the red spectral range, mainly the spectroscopy of objects fainter than the foreground sky emission. Recent examples of this include the measurement of CaII triplet velocities at 845–866 nm in the dwarf galaxy KKH 22 [27] and spectroscopy of the distant quasar at redshift $z = 5.47$ [28].

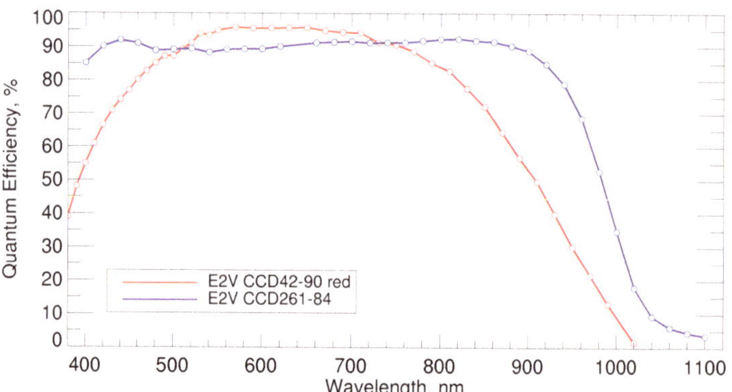

Figure 9. Quantum efficiency (QE) of CCD261-84 in comparison with CCD42-90. QE was obtained using an experimental setup which included the explored CCD System, a monochromator, and a rotating mirror. The measurement method is described in [29]. The increased sensitivity in the blue wavelength range of the blue curve is due to Multi-2 AR coating.

The cost of using this high-efficiency CCD is a higher rate of cosmic ray hits (CHs) than with a thin back illuminated detector. Figure 10 shows an example of the CH pattern collected on the small fragment of the CCD261-84 chip during 40 min of exposure: the length of some cosmic ray tracks can reach several hundreds (in some cases, even several thousands) of pixels. Fortunately, most of these high-contrast CHs are easy to remove with standard algorithms based on Laplacian edge detection (see L.A.COSMIC program in [30]). However, some cosmic rays trapped during the readout process introduce a more serious problem because their tracks are smoothed and seem like images of stars or emission knots (the red arrows in Figure 10, middle). To remove these sorts of CHs in SCORPIO-2, a sigma-clipping procedure to multiple (3–5) frames of the same objects is employed in SCORPIO-2 data reduction after subtraction of the sky emission in individual frames (Figure 10, bottom). Our first experience showed that new types of algorithms need to be developed in this area, different from the standard ones.

Figure 10. Long-slit spectrum obtained with SCORPIO-2 exposed by CCD261-84: 800×300 pixel fragment near the [OIII] and Hβ emission lines of the ionized gas cloud between the galaxies NGC 235 and NGC 232 (for details, see [31]). The sum of 4×600 s exposures is shown. From top to bottom: the raw spectrum including airglow emission and cosmic ray hits; sky emission is subtracted, the red arrows show hits after cosmic ray particles are trapped during the readout; the 'cleaned' spectrum after sigma-filter comparison of individual frames.

10. Results

We have presented the characteristics and performance of the CCD261-84 camera system installed on the multimode focal reducer SCORPIO-2 at the 6 m telescope of the SAO RAS. The main performance characteristics of the CCD system with the CCD261-84 are shown in Table 1.

Table 1. Characteristics of the CCD261-84 camera system.

Parameter	Value	
	Low Gain	High Gain
CCD coating type	Astro Multi-2	
Image size (mm)	30.7×61.6	
Pixel pitch (μm)	15×15	
Readout rates (kpixel/s)	65 (slow), 185 (normal), 335 (fast)	
Readout noise (e^-)	2.62 @ 65 kpixel/s	2.18 @ 65 kpixel/s
Gain factor (e^-/ADU)	3.2	0.6
Full well capacity (ke$^-$/pixel)	165	35
Dark current (e^-/s/pixel)	0.001	
Operating temperature ($^\circ$C)	-130	
Thermal stability ($^\circ$C)	±0.1	
Linearity (%)	>99.3 (0–140 ke$^-$)	

The operating spectral range of the SCORPIO-2 multimode focal reducer, which provides a significant part of the photometric and spectral observational programs of the BTA, has been extended to the red region (up to 1000 nm).

For the designed CCD system, a deviation from the Poisson statistics is observed for fluxes weaker than 250 e$^-$ with a readout noise of 2.18 e$^-$. The CCD system can be regarded as almost 'ideal' in the flux range of 250–40,000 e$^-$. The dark current decreased by a factor of three compared to the previous detector (CCD42-90). The amplitude of the fringes decreased five times.

An image filtering procedure has been developed to remove traces of cosmic particles in relation to detectors with a thick substrate.

The features and size of fringes in a thick substrate have been studied, and recommendations have been developed for taking them into account in observations.

11. Discussion

This article presents a finished working CCD system based on a fully depleted scientific detector CCD261-84 with substrate thickness of 200 μm. Descriptions of implemented CCD systems with such a photodetector have not been published before; there are only data on the study of the photodetectors themselves by the manufacturer [8–10] and measurements of the detector characteristics to substantiate the charge transfer model in such detectors, as well as to determine the expediency of using detectors in planned projects [32]. It is shown that the thoughtful design of the camera and the use of a universal CCD controller developed at the SAO RAS made it possible to achieve high photometric characteristics that are inherent in the CCD detector itself. Primarily, it has low readout noise and low dark current, which makes it possible to effectively use the device for spectral observations of faint objects with exposures up to 1 h. The achieved readout noise was 2.18 e$^-$ @ 65 kpixel/s at the declared value of 2.8 e$^-$ @ 50 kpixel/s in the manufacturer datasheet. Such a level of readout noise is inherent in a few working astronomical CCD systems.

To characterize the degree of charge spreading in the substrate, we used the method for finding the two-dimensional autocorrelation function of the difference between two flat-field frames. The application of this method does not require the construction of a complex stand with a movable point light source followed by obtaining a point spread function (PSF); rather, only a flat-field stand is needed. In this case, the method makes it possible to estimate the degree of charge spreading for a given wavelength and a known charge value in a pixel. This article shows that in the operating range of substrate voltage (-60 V...-100 V), the amount of charge spreading varies slightly depending on the substrate voltage. This is consistent with the data of Figure 4 in [8], which shows that the PSF changes insignificantly in this range of substrate voltages for samples of devices with varying degrees of silicon depletion. When choosing the operating voltage of the back substrate, it is also necessary to consider the fact that, with an increase in the potential difference, the generation of a parasitic charge in hot pixels increases (Figure 8 in [8]). We also demonstrated a pronounced direct dependence of the autocorrelation coefficients on the signal level, which is consistent with theory. We can compare our data with the data from Figure 6.9 in [32]. In the range of signals from 20 to 1000 ke$^-$ at a substrate voltage of -70 V, the author does not observe a clear dependence, although the order of the coefficients is the same. This may indicate that the quality of the video channel in the data acquisition system used.

The Fano factor also indicates the quality of the video signal processing channel. The graph shown for this system (Figure 7, bottom) testifies to the high quality of the video channel, evidenced by the absence of interference and distortion of statistics in various input signals. Since this parameter is practically not given in the description of CCD systems, it is difficult to compare it with other projects.

The characteristic quantum sensitivity curve for detectors with depleted silicon and the Multi-2 anti-reflection coating is also given in Figure 14 in [8]. The measured values of the quantum efficiency for our CCD system are in good agreement with the curve from this publication as well as with the typical curve presented in the datasheet of the detector, considering some losses in the camera optical window. Thanks to the Multi-2 coating, the detector has very high sensitivity, not only in the red, but also in the blue region of the spectrum.

12. Conclusions

In this paper, we describe the design, implementation and operating principles of an astronomical camera system based on a large-format CCD261-84 detector. To use the new BSI CCD with a very thick substrate, a CCD controller with the possibility of generating a controlled high-voltage level has been developed at SAO RAS. The CCD System with CCD261-84 is now used as a detector on the multimode focal reducer SCORPIO-2 at the 6 m telescope of the SAO RAS. Studies of the photometric characteristics of the system as well as the effect of charge spreading in the substrate and the formation of fringes have been carried out. Low readout noise and dark current were achieved. Based on the images obtained in the observations, new methods of processing and reduction of spectral images have been developed to minimize the increased impact of cosmic ray particles. The developed CCD System has significantly higher sensitivity, with a low contrast of fringes at a wavelength of more than 800 nm, which makes it possible to use the SCORPIO-2 at the SAO RAS 6 m telescope to solve new observational tasks in the red spectral range.

Author Contributions: Conceptualization, V.M., V.A., I.A. and N.I.; methodology, all authors; software, I.A., V.M. and A.M.; validation, V.M., I.A., V.A., N.I. and M.P.; data curation, V.M., I.A., A.M. and E.M.; writing—original draft, V.A.; writing—review & editing, V.A., I.A., V.M., A.M., E.S. and E.M.; visualization, I.A., V.A., A.M. and E.M.; supervision, I.A. and V.M.; project administration, I.A. All authors have read and agreed to the published version of the manuscript.

Funding: The observational data were collected using the unique scientific facility Big Telescope Alt-azimuthal of SAO RAS, and data processing was performed with the financial support of grant No075-15-2022-262 (13.MNPMU.21.0003) of the Ministry of Science and Higher Education of the Russian Federation.

Institutional Review Board Statement: Not applicable.

Informed Consent Statement: Not applicable.

Data Availability Statement: The data are available from the first author upon reasonable request.

Acknowledgments: We express our gratitude to the staff of the Laboratory of Spectroscopy and Photometry of Extragalactic Objects (SAO RAS): Perepelitsyn, A.E.; Uklein, R.I.; Oparin, D.V.; and Kotov, S.S. for their assistance in laboratory measurements of CCD characteristics.

Conflicts of Interest: The authors declare no conflict of interest.

References

1. Rieke, G.H. History of infrared telescopes and astronomy. *Exp. Astron.* **2009**, *25*, 125–141. [CrossRef]
2. Vincent, J.D.; Hodges, S.E.; Vampola, J.; Stegall, M.; Pierce, G. *Fundamentals of Infrared and Visible Detector Operation and Testing*, 2nd ed.; John Wiley & Sons: New Jersey, NB, Canada, 2016. [CrossRef]
3. McLean, I. *Electronic Imaging in Astronomy: Detectors and Instrumentation*, 2nd ed.; Springer: New York, NY, USA, 2008. [CrossRef]
4. Teledyne-E2V. Scientific CCD Image Sensors. Available online: https://www.teledyne-e2v.com/en/solutions/scientific/scientific-ccd-image-sensors (accessed on 18 May 2023).
5. Hayes, P.S.; Pool, P.J.; Holtom, R. A new generation of scientific CCD sensors. *Proc. SPIE* **1997**, *3019*, 201–209. [CrossRef]
6. Downing, M.; Baade, D.; Deiries, S.; Jorden, P. Bulk Silicon CCDs, Point Spread Functions, and Photon Transfer Curves: CCD Testing Activities at ESO. 2009 Workshop Detectors for Astronomy. Available online: https://scholar.google.com/scholar?hl=ru&as_sdt=0%2C5&q=Bulk+Silicon+CCDs%2C+Point+Spread+Functions%2C+and+Photon+Transfer+Curves&btnG=1 (accessed on 18 May 2023)
7. Jorden, P.R.; Ball, K.; Bell, R.; Burt, D.; Guyatt, N.; Hadfield, K.; Jerram, P.; Pool, P.; Pike, A.; Holland, A.; et al. Commercialisation of full depletion scientific CCDs. *Proc. SPIE* **2006**, *6276*, 627604. [CrossRef]
8. Robbins, M.S.; Mistry, P.; Jorden, P.R. Detailed characterisation of a new large area CCD manufactured on high resistivity silicon. *Proc. SPIE* **2011**, *7875*, 787507. [CrossRef]
9. Jorden, P.R.; Downing, M.; Harris, A.; Kelt, A.; Mistry, P.; Patel, P. Improving the red wavelength sensitivity of CCDs. *Proc. SPIE* **2010**, *7742*, 77420J. [CrossRef]
10. Jorden, P.R.; Jordan, D.; Jerram, P.A.; Pratlong, J.; Swindells, I. E2V new CCD and CMOS technology developments for astronomical sensors. *Proc. SPIE* **2014**, *9154*, 91540M. [CrossRef]
11. Afanasiev, V.L.; Moiseev, A.V. Scorpio on the 6 m Telescope: Current State and Perspectives for Spectroscopy of Galactic and Extragalactic Objects. *Balt. Astron.* **2011**, *20*, 363–370. [CrossRef]
12. Markelov, S.V.; Murzin, V.A.; Borisenko, A.N.; Ivaschenko, N.G.; Afanasieva, I.V.; Ardilanov, V.I. A high-sensitivity CCD camera system for observations of early universe objects. *Astron. Astrophys. Trans.* **2000**, *19*, 579–583. [CrossRef]
13. Murzin, V.A.; Markelov, S.V.; Ardilanov, V.I.; Afanasieva, I.V.; Borisenko, A.N.; Ivashchenko, N.G.; Pritychenko, M.A. Astronomical CCD systems for the 6-meter telescope BTA (a review). *Adv. Appl. Phys.* **2016**, *4*, 500–506. (In Russian)
14. Ardilanov, V.I.; Murzin, V.A.; Afanasieva, I.V.; Ivaschenko, N.G.; Pritychenko, M.A. Development of Large-Format Camera Systems Based on the Latest Generation Sensors for the 6-m Telescope. In Proceedings of the All-Russian Conference "Ground-Based Astronomy in Russia, 21st Century", Arkhyz, Russia, 21–25 September 2020; Romanyuk, I.I., Yakunin, I.A., Valeev, A.F., Kudryavtsev, D.J., Eds.; Special Astrophysical Observatory of RAS: Nizhnii Arkhyz, Russia, 2020; pp. 115–118. [CrossRef]
15. IEEE 802.3 Ethernet Working Group. Available online: https://www.ieee802.org/3/ (accessed on 18 May 2023).
16. Li, J.; Wu, C.; Ye, J.; Ding, J.; Fu, Q.; Huang, J. The Comparison and Verification of Some Efficient Packet Capture and Processing Technologies. In Proceedings of the 2019 IEEE International Conference on Dependable, Autonomic and Secure Computing, International Conference on Pervasive Intelligence and Computing, International Conference on Cloud and Big Data Computing, International Conference on Cyber Science and Technology Congress (DASC/PiCom/CBDCom/CyberSciTech), Fukuoka, Japan, 5–8 August 2019; pp. 967–973. [CrossRef]
17. Association for Advancing Automation. *GigE Vision—Video Streaming and Device Control Over Ethernet Standard*; Association for Advancing Automation: Ann Arbor, MI, USA, 2018; Rel. 2.1.
18. Afanasieva, I.V. Data Acquisition and Control System for High-Performance Large-Area CCD Systems. *Astrophys. Bull.* **2015**, *70*, 232–237. [CrossRef]
19. Pence, W.D.; Chiappetti, L.; Page, C.G.; Shaw, R.A.; Stobie, E. Definition of the Flexible Image Transport System (FITS), version 3.0. *Astron. Astrophys.* **2010**, *524*, A42. [CrossRef]
20. Downing, M.; Baade, D.; Sinclaire, P.; Deiries, S.; Christen, F. CCD Riddle: a) Signal vs. Time: Linear; b) Signal vs. Variance: Non-Linear. *Proc. SPIE* **2006**, *6276*, 627609. [CrossRef]
21. Howell, S.B. *Handbook of CCD Astronomy*, 2nd ed.; Cambridge University Press: New York, NY, USA, 2006; pp. 36–64. [CrossRef]
22. Fano, U. Ionization Yield of Radiations. II. The Fluctuations of the Number of Ions. *Phys. Rev.* **1947**, *72*, 26. [CrossRef]
23. Afanasieva, I.V. Study of distortions in statistics of counts in CCD observations using the Fano factor. *Astrophys. Bull.* **2016**, *71*, 366–370. [CrossRef]
24. Afanasiev, V.L.; Moiseev, A.V. The SCORPIO Universal Focal Reducer of the 6-m Telescope. *Astron. Lett.* **2005**, *31*, 194–204. [CrossRef]
25. Afanasiev, V.L.; Amirkhanyan, V.R.; Moiseev, A.V.; Uklein, R.I.; Perepelitsyn, A.E. SCORPIO-2 guiding and calibration system in the prime focus of the 6-m telescope. *Astrophys. Bull.* **2017**, *72*, 458–468. [CrossRef]
26. Afanaisev, V.L.; Egorov, O.V.; Perepelitsyn, A.E. IFU Unit in Scorpio-2 Focal Reducer for Integral-Field Spectroscopy on the 6-m Telescope of the Special Astrophysical Observatory of the Russian Academy of Sciences. *Astrophys. Bull.* **2018**, *73*, 373–386. [CrossRef]
27. Karachentsev, I.D.; Makarova, L.N.; Tully, R.B.; Anand, G.S.; Rizzi, L.; Shaya, E.J.; Afanasiev, V.L. KKH 22, the first dwarf spheroidal satellite of IC 342. *Astron. Astrophys.* **2020**, *638*, A111. [CrossRef]

28. Khorunzhev, G.A.; Meshcheryakov, A.V.; Medvedev, P.S.; Borisov, V.D.; Burenin, R.A.; Krivonos, R.A.; Uklein, R.I.; Shablovinskaya, E.S.; Afanasiev, V.L.; Dodonov, S.N.; et al. Discovery of the Most X-ray Luminous Quasar SRGE J170245.3+130104 at Redshift $z \approx 5.5$. *Astron. Lett.* **2021**, *47*, 123–140. [CrossRef]

29. Jähne, B. Release 4 of the EMVA 1288 standard: Adaption and extension to modern image sensors. In *Forum Bildverarbeitung 2020*; Heizmann, M., Längle, T., Eds.; KIT Scientific Publishing: Karlsruhe, Germany, 2020; pp. 13–24. Available online: https://scholar.google.com/scholar?hl=ru&as_sdt=0%2C5&q=Release+4+of+the+EMVA+1288+Standard%3A+Adaption+and+Extension+to+Modern+Image+Sensors&btnG= (accessed on 20 June 2023)

30. van Dokkum, P.G. Cosmic-Ray Rejection by Laplacian Edge Detection. *Publ. Astron. Soc. Pac.* **2001**, *113*, 1420–1427. [CrossRef]

31. Keel, W.C.; Moiseev, A.V.; Kozlova, D.V.; Ikhsanova, A.I.; Oparin, D.V.; Uklein, R.I.; Smirnova, A.A.; Eselevich, M.V. The TELPERION survey for distant [O III] clouds around luminous and hibernating AGN. *Mon. Not. R. Astron. Soc.* **2022**, *510*, 4608–46250. [CrossRef]

32. Weatherill, D. Charge Collection in Silicon Imaging Sensors. Ph.D. Thesis, The Open University, Milton Keynes, UK, 2016. Available online: https://scholar.google.com/scholar?hl=ru&as_sdt=0%2C5&q=+Charge+collection+in+silicon+imaging+sensors&btnG= (accessed on 20 June 2023)

Article

Optical Design of a Slitless Astronomical Spectrograph with a Composite Holographic Grism

Eduard Muslimov [1,2,3,*], Damir Akhmetov [2,4], Danila Kharitonov [2,4], Erik Ibatullin [2], Nadezhda Pavlycheva [2], Vyacheslav Sasyuk [5] and Sergey Golovkin [6]

[1] NOVA Optical IR Instrumentation Group, ASTRON—The Netherlands Institute for Radio Astronomy, Oude Hoogeveensedijk 4, 7991 PD Dwingeloo, The Netherlands
[2] Optical and Electronic Systems Department, Kazan National Research Technical University Named after A.N. Tupolev KAI, 10 K. Marx, 420111 Kazan, Russia
[3] LAM, CNES, CNRS, Aix Marseille Univ, 13388 Marseille, France
[4] Scientific and Production Association "State Institute of Applied Optics", 2 Lipatov, 420075 Kazan, Russia
[5] Engelhardt Observatory, Kazan Federal University, 420008 Kazan, Russia
[6] Leonov Planetarium, Kazan Federal University, 7 AOE Street, 422526 Oktyabrskiy, Russia
* Correspondence: muslimov@astron.nl or eduard.muslimov@lam.fr

Citation: Muslimov, E.; Akhmetov, D.; Kharitonov, D.; Ibatullin, E.; Pavlycheva, N.; Sasyuk, V.; Golovkin, S. Optical Design of a Slitless Astronomical Spectrograph with a Composite Holographic Grism. *Photonics* **2023**, *10*, 385. https://doi.org/10.3390/photonics10040385

Received: 16 January 2023
Revised: 3 March 2023
Accepted: 29 March 2023
Published: 31 March 2023

Abstract: In the present work, we consider an optical design of a slitless spectrograph for an existing 0.5 m-class telescope. This design concept has a number of advantages such as compact size, simplicity, and simultaneous coverage of a large field of view. A challenge with this design is correcting aberrations caused by placing a dispersing element in a converging beam. To overcome this issue, we propose to use a composite grism, which represents a combination of a prism and a volume-phase holographic grating, the latter which is split into zones with independently optimized parameters. We demonstrate two designs of such a grism. In both designs, the spectrograph operates in the range of 450–950 nm in an $F/6.8$ beam and covers a field of view of $35.6' \times 7.2'$. Through advanced modeling, it is shown that a composite grism having four rectangular zones with different thickness and index modulation depth of the hologram and recorded with an auxiliary deformable mirror decreases the astigmatic elongation by a factor of 85, increases the spectral resolving power by 4.4 times, and reaches $R1389$ while increasing the average diffraction efficiency by a factor of 1.31. If we reduce the number of zones to only two, replace the deformable mirror with two static corrector plates, and fix the hologram thickness, the corresponding performance gains still remain high: the astigmatism is reduced by a factor of 61, the spectral resolving power is up to 1.7 times higher, reaching $R1067$, and the efficiency is increased by a factor of 1.27. This shows that the proposed design allows the construction of a simple and compact instrument, providing high performance over the entire field of view and spectral range.

Keywords: slitless spectroscopy; astronomical spectrograph; grism; volume-phase hologram; aberration correction; diffraction efficiency; composite hologram

1. Introduction

Spectroscopy is a key method in astronomy, as it allows one to determine the temperature, chemical composition, and kinematics of celestial objects. Therefore, most professional astronomical telescopes are equipped with spectroscopic instruments of different kinds. Usually, such an instrument projects a static, spectrally dispersed image of a source onto a two-dimensional photo-detector array without scanning and therefore is classified as a spectrograph. One dimension of the detector array corresponds to the spectral coordinate, but there are different approaches to handling the spatial coordinate of the object. Depending on the design solution regarding this second coordinate, one can distinguish such classes as slitless, longslit, multislit, and integral-field spectrographs [1].

Slitless spectroscopy [2] has a number of well-known shortcomings such as a high background level, which reduces the signal-to-noise ratio and contamination of the spectral image by the image features of the object of interest itself and surrounding objects. However, this technique has a number of advantages, which result in a growing interest in its astronomical applications. First, it can be implemented with a relatively simple instrument, both in terms of design and operation. Second, slitless spectroscopy has a very high multiplexing capability, allowing one to obtain a large number of spectra at once. Third, it has the potential to provide flux-limited surveys with high spectro-photometric accuracy, insensitive to fiber or slit losses [3].

One field that can obviously benefit from the advantages of slitless spectroscopy is space optics: compact and simple instruments capable of detecting multiple spectra in a relatively wide field of view at once are mounted on board space telescopes. Flagship missions payload such as Advanced Camera for Surveys on board the Hubble Space Telescope [4] and the NIRISS instrument on the James Webb Space Telescope [5] provide perfect examples for this type of design and the corresponding observation technique. Another example is the UV (ultraviolet) slitless spectroscopy mode of the GALEX space telescope [6]. The latter example is of specific interest since for operating in the UV band, where the choice of materials and coatings is limited, it is very desirable to reduce the number of optical components and thicknesses of transmission ones, whereas a single dispersive element should form spectral images in two sub-bands on two separate detectors. This urges mounting a single dispersive element in a converging beam and fine-tuning its diffraction efficiency curve.

Another application for slitless spectrographs is represented by small ground-based telescopes. Examples such as [7,8] show how the compact size, reduced pointing accuracy requirement, and design simplicity of a slitless spectrograph can become crucial for this class of instruments.

Another complication with slitless spectroscopy is the optical design of the systems. In particular, such an optical system is often mounted in a converging beam and operates with a wide field of view (FoV). Therefore, the aberrations and throughput of the optics will vary both across the aperture and FoV. In [9] these, issues were noted, and an approach to correct the aberrations of a grating mounted in a converging beam was proposed.

In previous work, we considered the design of a slitless spectrograph for a fast and wide-field small telescope [10]. We also had demonstrated that by using the concept of a composite holographic element [11], one can substantially increase the spectral resolution and the diffraction efficiency (DE) of a holographic grating working in such a non-conventional setup. The composite element represents a volume-phase holographic (VPH) grating split into several zones. In each of the zones, the grooves pattern and their profile can be varied independently to maximize the overall performance. It is convenient to combine such a grating with a prism to moderate the chief ray deviation. The resulting optical component is referred to hereafter as a composite grism.

We consider the design of a slitless spectrograph as a good test case for a comparative study of a composite holographic element's performance. On the one hand, the hologram operating conditions vary across the aperture because the element is mounted in a converging beam, across the field of view, because the spectrograph works with an extended two-dimensional field of view, and across the spectrum since it operates in a wide spectral range. These features distinguish a slitless spectrograph design from those considered before in the context of the composite holograms studies as a waveguide holographic display, a flat-field spectrograph with concave grating, or an imaging spectrograph with a grism in a collimated beam. Once the hologram working conditions are more challenging, the performance difference between the composite element and a classical one will be easier to demonstrate, both in modeling and in a real experiment. On the other hand, the slitless spectrograph design is easier to implement in practice, since it does not require dedicated pre-optics or a detector and can consist of a relatively simple opto-mechanical assembly with a single optical component, namely the composite grism. However, if we presume that the design under consideration should eventually be turned into a real device and become a proof of concept, we have to apply some additional limitations and perform a more detailed analysis. Thus, in contrast with [10], we simplify the grism surfaces' shapes. Because of the large influence of conical diffraction, we should use a precise numerical method instead of analytical scalar diffraction theory to optimize and analyze the diffraction efficiency. As a result, we should explicitly compute the instrument functions for a finite-width virtual slit, taking into account aberrations of individual zones of a composite hologram and not limit the performance comparison to just spot diagrams analysis. Both of these points require some work on custom modeling tools.

The goals of this research are to develop and compare slitless spectrograph designs for an existing small telescope, providing a comprehensive performance analysis in each case in order to prepare for a future proof-of-concept experiment. The paper is organized as follows: Section 2 presents the optical design versions; in Section 3 we provide the image quality and spectral resolution analysis; Section 4 provides the DE optimization results and analysis; Section 5 discusses the analysis; and Section 6 contains the main conclusions and the plan of future work.

2. Optical Design

As the target hosting telescope we chose a CDK500 [12] from *PlaneWave Instruments^{TM}*, (Adrian, MI, USA) an instrument owned by the Institute of Physics of Kazan Federal University. It is a Modified Dall–Kirkham telescope with a lens corrector and a primary mirror of 508 mm in diameter and central obscuration of 0.39. The focal length is 3454 mm, resulting in $F/\# = 6.8$. One of our main goals was to re-use the existing CCD camera of the telescope for the detection of the spectral images. Therefore, the key parameters of the spectrograph under development were chosen accordingly.

The CCD represents an array of 4096×4096 pixels having linear dimensions of 36.8×36.8 mm^2. Taking the telescope focal length into account, we set the target FoV equal to $35.6' \times 7.2'$ and the spectral image length (i.e., the distance between the centers of monochromatic images formed at the edges of the working spectral range) of 29 mm. The working spectral range is 350–950 nm, which corresponds to the sensitivity region of a typical Si-based sensor.

Accounting for the actual mechanical interfaces of the telescope, we limit the distance between the first surface of the composite grism and the incoming beam focus to 160 mm.

Considering the basic Dall–Kirkham design [13], we can restore a presumable optical layout of the CDK500 and design a slitless spectrograph coupled to it (see Figure 1). The main optical system of the telescope, consisting of the primary and secondary mirrors (1, 2, resp.) and lens corrector 3, forms a converging beam which is focused on the nominal focal plane 4. The grism is mounted in a pre-focal position. It represents a transmission VPH grating 5, recorded on a 3 mm-thick substrate made of *LZOS^{TM}* (Lytkarino, Russian Federation) LK7 glass and glued together with a prism 6. The composite grating represents a mosaiced element consisting of a few rectangular parts, hereafter called zones. The number of zones and their sizes change for different design versions. In practice, the zones can be formed by imposing the photosensitive layer and recording the fringes pattern consequently through a few masks on the same substrate. It was demonstrated that parts of a mosaiced VPH can be co-aligned with high precision and act as a single dispersive element in an astronomical spectrograph [14]. We propose to introduce additional degrees of freedom in the current design, defined for each of the zones separately. The prism is made of *LZOS^{TM}* TK8 glass with an axial thickness of 25 mm and its facets having tilt angles of $4.47°$ and $6.85°$ with respect to the telescope optical axis. These parameters are constant for all the design versions discussed below to facilitate the comparison. Note that a solution with a single prism does not allow the creation of an exact zero-deviation geometry, but makes the component much simpler in manufacturing while the deviation angle is moderated. The spectral image is detected by the same CCD array mounted in position 7.

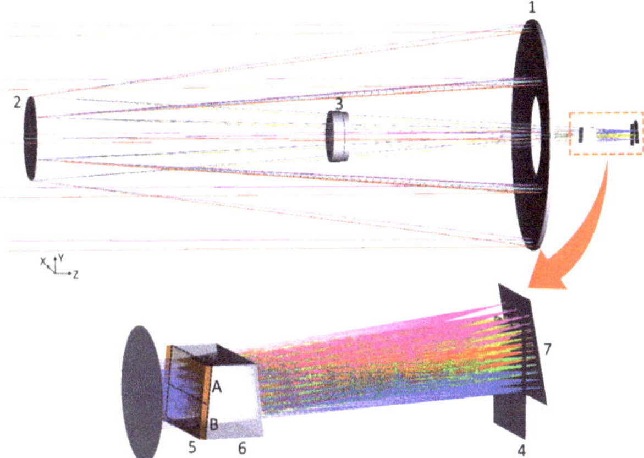

Figure 1. General view of the slitless spectrograph optics coupled with a Dall–Kirkham telescope: 1—primary mirror, 2—secondary mirror, 3—lens corrector, 4—focal plane position in the imaging mode, 5—composite holographic grating, split into zones with independently optimized parameters as A and B, 6—prism, 7—spectral image plane position.

The telescope provides nearly diffraction-limited image quality across the entire FoV and an almost telecentric output beam with the exit pupil position at −1419 mm. For these reasons, we exclude the telescope system from the model and replace it by a perfect lens with the same field and aperture.

As can be seen in Figure 1, adding the spectrograph does not significantly change the telescope's overall dimensions, and the optical design includes only one optical component. The key functionality is performed by the VPH grating and we consider the following versions of that.

- Classical grating. This is a default option, which can be produced by most manufacturers or even found as an off-the-shelf component. It represents a grating with straight equidistant fringes having a profile perpendicular to the substrate surface. Such a grating can be recorded in a setup shown in Figure 2a, where the laser beam forms two sources 1 and the beams are collimated by two off-axis parabolic (OAP) mirrors 2, which interfere in a symmetric geometry on the substrate 3. The holographic layer thickness is equal to a standard value and the modulation depth is optimized for a chief ray using scalar diffraction theory.

- Composite grating. This version represents all the capacities of the concept shown in [11]. The recording beams in this setup Figure 2b are formed by the same sources 1 and OAP's 2, but the angles of incidence onto the substrate 3 are not symmetric. A deformable mirror (DM) 4 is introduced in one of the interferometer's branches to control the wavefront and introduce the aberration correction. The mirror surface shape is described by the Zernike polynomials [15] up to the 2nd order (we use only the YZ-symmetric terms):

$$z = \frac{r^2/R}{1 + \sqrt{1 - (1+k)r^2/R^2}} + \sum_{i=4}^{25} A_i Z_i(\rho, \phi); \tag{1}$$

where R, mm is the vertex radius of curvature, k is the conic constant, Z_i are Zernike polynomials, and A_i are corresponding coefficients, r, mm is radial coordinate and (ρ, ϕ) are normalized polar coordinates of a point on a surface. We presume that the grating is split into four equal horizontal stripes and the deformable mirror shape is optimized for every stripe independently, whereas the rest of the setup arrangement remains the same. In addition to this, we assume that the thickness of the deposed photosensitive layer can be chosen for the stripes independently as well as the exposure time, which defines the index modulation depth. The latter condition can be relatively easily implemented if the grating is recorded through a movable rectangular mask.

- Simplified composite grating. Producing a recording setup as described in the previous item, with all its degrees of freedom, may be very challenging. Therefore, we consider a simplified version. In this setup, the aberrated wavefront is formed by a tilted corrector plate 4 with an ordinary axisymmetric asphere on its first surface:

$$z = \frac{r^2/R}{1 + \sqrt{1 - (1+k)r^2/R^2}} + \alpha_2 r^2 + \alpha_4 r^4, \tag{2}$$

where α_2, mm^{-1} and α_4, mm^{-3} are the asphericity coefficients. In this case, we use only two rectangular zones, and the corresponding corrector plates are substituted during the recording. The rest of the recording setup remains the same, including the branches' asymmetry. Finally, we assume that the thickness of the hologram structure is fixed and equal to a standard value, whereas the modulation depth can be optimized separately for each zone.

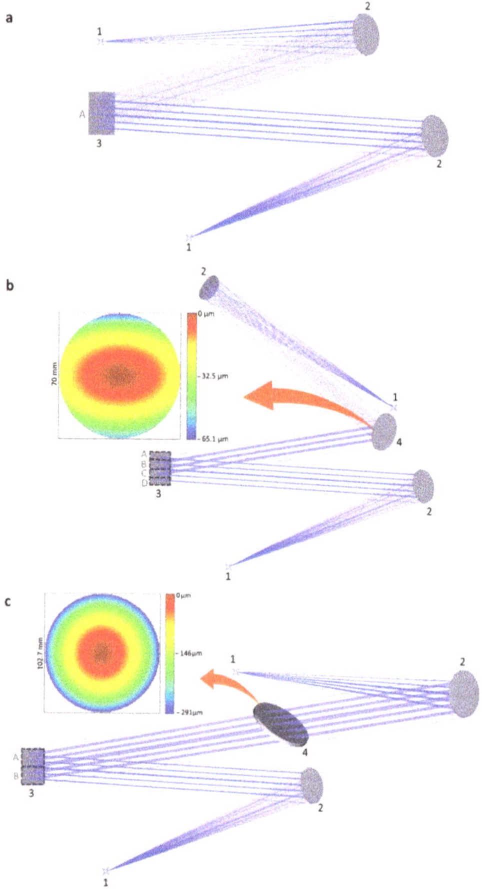

Figure 2. Holographic grating recording setups: (**a**)—classical grating, (**b**)—composite grating, (**c**)—simplified composite grating. The setup elements are: 1—laser point sources, 2—collimating mirrors, 3—hologram substrate, 4—auxiliary optics for the wavefront control; A, B, C and D—composite hologram zones.

We consider the classical grating as a good reference case, since the majority of dispersive units in spectral instruments is represented by such elements and it would be easy to produce and test in practice. The composite element corresponding to Figure 2b is intended to show the capabilities of the proposed solution in terms of image quality and diffraction efficiency improvement once all the currently available technological facilities are employed. The simplified version shown in Figure 2c may be used for early experimental proof of concept since it does not require the use of customized photosensitive layers or active optics. In each case, the recording geometry directly defines the fringes' shapes and their section profiles, so it should be directly included in the modeling and optimization process. In the following sections, we show the optimization and analyze the performance of each of the presented design options.

3. Image Quality Optimization and Analysis

The primary tool used to assess the spectrograph's image quality is spot diagrams obtained by ray tracing a set of control wavelengths and FoV points. The diagrams for the classical grism are shown in Figure 3, where the X axis corresponds to the spatial direction and the Y axis to the spectral dimension. Note that the optical design is symmetrical with respect to the tangential (ZY) plane and the aberrations vary gradually across the FoV and spectral range. The spot diagrams corresponding to the center and four corners of the FoV at three reference wavelengths characterize the image quality across the entire spectral image. It is assumed that the detector plane is perfectly aligned in its nominal position. It is clear from the plots that the classical grism introduces large aberrations, mainly astigmatism. These cannot be corrected because of a lack of free design variables. In this case, we can only find a focal plane position that minimizes the spot size in the dispersion direction for the central wavelength and a tilt angle for which the spots' blurring is moderated at the edges of the spectral range. With the given aberrations the spectra of neighboring objects may intersect, limiting the number of simultaneously observable targets drastically. A positive aspect is that the image quality remains relatively stable across the FoV, slightly simplifying the task of aberration correction.

In addition to the spot diagrams, we use instrument functions (IF). An IF represents a transverse illumination distribution in an image of a slit of finite width, or that of a middle section of a finite-size spot image. By definition, this distribution can be obtained as a convolution of the object luminosity function and the optical system line spread function (LSF). This approach works for the classical grism, but not for a composite one, since the LSF is usually computed using a Fourier transform, which ignores splitting of the grating into independent zones. Therefore, for this work, we had to use another, more straightforward method as described in [16]. It relies on tracing a large number of rays covering the full system aperture and the full object width and counting ray intercepts in a number of narrow stripes at the image plane. This technique was used in the past for the IF computation of spectrographs with aberration-corrected holographic gratings and has shown a good match with experimental data [17]. Before applying it for analysis and optimization of the current design, we performed a comparison and verification. In Figure 4, we show two IF plots computed at 700 nm for the FoV center in the case of a single holographic grating—which is not split into zones—computed with two methods. In both computations, we assume that the object luminosity represents a rectangular function with a width of 59 µm, which corresponds to 3.5'' seeing. The first curve is computed using a fast Fourier transform (FFT) LSF and the rectangular function. The second one employs the rays-counting approach described above. As one can see, the curves are very similar, with a deviation of 8.6% in width, indicating that the two methods converge well. To illustrate the effects of seeing, aberrations, and the grating mosaicing, we also show the initial LSFs in the same axes. We consider two cases: a single rectangular aperture and an aperture split into four identical rectangular zones. The second case corresponds approximately to Figure 2b but neglects the difference in aberrations for the zones. The zones are separated by thin opaque border lines, corresponding to possible defects at the edges of the recording masks. To emphasize the difference, we set the lines' width equal to 1 mm. Even with these exaggerated borders' widths, the LSFs are almost identical. Part of the energy of the LSF peak is diffracted to its wings, but the overall shape and width remain almost the same. Thus, we can conclude that the actual instrument function shape will be driven by the object width, i.e., the seeing, and affected by the optical system aberrations, whereas the influence of diffraction at the apertures is negligibly small.

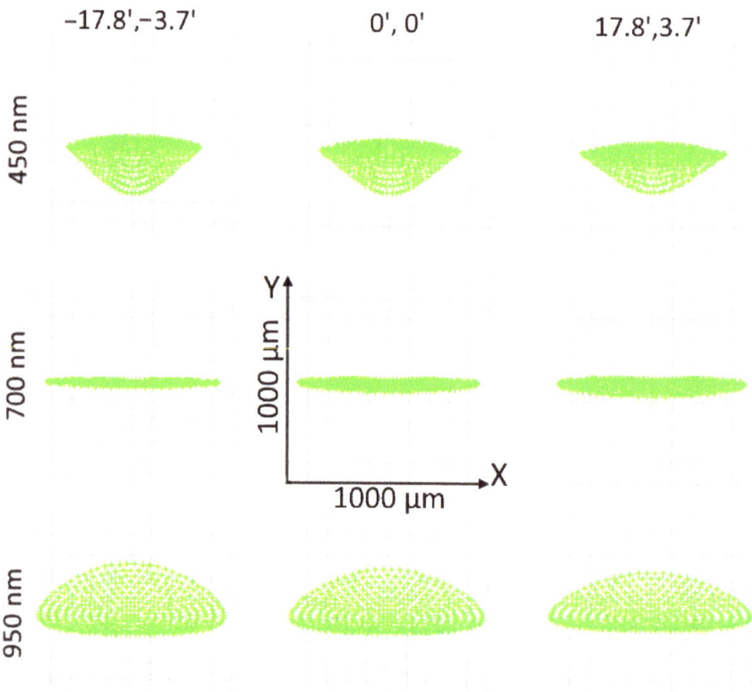

Figure 3. Spot diagrams of a spectrograph with a classical grism.

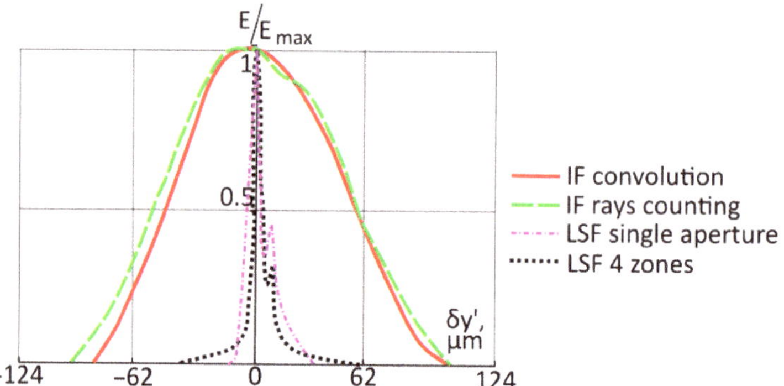

Figure 4. Instrument function computation verification.

This technique allows to account for different zones by introducing vignetting and switching between the optical system configurations in a loop. Another point, which usually is not considered when analyzing a spectrograph's performance, is the object luminosity function. For most instruments intended for laboratory applications it is sufficient to represent it as a simple rectangular function. However, for a slitless spectrograph the object represents a spot already blurred by the atmospheric turbulence and the telescope optics, which does not pass any spatial filter, as an entrance slit, optical fiber, or integral-field unit. Therefore, in the analysis below, we use a conservative estimate of $3.5''$ seeing, which corresponds to a 59 µm spot in the telescope focal plane and consider two cases:

- a rectangular function of the given width

$$rect(y') = \begin{cases} 1, -b_1'/2 \leq y' \leq b_1'/2, \\ 0, |y'| \geq b_1'/2. \end{cases} \tag{3}$$

where b_1' is the slit image width, taking into account the magnification.
- a Gaussian distribution with the given width at level of 0.1

$$\sigma = \sqrt{\frac{-0.5}{ln0.1}}, \tag{4}$$

$$G(y') = e^{-0.5y'^2/\sigma^2}. \tag{5}$$

where σ is the Gaussian distribution dispersion.

The described computations were implemented as a macro for *Zemax Optics Studio^{TM}* (Ansys, Canonsburg, PA, USA) and shown good convergence with the LSF-based technique for a few test cases. The IFs for the classical grism are shown in a summary comparative diagram in Figure 7. The main outcome of the IFs is their full width at half maximum (FWHM). This value, multiplied by the reciprocal linear dispersion of the spectrograph, gives us its spectral resolution $\delta\lambda$ and from that the spectral resolving power $R = \lambda/\delta\lambda$. All the numerical values, characterizing the spectrograph imaging performances, are shown in a summary in Table 2.

Considering aberration correction, the main difference between the classical grism and the composite ones consists of the presence of a deformable mirror or corrector plates. In the first case, the deformable mirror is mounted with a fixed angle of incidence of $10°$, 550 mm away from the substrate. Its shape is defined by automated optimization of the root mean square (RMS) spot sizes at the mentioned control wavelengths and FoV points, whereas the aberrations in the spectral direction Y have their weight coefficient increased by a factor of 10 with respect to that for the aberrations in the spatial direction X. Furthermore, the merit function includes boundary conditions, which require maintenance of the image size along the spectral axis, limiting the global Y position of the image edges, the distance between the grism and the detector plane as well as the angle of incidence onto the detector plane at the central wavelength. We may note here that the grating frequency is large enough to keep the zeroth order of diffraction far from the detector, whereas the intensities of higher order diffraction orders for a thick VPH are negligibly low. The resulting profile is dominated by defocusing and astigmatism as can be seen in Figure 2b. The overall deformation of the auxiliary mirror surface found after optimization for the composite hologram is quite large (61–784 µm). It can be implemented in practice in two different ways. One approach is to make use of an active thin-shell mirror, similar to that described in [18], which is capable of generating a peak-to-valley (PTV) deformation of 3.4 mm over a 160 mm aperture and create Zernike modes up to and including the 2nd order. This solution is straightforward but requires a customized active component. Another approach is to introduce the dominating defocus mode by re-focusing the collimator and using the DM only for higher modes. In this case, an existing MEMS (micro-electromechanical system)-based DM with a maximum stroke of $\approx$27 µm [19] or even less can introduce the required wavefront deformation. In Table 1 we specify separately the beam defocus in millimeters and the PTV deformation of the mirror surface sag corresponding to the rest of the modes (the tip, tilt, and piston are also excluded). In the case of a simplified hologram, the aspheric corrector can be relatively easily polished to the desired surface, but we keep the modes separation to facilitate the comparison. The values indicate that the sag deviation is relatively small, whereas the defocus is large in comparison with the grism size, which is 60×34 mm.

Table 1. Optimized parameters of the recording setup.

Zone	Composite		Simplified Composite	
	z'_{def}, mm	$\delta z'_{PTV}$, µm	z'_{def}, mm	$\delta z'_{PTV}$, µm
A	−2168.9	13.7	−3844.8	2.67
B	−3575.9	24.5	−3744.4	2.77
C	−1074.3	22.4	-	-
D	−2984.8	20.9	-	-

Similarly to the design using a classical grism, we first investigate the spot diagrams—see Figure 5 and Table 2. The diagrams clearly show the contribution of each zone and the gain in the image quality. If we compare the aberrations in the spectral $\delta y'$ and spatial $\delta x'$ directions, it becomes clear that for the spectral direction at the central wavelength, the aberrations are comparable, whereas at the edges of spectrum, they are reduced by a factor of 2.3. Meanwhile, the aberrations in the spatial direction are decreased significantly with a gain up to 85 times at the red edge of the spectrum. Therefore, we can expect that the spectra of adjacent objects will be resolved separately and that of extended objects will be free of self-contamination.

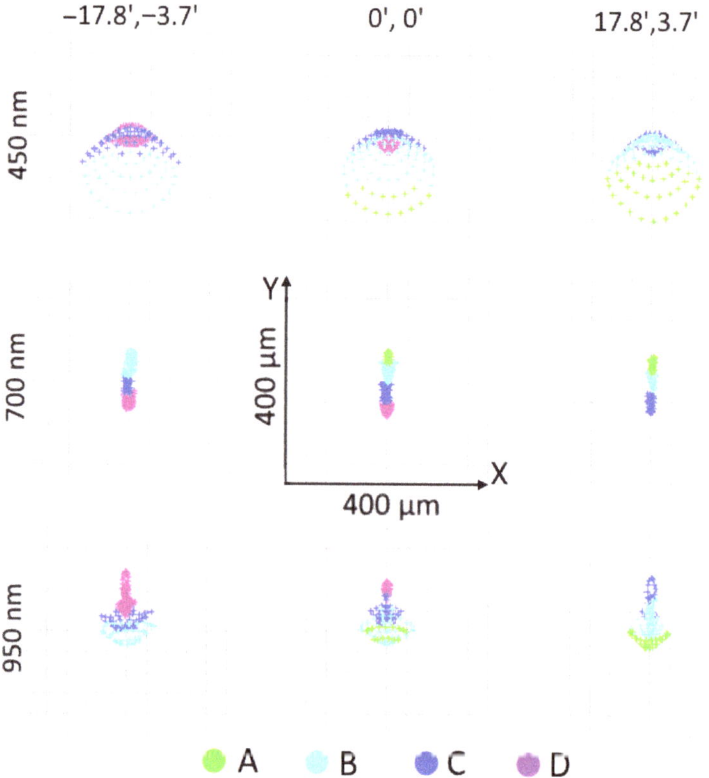

Figure 5. Spot diagrams of the spectrograph with composite grism.

Table 2. Image quality metrics of the slitless spectrograph.

Metric	Classical			Composite			Simplified Composite		
Wavelength, nm	450	700	950	450	700	950	450	700	950
FWHM Rect., μm	73.4	62.8	208.8	61.3	62	60.8	66.1	63.7	122.4
$\delta\lambda$ Rect.	1.28	1.10	3.64	1.06	0.07	1.05	1.15	1.11	2.13
R Rect., nm	352	639	260	425	654	905	392	632	447
FWHM Gauss., μm	61.8	41.4	172.8	35.9	34.8	39.6	49.6	37.8	118.8
$\delta\lambda$ Gauss.	1.08	0.72	3.02	0.62	0.60	0.68	0.86	0.66	2.06
R Gauss., nm	417	967	315	726	1164	1389	523	1067	460
$\delta y'_{best}$, μm	226	35	223	147	58	95	138	40	128
$\delta x'_{best}$, μm	309	425	441	74	5	15	61	7	22
$\delta y'_{worst}$, μm	257	69	259	156	65	114	177	77	170
$\delta x'_{worst}$, μm	340	457	490	85	9	41	95	27	44

The instrument functions of this version of the spectrograph are also shown in Figure 7. They clearly demonstrate the spectral resolution increase for the entire spectrum, but the gain obtained at its long-wavelength edge is very notable. Some high-frequency features can clearly be seen in the plots, which can be attributed to the zone edges. Comparing the numerical values in Table 2 we see that the spectral resolving power growth reaches a factor of 4.4. It is also notable that the gain is higher by ≈18% compared to the Gaussian distribution assumption. In general, the model predicts a spectral resolving power of up to $R1389$, which is quite high for this class of instruments.

If we step back to the simplified version of the composite grism, the deformable mirror is replaced by two interchangeable corrector plates. Each of them is a 7 mm-thick glass substrate made of $LZOS^{TM}$ K8 glass and placed 550 mm away from the hologram substrate. For the zone A recording the tilt angle is 40.604° and for the zone B it is 41.571°. The key values, describing the first aspherical surface of each corrector are given in Table 1. Flat aspheres with such parameters are definitely manufacturable.

The spot diagrams obtained in this design version are shown in Figure 6. In general, the effect of aberration correction with a composite grism is similar to that seen in Figure 5, although the correction is less uniform across the FoV, and not that efficient at the central wavelength and the red edge. However, in comparison with the classical grism aberrations (Table 2), we still gain a factor of 1.7 in the spectral direction and 60.7 in the spatial direction. Thus, the main goal of separating the spectra from different objects can be sufficiently reached even with all the simplifications.

Finally, let us consider the IFs and the spectral resolving power of the simplified grism. The corresponding data are given in Figure 7 and Table 2. Comparing the results with that for a classical grism, one can see that the spectral resolution remains nearly the same at the central wavelength, whereas at the spectrum edges it increases by a factor of 1.12–1.71. The most significant gain is observed for 950 nm with the Gaussian distribution assumption.

In general, both composite grism versions provide good astigmatism correction, which is a key requirement for a slitless spectrograph . This enables the detection of spectral images of multiple objects without intersections and contamination. Furthermore, using auxiliary correcting optics for the grating recording improves the spectral resolving power. If we use four zones of the grating and a deformable mirror, it becomes possible to reach a high and uniform spectral resolution for the entire working range. With two zones and static tilted corrector plates the results are more modest, but the gain at the long-wavelength edge is still significant.

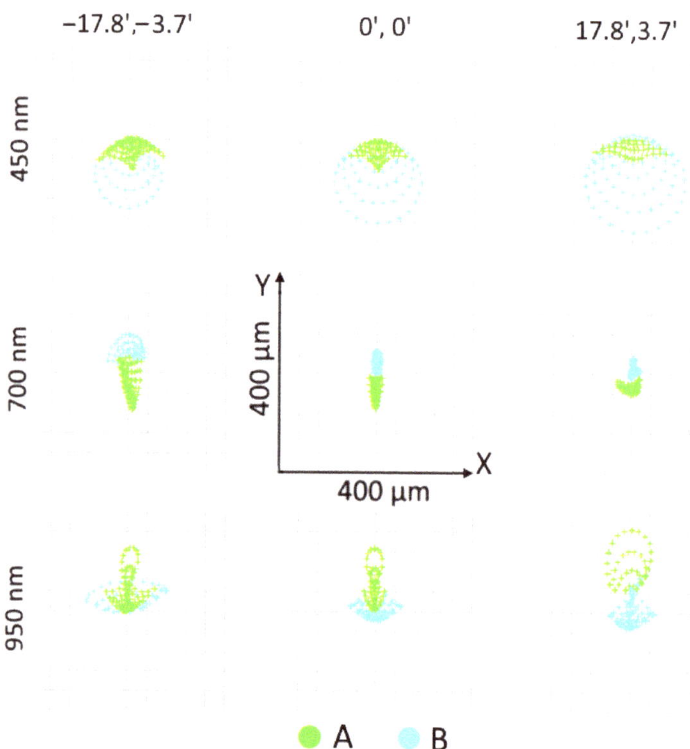

Figure 6. Spot diagrams of the spectrograph with simplified composite grism.

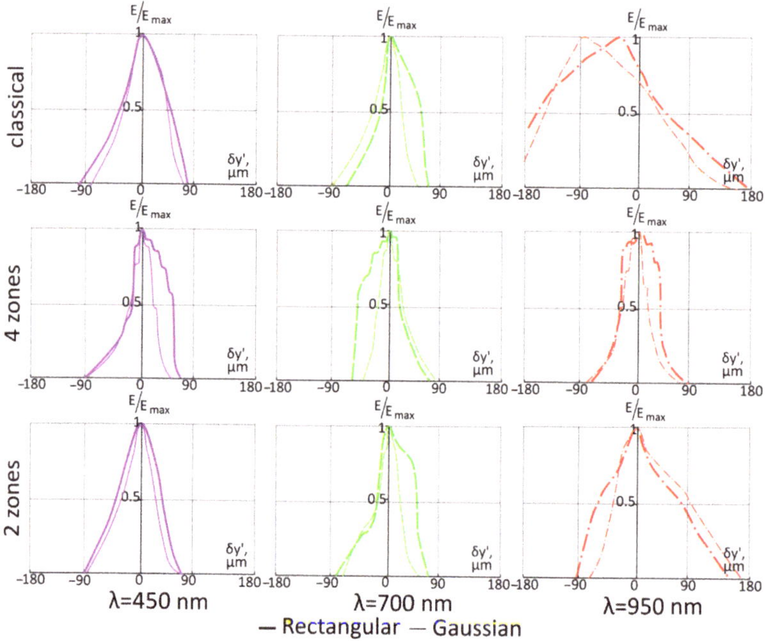

Figure 7. Instrument functions of the slitless spectrograph for a 3.5″ object in the FoV center.

All of the results shown above correspond to the perfect nominal values of the parameters describing the recording and operation setups. In practice, there will be unavoidable errors of manufacturing and assembly. To estimate the tolerances on individual parameters, we perform the following analysis: we introduce small deviations into every parameter p and measure the corresponding change of the weighted RMS of the aberrations over the FoV and aperture at the central wavelength in order to compute the corresponding sensitivity $\partial\sqrt{\sum(w_x x'^2 + w_y y'^2)}/\partial p$. Here, x' and y' are the transverse aberrations in two directions and w_x and w_y are the corresponding weight coefficients. For a classical grism, we use $w_x = 0$ and $w_y = 1$, since it has no astigmatism correction, whereas for the composite grisms, we use $w_x = 0.1$ and $w_y = 1$ as it was during the optimization. Then, presuming the errors for all the S parameters in the system sum up quadratically, we find the individual parameter tolerance as

$$\Delta p = \frac{\Delta\sqrt{\sum(w_x x'^2 + w_y y'^2)}}{\partial\sqrt{\sum(w_x x'^2 + w_y y'^2)}/\partial p}\frac{1}{\sqrt{S}}, \tag{6}$$

where the acceptable change of the aberrations' RMS is set to 5% of its nominal value.

The computation results are given in Table 3. All the tolerances, which appear to be too wide, are truncated down to a technologically feasible value. The radii of curvature R and irregularities ΔN are measured in interference fringes at $\lambda = 632.8$ nm. t denotes the axial thickness. The tip and tilt angles are denoted as α_x and α_y, respectively. n is the index of refraction and ν is the Abbe number. Note that the latter matters only in the operation setup. i_{rec} is the incidence angle in the recording setup and f_{rec} denotes defocusing of the recording beam. Finally, the indices are as follows: "sub"—grating substrate first surface, "grat"—grating surface, "prism"—second surface of th prism, "grism"—the entire grism, "corr"—the recording wavefront corrector, i.e., the DM or the corrector plate, and "air"—the airgap before the grism.

These values provide approximate estimates of the individual parameter tolerances, but they indicate some notable points regarding manufacturing and alignment:

- Most of the values are feasible to accomplish in practice. It appears that the classical grism is sensitive to decenters and tilts. This can be explained by the large uncompensated aberrations, which grow rapidly when deviating from the found best-fit position. However, the requirements for the individual surfaces' tilts can be met by defining proper specifications in the drawings of corresponding elements. An angular precision at the level of $\approx 1'$ is definitely possible for both the prism and grating.
- The composite grism is more sensitive to the prism refractive properties, but the absolute values are high enough to reach the required precision.
- The simplified version of the composite hologram recording setup does not have wider tolerances for the auxiliary components' parameters and alignment. The reason for this is that the quality of both surfaces, their position, and material refraction index contribute to the recording of wavefront aberrations.
- The tightest tolerance for a composite gratings' recording setup corresponds to the corrector plate surface irregularity. However, it should still be technologically feasible.

Table 3. Tolerances based on the image quality criterion.

Parameter	Classical	Comp.	Simpl. Comp.
R_{sub}, $fr.$	8	8	8
R_{grat}, $fr.$	7	8	8
R_{prism}, $fr.$	8	8	8
t_{air}, mm	0.68	1	1
t_{grat}, mm	0.22	0.5	0.5
t_{prism}, mm	0.20	0.5	0.5
dy_{grism}, mm	0.06	0.2	0.2
$\alpha_{x,grism}$, $^\circ$	0.6	1	1
$\alpha_{y,grism}$, $^\circ$	1	1	1
$\alpha_{x,grat}$, $^\circ$	0.013	0.28	0.27
$\alpha_{y,grat}$, $^\circ$	0.26	0.5	0.5
$\alpha_{x,prism}$, $^\circ$	0.023	0.1	0.1
$\alpha_{y,prism}$, $^\circ$	0.94	1	1
ΔN_{grat}, $fr.$	0.52	1	1
ΔN_{prism}, $fr.$	0.52	0.9	0.9
$n_{prism} \times 10^{-3}$	2.22	1.82	1.76
ν_{prism}	10	6.1	5.9
z_{corr}, mm	-	2	2
$\alpha_{x,corr}$, $^\circ$	-	1	0.3
t_{corr}, mm	-	-	0.5
ΔN_{corr}, $fr.$	-	5.7	1.6
$n_{corr} \times 10^{-3}$	-	-	6.8
i_{rec}, $^\circ$	0.203	0.145	0.141
f_{rec}, mm	-	0.79	0.77

4. Diffraction Efficiency Optimization and Analysis

Further, we discuss the diffraction efficiency of the grisms in a similar way. As we stated before, it is presumed that the classical grating is recorded in a symmetric setup, so its fringes are perpendicular to the surface. Let us assume that the VPH grating is recorded in dichromated gelatin (DCG) with an Ar laser at a wavelength of 514.5 nm, which is a very typical combination. Then the angles of incidence in the recording setup are 5.087° and −5.087°, resulting in a spatial frequency of $N = 344.7$ mm^{-1}, which corresponds to the required linear dispersion at the image plane. We also assumed that the hologram structure thickness is equal to a standard value of 20 μm provided by DCG manufacturers. If all of these parameters are fixed, then the only variable we can use to optimize the DE spectral distribution is the index modulation depth. We use the analytical equations of Kogelnik's coupled wave theory [20] to compute the diffraction efficiency and perform this computation in a loop to maximize the sum of DE at five control wavelengths covering the working range uniformly. Note, that we make these calculations only for the FoV center and the chief ray, so the distribution across the field and aperture is ignored. The resulting parameter values are given in the summary table Table 4 and the corresponding DE curve is shown in a comparative plot (Figure 11).

Table 4. Optimized parameters of the hologram structure.

Zone	Classical		Composite		Simpl. Composite	
	t, µm	δn	t, µm	δn	t, µm	δn
A	20.0	0.012	18.5	0.022	20.0	0.019
B	-	-	17.5	0.023	20.0	0.022
C	-	-	22.9	0.018	-	-
D	-	-	20.0	0.028	-	-

This simplified approach provides us with a starting point and a reference for the hologram structure optimization. However, it is obviously not sufficient to find the best possible design. The grism is mounted in a converging beam and works with an extended FoV; whereas the angle variation across the FoV was shown to be relatively small, the changes across the aperture with $F/\# = 6.8$ are significant. Both the change of the angle of incidence itself and that of the conical diffraction angle take place here. The latter is not taken into account by Kogelnik's scalar theory. It is necessary that we should use another method to compute the DE precisely. In the present work, we use the rigorous coupled wave analysis (RCWA) [21] implemented in the Reticolo software for these computations. This is a MatLab-based code that computes the diffraction efficiencies and the diffracted amplitudes of gratings composed of stacks of lamellar structures. It incorporates routines for the calculation and visualization of the electromagnetic fields inside and outside the grating. Reticolo implements the RCWA method or frequency–domain modal method for diffraction using 1D structures and is also capable of computing Bloch modes and analyzing stacks of arbitrarily anisotropic multilayered thin films. More details about the RCWA method can be found in [22] and the software description is available at [23].

We use the exact ray tracing results as an input for the diffraction problem to be solved with RCWA across an array of probing points (or elementary gratings), as it was made in [24].

The DE distribution obtained with RCWA for the spectral range center and edges is shown in Figure 8. The corresponding spectral distribution for the FoV center is shown in Figure 11. Note that due to the beam convergence, the DE varies significantly across the aperture, so we plot the average and maximum values in each case separately.

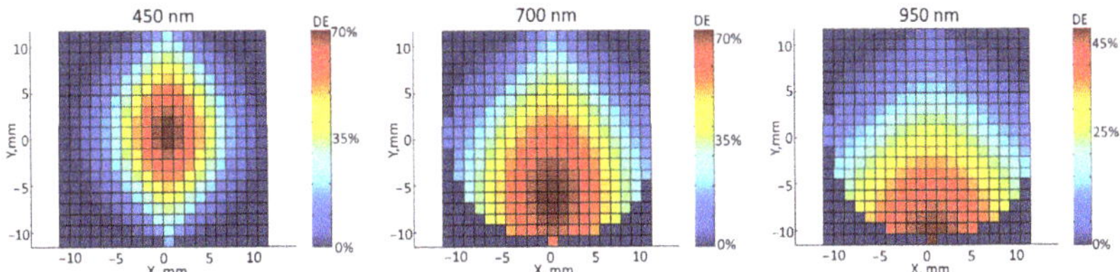

Figure 8. Spatial distribution of the DE across the beam footprint for the classical grism.

We begin the composite hologram structure design by breaking the recording setup symmetry. We compute the desired fringes' tilt angle according to the Bragg condition at the central wavelength for the chief ray in the FoV center and obtain angles of incidence of $2.533°$ and $-7.141°$. We keep these values fixed whilst performing an automated search of the thickness and modulation depth values to minimize the following function:

$$f(t, \delta n) = \sum_{\omega_x = -17.8'}^{17.8'} \sum_{\omega_y = -3.7'}^{3.7'} \sum_{\lambda = 450\text{nm}}^{950\text{nm}} 1 - \eta_{av}(t, \delta n), \tag{7}$$

here t is the VPH thickness, δn is the index modulation depth, and each of them represents a vector of four elements for the A-D zones; (ω_x, ω_y) are the angular coordinates of the point in the FoV and η_{av} is the average DE across the aperture including all the zones of composite grating.

We use a standard simplex method [25] to minimize this function. Since the achievable values of the hologram structure parameters have some technological limits, we apply the following boundary conditions typical for DCG [26]: $10\ \mu m \leq t \leq 30\ \mu m$ and $0.005 \leq \delta n \leq 0.05$.

The optimized parameter values are given in Table 4 and the DE distributions are shown in Figure 9. Note that in this case we plot data for points covering the entire FoV since the zone coverage differs across it. One can see that the distribution is more uniform and the values at the red edge are notably higher. Furthermore, these diagrams carry clear footprints of the zone boundaries. The DE at 450 nm represents an illustrative example of the composite hologram concept—the maximum is shifted to the top side of the grating, but the parameters of the lower zone D are optimized locally to partially compensate for this shift. It is necessary to state here that for the RCWA modeling, we introduce some simplification of the recording geometry. When calculating the recording beams' incidence angles we take into account only the chief ray incidence and the defocus given in Table 1. This allows us to simplify the calculation significantly and avoid the issue of sampling difference in the recording and operation setups, whereas the recording angle deviation does not exceed $0.16°$.

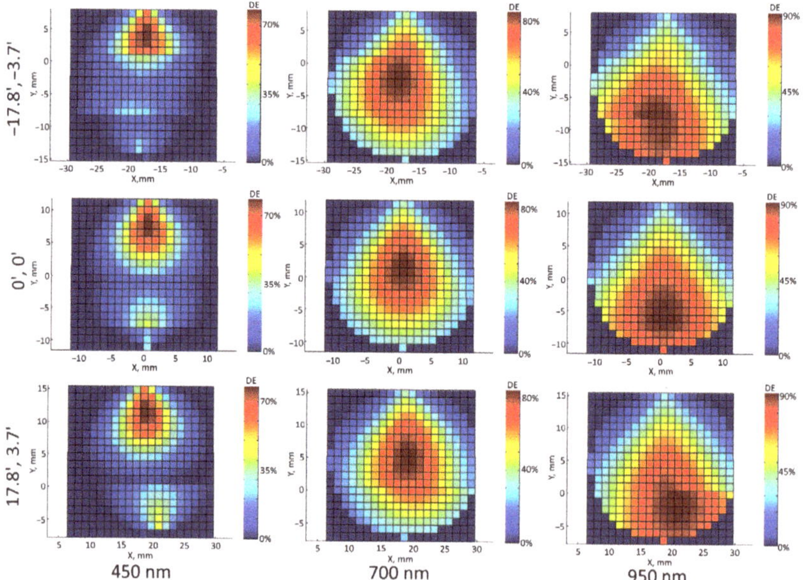

Figure 9. Spatial distribution of the DE across the beam footprint for the composite grism.

Similarly, the spectral distribution for the center is shown in Figure 11. Generally, the main qualitative changes are confirmed, but in order to quantify the gain we propose the following metric:

$$Q = \frac{\int_{450nm}^{950nm} \eta_{av}(\lambda)}{\int_{450nm}^{950nm} rect(\lambda)},\qquad(8)$$

i.e., how much energy is diffracted into the working spectral order on average in comparison with an ideal 100% case. After evaluation, the composite grism provides $Q = 0.396$, whereas the classical grism had $Q = 0.303$. All the values of the Q metrics are summarized in Table 5. On top of these, we see a very notable increase in the maximum DE.

Table 5. Integral diffraction efficiency metrics.

Field	Classical		Composite		Simpl. Composite	
	max	av	max	av	max	av
$-17.8', -3.7'$	0.690	0.303	0.843	0.383	0.846	0.394
$0', 0'$	0.690	0.303	0.881	0.386	0.853	0.396
$17.8', 3.7'$	0.690	0.303	0.878	0.387	0.845	0.386

Next, we perform a similar optimization and analysis for the simplified composite grism. In this case, the hologram thickness is fixed and set equal to 20 μm and the hologram consists only of two zones. Therefore, the number of free optimization variables is limited to just two modulation depth values. It would not be enough to provide correction for the entire field of view, so Equation (7) is rewritten for only the FoV center:

$$f^*(t, \delta n) = \sum_{\lambda=450nm}^{950nm} 1 - \eta_{av}(t, \delta n), \tag{9}$$

We minimize this function using the same method and the same boundary conditions and obtain the values listed in Table 4.

The corresponding DE distributions are given in Figure 10. The general distribution patterns are the same, so in comparison with the classical grism, we obtain a more uniform distribution with a better performance at the long-wavelength edge, whereas the specific features of the composite grating are used to recompensate the DE losses at the short-wavelength edge.

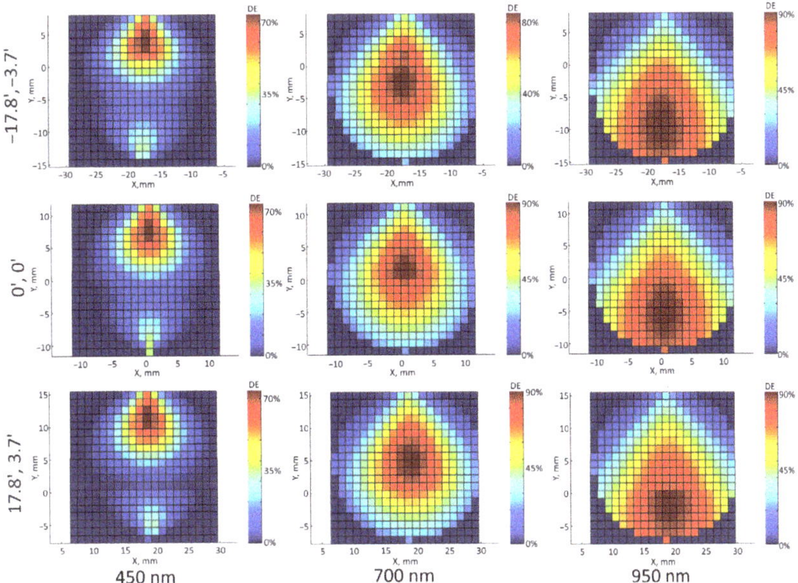

Figure 10. Spatial distribution of the DE across the beam footprint for the simplified composite grism.

The spectral dependencies of the average and maximum DE for the simplified composite grism are shown on the comparative plot Figure 11 together with the other designs.

Here the maximum and average values are computed across the beam footprint, i.e., the 2D distribution in every sub-plot in Figures 8–10. One can see that despite the simplifications, the performance is superior to that of a classical grism and very close to the case of the initial composite element. We can confirm this using the metric according to Equation (8): $Q = 0.386$. From this, it can be seen that in terms of the DE improvement, further complication of the design brings little gain. One possible explanation for this is the grism position. It is located far from the focal and pupil planes, so beams corresponding to different FoV points and aperture parts are mixed, which limits the efficiency of local parameter optimization.

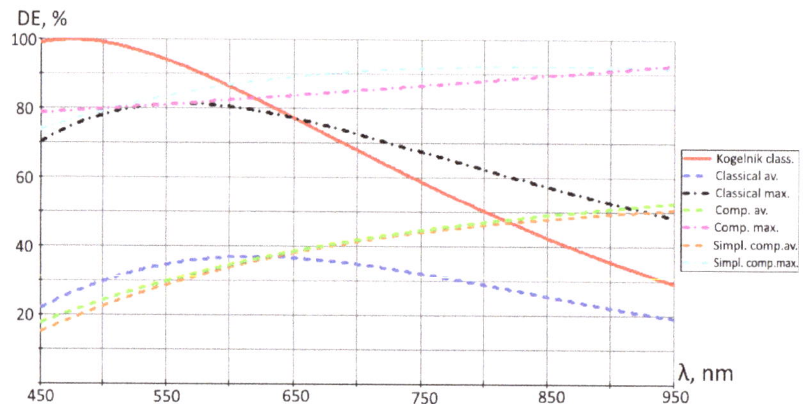

Figure 11. Comparison of the DE spectral dependence in the FoV center.

In general, the use of a composite grism will allow us to obtain brighter spectral images and improve the instrument sensitivity, especially in the red part of the spectrum.

Similar to the image quality analysis, we estimate the tolerances for the parameters which define the diffraction efficiency. We use the median value of DE over the working spectral range measured at the FoV center η_m as the merit function and assume that its acceptable change is $\Delta\eta_m = 5\%$ in the absolute DE value. Then, we use the following relation, similar to Equation (6):

$$\Delta p = \frac{\Delta\eta_m}{\partial\eta_m/\partial p} \frac{1}{\sqrt{S}},$$ (10)

In this case, we only consider the hologram structure thickness t, its index modulation depth δn, and incidence angles in the recording setup $i_{rec}1$ and $i_{rec}2$. Even for the slanted fringes, the influence of the two incidence angles on the DE is almost identical. Furthermore, both the frequency and tilt of the fringes change with incident angle deviation. The sensitivity of the DE to the fringes tilt angle change is an order of magnitude higher, so we use the values derived from the fringes tilt angle. All the tolerance estimates are summarized below in Table 6.

Table 6. Tolerances based on the diffraction efficiency criterion.

Parameter	Classical	Comp.	Simpl. Comp.
i_{rec}, deg	0.082	0.046	0.067
t, μm	1.6	0.9	1.3
$\delta n \times 10^{-4}$	6.3	3.5	5.1

The values in Table 6 are only estimates, but they indicate a high sensitivity of the DE to the hologram structure parameters. The tolerances on recording angles are tighter than those found with the image quality criterion, but remain at a feasible level >2′. The tolerance on thickness is relatively wide—for a 20 μm layer a 4% change appears to be acceptable. The highest sensitivity was found for the index modulation depth, but even in the worst case, the tolerance is 2.9%. Finally, we can note that the increase of the design variable number obviously causes some tightening of the tolerances, but the difference is moderate.

5. Discussion

The analysis presented above demonstrates that the developed designs of slitless spectrographs based on composite grisms exhibit a clear and measurable gain in performance in terms of both resolution and throughput. Furthermore, these designs, especially the simplified one, are feasible and can be used in the future for an experimental proof-of-concept. From the standpoints of design technique and holographic technology, it introduces some novelty to the field since, to the best of our knowledge, the holographic gratings used in astronomical spectrographs so far could be split into sub-zones [14,27], but did not have local optimization of both aberrations and diffraction efficiency at the same time. However, besides being a demonstrative design, the proposed spectrograph can take its own niche among the astronomical instruments and be useful for real observations. Once we have the numerical estimates of the key performance metrics, it would be useful to compare them with those of some other instruments in order to show the most prospective application areas and the key advantages.

There are a few low-resolution spectrographs working with large ground-based telescopes. They use VPH-based grisms as dispersive elements and operate over an extended field of view. FOCAS [28] and EFOCS [29] can serve as examples. These instruments also try to cover the entire waveband available with a standard CCD detector and utilize the entire field provided by the telescope. The field of view in these cases is smaller than in the proposed design whilst the resolving power and the throughput is comparable or higher than the values we obtained. However, there are notable differences in the optical architectures. In contrast to our design, these spectrographs still use some spatial filters like multi-slit setups in the focal plane and then use auxiliary relay optics and interchangeable dispersers to form the spectral image. This also implies the use of a dedicated detector for the spectral images. Thus, our spectrograph also can be used for observations to take spectra of multiple or extended faint objects. On the one hand, the maximum observable magnitude will be relatively modest, taking into account the difference in collecting area between the telescopes and the optical system transmission. On the other hand, the proposed instrument will benefit from observing a notably larger FoV and the entire waveband at once, whereas its design is much simpler, more compact, and less expensive. The key parameters of the optical designs are compared in Table 7.

Table 7. Spectrograph characteristics comparison.

Instrument	Comp. Grism	FOCAS	EFOCS	MAO	AFA	Galex-NUV
Wavelengths	450–950 nm	365–900 nm	350–1000 nm	400–1000 nm	350–700 nm	177–283 nm
Spectral resolving power	R425–905	R250–2000	R250–7140	R100	R103–147	R90
Throughput	18–51%	57–82%	18–65%	25–75%	25–75%	10–82%
FoV	35.6′ × 7.2′	6′ × 6′	3.5′ × 5.7′	10.8′ × 10.8′	14′ × 14′	⌀1.24°
Number of elements	1	15	7	1	1	2
Hosting telescope	CDK500 0.5 m ground	Subaru 8.2 m ground	ESO 3.6 m ground	Zeiss-600 0.6 m ground	DFM 0.4 m ground	Galex 0.5 m space

Another group of instruments that could be compared with our design consists of single-grating slitless spectrographs for small ground-based telescopes. For instance, such an instrument was built at MAO NASU [7,30] for the detection of multiple stellar spectra in a single observation and flare star observations. A similar setup was developed at US AFA [31] and used for the detection of spectra of geosynchronous satellites. In both of these cases, the spectrographs are based on a single transmission grating mounted in a converging telescope beam. Their main parameters are also shown in Table 7. As one can see, the proposed design has a larger FoV and the spectral resolving power is notably higher. For the indicated cases, there are no exact values of transmission available, so we assume that it is close to that of a typical commercial grating [32]. Even with this optimistic assumption, our design shows a comparable performance. This comparison is useful not only to demonstrate the advantages in performance but also to indicate potential applications such as flare star search and spectral characterization as well as the observations of satellites or space debris.

Finally, the developed design can be compared to some space instruments. It is difficult to find a direct analog, but the Galex-NUV channel [6] has a very similar optical design. In comparison with it, our design has a wider working range, higher spectral resolution, and higher minimum throughput. We believe that these advantages can be at least partially maintained if we repeat a composite grism-based design for the NUV domain. We keep this example in the comparative table to emphasize the prospects of our solution for space-based spectrographs as more and more projects of spectral instruments for small satellites like [33] appear.

6. Conclusions and Future Work

In the present work, we have demonstrated the advantages of a composite holographic element being applied in a slitless spectrograph design. It allows for the creation of an extremely compact and simple instrument, which can be mounted in a converging beam at the telescope output, and use the nominal CCD camera of the telescope to detect the spectral image. The key advantage of the composite grism is the astigmatism correction, which allows for decreasing the spot diagram elongation by almost two orders of magnitude. This will make it possible to observe spectra from different objects or different points of an extended object without overlapping, thus implementing the key feature of slitless spectroscopy. On top of this, we can increase the spectral resolving power by a factor up to 4.4 and improve it significantly at the long-wavelength edge of the spectrum.

In addition to this the composite, VPH has a higher diffraction efficiency that is also distributed more uniformly across the spectral range. Thanks to hologram structure parameters optimization, we can increase the average DE value by a factor of 1.31. One should keep in mind that when the DE defines the flux power directed to the working diffraction order, the aberrations correction decreases the area in which this flux is focused. Thus, we expect a very significant increase in the spectral image illumination and, therefore, the overall sensitivity of the instrument.

We have considered two versions of the composite grism. The modeling has shown that even with some simplifications, like decreasing the number of zones, using static corrector plates in the holographic recording setup, and using a standard thickness of the holographic layer, it is still possible to obtain a significant gain in performance. Moreover, in terms of DE optimization, a further complication of the design appears to be unnecessary.

The next stage of our research implies an experimental proof of concept. The plan is to turn the simplified composite grism design into practice and test it experimentally. First, it should be tested in the laboratory, including measurements of the resolution and throughput. Second, we plan to make trial observations with the spectrograph coupled with the CDK500 telescope owned and operated by Kazan Federal Univ. Figure 12 shows the optomechanical design of the slitless spectrograph unit to be mounted on the telescope.

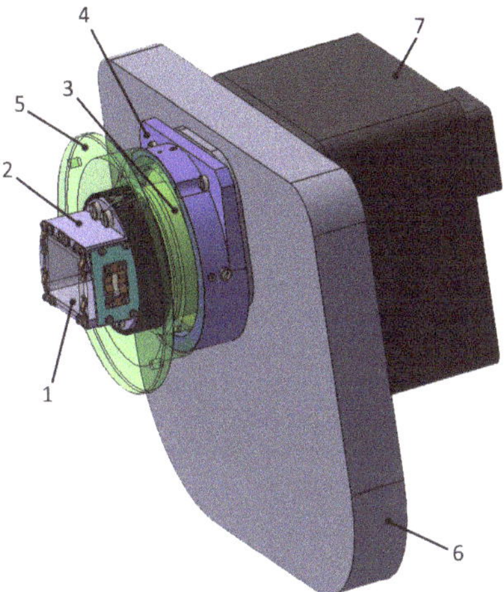

Figure 12. Opto-mechanical design of the slitless spectrograph unit: 1—grism, 2—grism mount, 3—connecting tube, 4—camera flange, 5—telescope mechanical interface, 6—telescope filter wheel, 7—telescope camera.

To the best of our knowledge, it would be the first application of a composite holographic dispersion element in an operational astronomical instrument. Thus, it should provide a strong confirmation of the very design concept advantages and to the corresponding design and modeling techniques. We hope that this instrument will open new prospects for spectral instrument design in astronomy and other fields.

Author Contributions: Conceptualization, E.M.; methodology, E.M.; software, D.K.; validation, N.P.; formal analysis, E.M. and N.P.; investigation, D.A.; resources, E.I. and S.G.; data curation, E.M. and D.A.; writing—original draft preparation, E.M.; writing—review and editing, E.M. and N.P.; visualization, E.M. and D.A.; supervision, E.M.; project administration, D.A. and V.S.; funding acquisition, E.M. All authors have read and agreed to the published version of the manuscript.

Funding: Work of E.M., D.A., and D.K. was funded by the Russian Science Foundation grant number 21-79-00082.

Institutional Review Board Statement: Not applicable.

Informed Consent Statement: Not applicable.

Data Availability Statement: No new data were created or analyzed in this study. Data sharing is not applicable to this article.

Acknowledgments: The authors would like to warmly thank Marjolein Verkouter from JIVE for correcting the manuscript and useful discussions. Furthermore, the authors are grateful to their colleague Ilya Guskov from the State Institute of Applied Optics for their help with the volume-phase grating modeling, and to the directorate of the Physics Institute of Kazan Federal University for their collaboration on the telescope interface.

Conflicts of Interest: The authors declare no conflict of interest.

References

1. Massey, P.; Hanson, M.M. Astronomical Spectroscopy. In *Planets, Stars and Stellar Systems*; Oswalt, T.D., Bond, H.E., Eds.; Springer: Dordrecht, The Netherlands, 2007; pp. 35–98.
2. Hoag, A.A.; Schroeder, D.J. "Nonobjective" Grating Spectroscopy. *Publ. Astron. Soc. Pac.* **1970**, *82*, 1141. [CrossRef]
3. Outini, M.; Copin, Y. Forward modeling of galaxy kinematics in slitless spectroscopy. *Astron. Astrophys.* **2020**, *633*, A43. [CrossRef]
4. Pasquali, A.; Pirzkal, N.; Larsen, S.; Walsh, J.R.; Kümmel, M. Slitless Grism Spectroscopy with the Hubble Space Telescope Advanced Camera for Surveys. *Publ. Astron. Soc. Pac.* **2006**, *118*, 270–287. [CrossRef]

5. Willott, C.J. The Near-infrared Imager and Slitless Spectrograph for the James Webb Space Telescope. II. Wide Field Slitless Spectroscopy. *Publ. Astron. Soc. Pac.* **2022**, *134*, 025002. [CrossRef]
6. Grange, R. GALEX UV grism for slitless spectroscopy survey. *Soc. Photo-Opt. Instrum. Eng. (SPIE) Conf. Ser.* **2017**,*10569*, 1056908. [CrossRef]
7. Zhilyaev, B.E.; Sergeev, A.V.; Andreev, M.V.; Godunova, V.G.; Reshetnik, V.N.; Taradii, V.K. Slitless spectrograph for small telescopes: First results. *Kinemat. Phys. Celest. Bodies* **2013**, *29*, 120–130. [CrossRef]
8. Ludovici D.A.; Mutel R.L. A compact grism spectrometer for small optical telescopes. *Am. J. Phys.* **2017**, *85*, 873–879. [CrossRef]
9. Moniez, M.; Neveu, J.; Dagoret-Campagne, S.; Gentet, Y.; Le Guillou, L.; LSST Dark Energy Science Collaboration. A transmission hologram for slitless spectrophotometry on a convergent telescope beam. 1. Focus and resolution. *Mon. Not. R. Astron. Soc.* **2021**, *506*, 5589–5605. [CrossRef]
10. Muslimov, E.; Sauvage, J.-F.; Lombardo, S.; Akhmetov, D.; Kharitonov, D. Low-resolution spectroscopy mode for CASTLE telescope with a composite grism. *SPIE Proc.* **2022**, *12184*, 121848G. [CrossRef]
11. Muslimov, E.; Pavlycheva, N.; Guskov, I.; Akhmetov, D.; Kharitonov, D.Y. Spectrograph with a composite holographic dispersive element. *SPIE Proc.* **2021**, *11871*, 1187112. [CrossRef]
12. CDK500 Observatory System. Available online: https://planewave.com/product/cdk500-telescope-system/ (accessed on 7 January 2023).
13. Jones, L. Reflective and catadioptric telescopes. In *Handbook of Optics*, 3rd ed.; Bass M., Ed.; McGraw-Hill: New York, NY, USA, 2010; p. 1568.
14. Blanche, P.A.; Gailly, P.; Habraken, S.; Lemaire, P.; Jamar, C. Mosaiced and high line frequency VPH gratings for astronomy. *Proc. SPIE* **2004**, *5494*, 208-216. . [CrossRef]
15. Lakshminarayanan, V.; Fleck, A. Zernike polynomials: A guide. *J. Mod. Opt.* **2011**, *58*, 545–561. [CrossRef]
16. Slyusarev, G.G. *Aberration and Optical Design Theory*; Taylor and Francis: Milton Park, UK, 1984; 674p.
17. Nazmeev M.M.; Pavlycheva N.K. New generation spectrographs. *Opt. Eng.* **2011**, *33*, 2777–2782. [CrossRef]
18. Lemaitre, G.R.; Wang, M. Active mirrors warped using Zernike polynomials for correcting off-axis aberrations of fixed primary mirrors. I. Theory and elasticity design. *Astron. Astrophys. Suppl. Ser.* **1995**, *114*, 373.
19. Dagel, D.J.; Cowan, W.D.; Spahn, O.B.; Grossetete, G.D.; Grine, A.J.; Shaw, M.J.; Resnick, P.J.; Jokiel, B. Large-Stroke MEMS Deformable Mirrors for Adaptive Optics. *J. Microelectromech. Syst.* **2006**, *15*, 572–583. [CrossRef]
20. Kogelnik, H. Coupled wave analysis for thick hologram gratings. *Bell Syst. Tech. J.* **1969**, *48*, 2909–2947. [CrossRef]
21. Lalanne, P. High-order effective-medium theory of subwavelength gratings in classical mounting: Application to volume holograms. *J. Opt. Soc. Am. A* **1998**, *15*, 1843–1851. [CrossRef]
22. Moharam, M.G.; Grann, E.B.; Pommet, D.A.; Gaylord, T.K. Formulation for stable and efficient implementation of the rigorous coupled-wave analysis of binary gratings. *J. Opt. Soc. Am. A*, **1995**, *12*, 1068–1076. [CrossRef]
23. Hugonin J.P.; Lalanne P. RETICOLO software for grating analysis. *arXiv* **2021**, arXiv:2101.00901.
24. Muslimov, E.R.; Ferrari, M.; Hugot, E.; Bouret, J.-C.; Neiner, C.; Lombardo, S.; Lemaitre, G.R.; Grange, R.; Guskov, I.A. Design and modeling of spectrographs with holographic gratings on freeform surfaces, *Opt. Eng.* **2018**, *57*, 125105. [CrossRef]
25. Lagarias, J.C. Convergence Properties of the Nelder-Mead Simplex Method in Low Dimensions. *SIAM J. Optim.* **1998**, *9*, 112–147. [CrossRef]
26. Newell, J.C.W. *Optical Holography in Dichromated Gelatin*; St. Cross College,University of Oxford: Oxford, UK, 1987; 270p.
27. Arns, J.; Wilson, J.C.; Skrutskie, M.; Smee, S.; Barkhouser, R.; Eisenstein, D.; Gunn, J.; Hearty, F.; Harding, A.; Maseman, P.; et al. Development of a large mosaic volume phase holographic (VPH) grating for APOGEE. *Proc. SPIE* **2010**, *7739*, 773913. [CrossRef]
28. Kashikawa, N.; Aoki, K.; Asai, R.; Ebizuka, N.; Inata, M.; Iye, M.; Kawabata, K.S.; Kosugi, G.; Ohyama, Y.; Okita, K.; et al. FOCAS: The Faint Object Camera and Spectrograph for the Subaru Telescope. *Publ. Astron. Soc. Jpn.* **2002**, *54*, 819–832. [CrossRef]
29. Buzzoni, B.; Delabre, B.; Dekker, H.; Dodorico, S.; Enard, D.; Focardi, P.; Gustafsson, B.; Nees, W.; Paureau, J.; Reiss, R. The ESO Faint Object Spectrograph and Camera (EFOSC). *Messenger* **1984**, *38*, 9B.
30. Zhilyaev, B.E.; Sergeev, O.V.; Andreev, M.V.; Godunova, V.G.; Reshetnyk, V.M.; Tarady, V.K. A slitless spectrograph for observing transient events with small telescopes. *Proc. SPIE* **2002**, *8446*, 84468I. [CrossRef]
31. Tippets, R.D.; Wakefield, S.; Young, S.; Ferguson, I.; Earp-Pitkins, C.; Chun, F.K. Slitless spectroscopy of geosynchronous satellites. *Opt. Eng.* **2015**, *54*, 104103. [CrossRef]
32. Visible Transmission Gratings. Available online: https://www.thorlabs.com/ (accessed on 21 February 2023).
33. Bowman, D.M.; Vandenbussche, B.; Sana, H.; Tkachenko, A.; Raskin, G.; Delabie, T.; Vandoren, B.; Royer, P.; Garcia, S.; Reeth, T.V.; et al. The CubeSpec space mission. I. Asteroseismology of massive stars from time-series optical spectroscopy: Science requirements and target list prioritisation. *Astron. Astrophys.* **2022**, *658*, A96. [CrossRef]

Article

Wavefront Sensing by a Common-Path Interferometer for Wavefront Correction in Phase and Amplitude by a Liquid Crystal Spatial Light Modulator Aiming the Exoplanet Direct Imaging

Andrey Yudaev *, Alexander Kiselev, Inna Shashkova, Alexander Tavrov, Alexander Lipatov and Oleg Korablev

Space Research Institute of Russian Academy of Science, Moscow 117997, Russia
* Correspondence: yudaev@phystech.edu

Abstract: We implemented the common-path achromatic interfero-coronagraph both for the wavefront sensing and the on-axis image component suppression, aiming for the stellar coronagraphy. A common-path achromatic interfero-coronagraph has its optical scheme based on a nulling rotational-shear interferometer. The angle of rotational shear can be chosen at a small angular extent of about 10 deg. Such a small angular shear maintains the coronagraphic contrast degradation known as the stellar leakage effect, caused by a finite stellar size. We study the phase and amplitude wavefront control by a liquid crystal spatial light modulator of reflection type which is used as the pixilated active adaptive optics unit. Therefore, adaptive optics perform a wavefront-correcting input toward a stellar interfero-coronagraph aiming at the direct exoplanet imaging. Presented here are both the numeric evaluations and the lab experiment stand to prove the declared functionality output.

Keywords: exoplanet direct imaging; adaptive optics; wavefront sensing and correction; phase-shifting interferometry; interfero-coronagraph; common-path; rotational-shear interferometer; power spectral density; point spread function; phase retrieval

Citation: Yudaev, A.; Kiselev, A.; Shashkova, I.; Tavrov, A.; Lipatov, A.; Korablev, O. Wavefront Sensing by a Common-Path Interferometer for Wavefront Correction in Phase and Amplitude by a Liquid Crystal Spatial Light Modulator Aiming the Exoplanet Direct Imaging. *Photonics* **2023**, *10*, 320. https://doi.org/10.3390/photonics10030320

Received: 26 January 2023
Revised: 25 February 2023
Accepted: 10 March 2023
Published: 16 March 2023

1. Introduction

The direct imaging of exoplanets orbiting their host star near the Solar System still remains a challenging engineering task due to the high contrast ratio and the small angular separation relative to the host star. Exoplanets have to be observed within 1 arcsec separation from, and between 6 and 10 orders of magnitude fainter than, the stars they orbit [1]. To meet these requirements, both space- and ground-based telescopes have become equipped with adaptive optics systems having high precision in a wavefront correction toward a stellar coronagraph. Both the adaptive optics (AO) and the coronagraph are optimized to acquire high-contrast images of the substellar surroundings (or a circumstellar companion), in the area of 1–20 λ/D (diffraction units), where D stands for telescope diameter and λ—central wavelength.

Aiming to image exoplanets in the Solar System vicinity and focusing on orbiting stars distanced at 10+ pc, both ground-based astronomical observatories [2] and space-based telescopes are equipped with stellar coronagraph instruments. For high-contrast imaging through the turbulent atmosphere, ground-based telescopes require AO units with kilohertz dynamic characteristics. Both space-based telescopes and ground-based telescopes require the ultimate precision of AO to control wavefront with a precision of about $\lambda/10^3$–$\lambda/10^4$.

Stellar coronagraph was originally planned by the NASA program TPF (Terrestrial Planet Finder) [3], and now it is split into three main project branches: (i) TPF-C dedicated to coronagraph, (ii) TPF-I dedicated to a multi-aperture interferometer, and (iii) TPF-O [4] dedicated to an external occulter. Additionally, the fourth project's *family* implements optical interferometers and steps from a single-dish telescope to multiple optically linked telescopes [5].

In particular, the space telescope WFIRST-AFTA [6], renamed the Nancy Grace Roman space telescope [7], will be equipped with modern stellar coronagraph instrumentation, including in-flight (space grade) adaptive optics (AO) for active wavefront correction. Vast R&D projects are completing important steps toward successful high-contrast imaging in the stellar vicinity.

Today, the most used star coronagraph optical schematics are various modifications of Lyot coronagraphs and apodization schemes. Roughly, we shall refer to these schemes to mask coronagraphs because of their working principles containing different masks in the focal, pupil, and intermediate optical planes [8]. Various modifications of the masks provide different properties of the coronagraph output. Conventionally, the optics of mask coronagraphs include several focusing lenses or mirrors for constructing focal- and pupil planes that are optically conjugated, where the main functional masks are placed. These optical planes are conjugated with wavefront sensors, deformable mirrors (DM) as wavefront correctors, field cameras, spectrographs, and additional analyzers. An opaque ring-form mask put in the plane, optically conjugated to the telescope's secondary (or successive) pupil, is historically called the Lyot mask, which presents the simplest version of apodization. Apodization by a graduate transmit-profile mask, together with more complex solutions, has functional advantages respective of the complexity of their implementations.

Alternatively, the working principle of an interfero-coronagraph [9] is to superimpose with anti-phase two pupil images being shifted, reversed, or rotated. Functionally, that results in destructive interference and, therefore, eliminates the starlight being received from the on-axis direction. Planetary (or companion) light does not interfere destructively because the off-axis tilt (or shift, or shear in a pupil) of the image component from a planet separates the two companion copies spatially. Interfero-coronagraphy is advantageous in terms of its broad spectral band achromaticity and, more generally, because of its small inner working angle (IWA). IWA characteristics can be considered the spatial resolution of a coronagraph instrument. The achromatic interfero-coronagraph (AIC) is known to have one of the smallest possible IWA ($0.38\,\lambda/D$), such as it was initially referred to in [8] with the fixed angle of a rotational shear of $180°$. Later, it was re-designed into a modified Sagnac scheme, implementing the common path (or cyclic path, or cavity) (CP-AIC) [10] aiming to relax the mechanic instability.

The severe functional disadvantage of an interfero-coronagraph is known as a stellar leakage effect. Due to this effect, the starlight cannot be completely suppressed because the apparent size of the star is not physically infinitesimal [8]. The observed size of a star is far beyond the spatial resolution of a *single-dish* optical telescope, but physically, the star size causes a non-fully coherent point-like source; spatially, an extended source. Observed by means of a two-meter diameter telescope, the Solar-type star at 10 parsecs has the apparent size of ~$0.02\,\lambda/D$ (for optical domain wavelength of $\lambda \approx 1$ micron and $D \approx 2$ m). This effect limits the destructive interference contrast to 5 orders of magnitude in terms of starlight suppression. To overcome this effect, an interfero-coronagraph with a variable rotational shear smaller than $180°$, abbreviated as ARC in [11–13], offers a possibility to adjust an arbitrary pupil rotation angle to achieve an optimal coronagraphic contrast considering the observing configuration. With a two-meter telescope class, the $10°$-shear interferometer offers the stellar leakage (due to the finite size of the stellar disk) for a Sun-like star at 10 pc almost below 8 orders of magnitude for a $5\,\lambda/D$ working angle of companion separation. The 8 orders in magnitude mentioned here are a theoretical raw coronagraphic contrast estimation without any aberrations of wavefront, i.e., any optical wavefront errors.

Optical wavefront errors of the order of 10^{-9} m from fabrication defects and misalignments introduce a bright starlight speckle halo around the theoretical stellar image that overwhelms the faint image of an orbiting planet or a circumstellar disk. This speckle halo evolves in response to minute changes in the thermal and mechanical state of the observatory. Stellar coronagraphs rely critically on closed-loop wavefront sensing and have to compensate for these aberrations in time. Therefore, space-based telescopes require on-orbit options for wavefront sensing and its control, where all computations associated with the sensing and control algorithms are carried out by the flight computer aiming to achieve the deepest possible coronagraphic contrast.

Coronagraphy-based wavefront control algorithms are known, such as stroke minimization (SM) [14] and electric field conjugation (EFC) [15]. These algorithms use a computer model of the coronagraph to solve an inverse problem using a wavefront control iteration process. An aberrated electric field is measured at the coronagraph detector. Internally, the algorithms evaluate a Jacobian matrix electric field response of each pixel within the focal-plane control region to each actuator or pixel of active optics. This transforms the wavefront control inverse problem to the number of actuators or pixels for a modern coronagraph, which is typically from 10^3 (to 10^6); this represents a substantial computational capacity. Here, fast linear least-squares coronagraph optimization (FALCO) [16] utilizes a coronagraph model highly optimized for speed showing practical improvement. Aiming at optimization, there are several more efforts known, such as reverse-mode algorithmic differentiation (RMAD) [17], pair-wise probing [18], Kalman filtering [19], and the self-coherent camera [20]. A coronagraph inverse problem can be solved to enable multi-wavelength wavefront control by adding intensity constraints for each control wavelength of interest [21].

Therefore, precise wavefront correction is ultimately required as pre-optics schematics for the coronagraph for its functionality. AO compensates for optical aberrations including optical defects given by the optical surfaces and apertures. Light radiation is collected by a telescope within or above the telluric atmosphere, and it is analyzed after a coronagraph by field camera, spectroscopy, or different instruments. It is important to note that the measurement of the wavefront has to be organized after the coronagraph. Otherwise, non-common-path wavefront errors caused by a different (or incomplete, similar) optical path generate additional phase and amplitudes errors that become magnified [22] by coronagraph. The AO-correcting techniques applied to Lyot and apodization coronagraphs are not automatically transferable if these techniques are applied to another type of coronagraph, e.g., interfero-coronagraph and such AO techniques have to be re-designed.

For wavefront sensing after a coronagraph, we have used an interferometer (the same interfero-coronagraph) to measure both the phase and the amplitude spatial distributions applying the techniques of phase shifting interferometry (PSI) [23]. PSI recovers the two-dimensional maps of phase and interference fringe visibility distributions in the plane optically conjugated with the detector (CCD) plane. In the optically conjugated plane, we mounted a liquid crystal spatial light modulator (LC SLM) [24] acting as the AO active element and having 1920×1080 pixels. Correcting the wavefront by LC SLM, we enable the control of wavefront both in amplitude and phase. Herewith, we aim for our unit toward practical coronagraphic contrasts.

In the present work, we propose and investigate a method for an AO-controlling wavefront for interfero-coronagraph, including simulations and a laboratory experiment.

2. Materials and Methods: Common-Path Rotational-Shear Interfero-Coronagraph for Wavefront Measurement

2.1. Common-Path Rotational-Shear Interfero-Coronagraph Principles

In an interfero-coronagraph, the stellar image component is suppressed as the result of destructive interference (with achromatic anti-phase). Stellar light interference occurs after a beam combiner by the on-axis superposition of two copies of a star image being mutually rotated by a certain angle of rotational shift (relative to the optical axis). Considering a point-like star as a light source, after beamsplitter, two point-spread-functions (PSFs) of the star image are set to be geometrically superimposed and, therefore, are suppressed (canceled) by destructive interference at dark port. The main energy of stellar PSF is re-directed to the bright port of interferometer. At the same time, off-axis image components—two PSFs from non-central regions of the image—(we assume an off-axis PSF is an exoplanet)—become geometrically separated (at least partially). Off-axis PSFs are non-superposed due to the rotational shift about the optical axis centered to stellar PSF. An exoplanet PSF cancelation or even its significant attenuation does not occur. This principle is explained in Figure 1a.

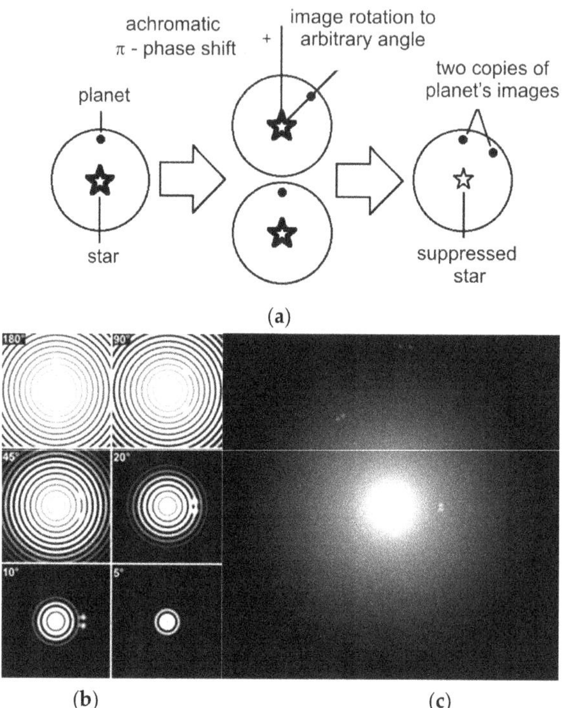

Figure 1. Rotation-shear interfero-coronagraph: principles and simulated output. (**a**)—anti-phase superposition on-axis for stellar PSF (asterisk denoted) and doubling off-axis planet's PSF (dot denoted). (**b**)—monochromatic starlight incomplete suppression, leakage considering the stellar size of $\Theta = 0.01 \lambda/D$ (@ $D = 2.4$ m, $\lambda = 500$ nm) by various rotational shears (the angles of rotation are noted), the shown planet PSF considered at the stellocentric separation of $\rho_0 = 5 \lambda/D$, and contrast $\varepsilon = 10^{-8}$. (**c**)—in polychromatic band ($\lambda = 350$–850 nm), picture of Solar System as observed from 10 pc with the telescope $D = 2.4$ m and the interfero-coronagraph having 3.6° angular shear; Jupiter, Saturn, Uranus, Neptune seen visualized toward periphery.

With an interfero-coronagraph, starlight suppression cannot be complete due to the physical size of a star. An apparent non-infinitesimal stellar disk size, as observed from a distance of several parsecs, effects an extended light-source. However, rather different suppression amounts of starlight can be achieved [13–15] if one varies the rotational shear by the angle; see Figure 1b. Here, the planet components with the *peak* contrast of $\varepsilon = 10^{-8}$ = (Stellar PSF peak)/(Planet PSF peak) are visually saturated (or overwhelmed) by starlight leaked at the rotational shears of $\psi = 180°, 90°$, and $45°$. A planet at a $5 \cdot \lambda/D$ separation with a $\varepsilon = 10^{-8}$ contrast can be clearly visualized at the rotational shear less than $\psi = 20°$; see corresponding panels. By the smaller angle of rotation shear (e.g., 5°), the planet images also become suppressed.

To illustrate an interfero-coronagraph applicability, in Figure 1c, we represent the simulated picture of the Solar System as observed from 10 pc in polychromatic band ($\lambda = 350$–850 nm) with a telescope having an ideal optics $D = 2.4$ m with an ideal optics interfero coronagraph attained at $\psi = 3.6°$ angular shear. We visualize Jupiter, Saturn, Uranus, and Neptune (double-dots images) when we trace them toward the periphery.

To verify, one can use Equation (1), which describes the intensity of residual light [15] because of the starlight leakage effect (because of apparent stellar size). Stellar PSF residual is in the first summand, and the intensity of planet light splits in two PSFs—in the second and third summands.

$$I(\alpha,\ \beta) = \frac{\Theta^2}{4}\sin^2\frac{\psi}{2}\left(\frac{J_2^2(\pi D\rho/\lambda)}{\rho^2}\right) + \varepsilon\frac{1}{4}\times\left[A\left(\alpha - \rho_0\cos\frac{\psi}{2},\ \beta - \rho_0\sin\frac{\psi}{2}\right) - A\left(\alpha - \rho_0\cos\frac{\psi}{2},\ \beta + \rho_0\sin\frac{\psi}{2}\right)\right]^2, \tag{1}$$

where

α and β—position angles on the celestial sphere $\alpha^2 + \beta^2 = \rho^2$:

$\alpha = 0$, $\beta = 0$—co-ordinates of a star (on-axis);

$\rho_0(\alpha_0,\ \beta_0)$—stellar-centric co-ordinates of a planet (off-axis);

Θ—apparent stellar size, e.g., $\Theta_\odot \approx 0.02\ \lambda/D$ (@ $\lambda = 500$ nm, $D = 2.4$ m);

ψ—angle of rotational shift;

$\varepsilon = (Stellar\ PSF\ peak)/(Planet\ PSF\ peak)$—peak-to-peak star-to-planet luminosity ratio;

$A(\alpha,\ \beta) = 2J_1(\pi D\rho/\lambda)/(\pi D\rho/\lambda)$—planet PSF assumed here having the equal luminosity to the stellar PSF;

J_1, J_2—first and second-order Bessel function of the first kind;

D—telescope diameter;

λ—sensing wavelength.

Without any image post-processing technique, a raw coronagraphic contrast (denoted *CC*) defines the ratio of planet light amount to residual light at the image area integrated over a spatial domain denoted by *S*. Area *S* can be chosen equal to the PSF core and for a central wavelength in $\Delta\lambda$ domain, Equation (2).

$$CC(\rho_0,\psi,\Theta) = \frac{\int_{\Delta\lambda}\iint_S[planet\ light\ (\lambda, D, \rho_0, \psi, \alpha, \beta)]d\lambda\,d\alpha\,d\beta}{\int_{\Delta\lambda}\iint_S[residual\ star\ light\ (\lambda, D, \Theta,\ \psi, \alpha, \beta)]d\lambda\,d\alpha\,d\beta}$$

$$= \frac{\int_{\Delta\lambda}\iint_S\frac{1}{4}\times\left[A\left(\alpha - \rho_0\cos\frac{\psi}{2},\ \beta - \rho_0\sin\frac{\psi}{2}\right) - A\left(\alpha - \rho_0\cos\frac{\psi}{2},\ \beta + \rho_0\sin\frac{\psi}{2}\right)\right]^2 d\lambda\,d\alpha\,d\beta}{\int_{\Delta\lambda}\iint_S\frac{\Theta^2}{4}\sin^2\frac{\psi}{2}\frac{J_2^2(\pi D\rho/\lambda)}{\rho^2}d\lambda\,d\alpha\,d\beta}. \tag{2}$$

Published in [15], the table of Equation (2) evaluates the *CC* contrasts versus the star-to-planet separation ρ_0, and the rotational shift ψ shows the possibility to visualize giant planets orbiting a Solar-type star from 10 pc with $D = 2.4$ m telescope at $\lambda = 500$ nm as Jupiter in the Solar System (SS) at $\rho_0 \approx 12.1\ \lambda/D$, with $\psi \approx 7.5°$ with $log(CC) \approx -8.8$. However, telluric planets such as the Earth in the SS can be observed at $\rho_0 \approx 2.3\ \lambda/D$ separation with $\psi \approx 40°$ and, by contrast, $log(CC) \approx -5.0$. Image processing can enhance one or two orders in *CC* contrast. We can optimistically visualize a Jupiter-like SS exoplanet and cannot visualize an Earth-like planet without processing about 5 orders of magnitude in the contrast that remains almost out of detector dynamic range.

Previously, without wavefront correction issue, we have experience with the nulling interferometers having different rotational shears [11,15]. We designed optical assemblies with out-of-plane principal ray propagation via successive reflections on plane mirrors. Several nulling optical schemes with the mutually balanced amplitudes of interfering waves by rotational-shear interferometers (RSI) are represented in Figure 2. The designed common-path nulling interferometers have the various numbers of reflections and the different amounts of rotational shear; moreover, the latter can be fixed or variable. Following the schemes in Figure 2, after the second beam splitter BS$_2$ (a) or after polarizing beam splitter PBS$_2$ in (b,c), we constructed a closed loop for a common optical path via successive mirror reflections. Indeed, two interferometer arms share the same path, are contrary directed, and, therefore, they have no optical path differences. Such a zero-difference in optical path lengths allows the achromatic functioning of the coronagraph. Rotational shear architecture achieves maximum transmission in the absence of pupil shielding. Shown in Figure 2, designs were tested under laboratory conditions, and their functionalities as achromatic coronagraphs were lab demonstrated [11,15].

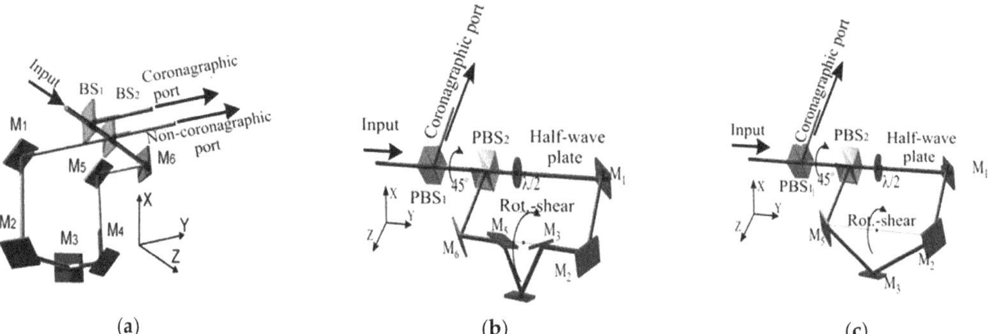

Figure 2. Optical schematics of common-path nulling interferometers with various rotational-shear angles acting as interfero-coronagraphs (on their dark port): (**a**) CP-AIC (RSI-180°) with rotational-shear angle fixed at 180°; (**b**) RSI-VAR with a variable rotational-shear angle; (**c**) RSI-10° rotational-shear angle fixed at 10°.

The interfero-coronagraph in Figure 2a has the 180° scheme, the Figure 2b scheme has a variable rotational shift, and the Figure 2c scheme has a fixed angle of rotation shear. The schemes of Figure 2b,c are characterized by increased sensitivity to polarization inhomogeneities across the aperture due to three successive polarization elements being installed in optical train: two polarization beam splitters and a half-wave plate. The wave plate is designed to be optimized to the central wavelength of a working spectral range, and its purpose is to increase the transmission of the coronagraph. Wave-plate has principal achromaticity, but it is not a critical issue for coronagraphic contrast at non-central wavelengths. To detail more functional and engineering aspects of these interfero-coronagraphs, we refer the reader to our former publications [11,15].

For simplicity, in the present work, we study an interfero-coronagraph with 180° only (shown in Figure 2a), with a function for wavefront sensing. With minor algorithm modifications, other nulling interferometers can perform similar wavefront measurement techniques.

2.2. Wavefront Measurement by Means Interfero-Coronagraph and Phase Shifting Interferometry

By means of an RSI-180° interfero-coronagraph, we implemented a PSI (phase shifting interferometry) technique to measure the interference fringes visibility $\gamma(u,v)$ (from which we can characterize the amplitude distribution $P(u,v)$) and the $\phi(u,v)$ phase distribution of the electric field in a pupil plane (u,v). Applying the PSI technique to coronagraphy, we found this approach is rather different from, e.g., EFC [17]. To find an incident wavefront characteristic, we solve the inverse problem substantially easier. We found, however, much closer similarity of PSI with [25,26].

In Section 2.1, we have already introduced angular sky coordinates set at (α, β) on a celestial sphere. Setting Cartesian coordinates, denoted as (x, y), in the image plane conjugated to the focal (or CCD) plane can be found by $(x = \sin \alpha \cos \beta, y = \sin \alpha \sin \beta)$. Complex amplitudes in the pupil (u,v) and in the image (x,y) are Fourier transformed:

$$\hat{P}(x,y)e^{i\,\hat{\phi}\,(x,y)} = F^{-1}\left\{ P(u,v)e^{i\phi(u,v)} \right\}.$$

To enable a wavefront measurement by PSI, we detect the intensities of several interference patterns obtained by a set of controlled phase shifts in the plane conjugated to the pupil. To measure the wavefront by interfero-coronagraph (with a 180° rotational shift), we perform sequentially the phase-step modulation in the half of pupil field. Before the interferometer, a pixelated AO element (LC SLM) was installed in a conjugated pupil. Then, using the PSI formulas, we recover the wavefront considered as referenced to a half view field via a 180° rotational shear interferometer.

For PSI details, we refer the reader to [23]. Just to briefly recall phase-shifting interferometry (PSI), let us fix the reference mirror position in one arm of the Michelson interferometer; in the second arm of the interferometer, the mirror is cyclically shifted along the optical path by a multiple of the phase shift $\Delta\phi_i$, and the corresponding intensity distributions $I_i(u,v)$ are measured. Three-, four-, and multi-step PSI techniques are therefore known [23].

The simplest three-step PSI technique assumes a constant phase step modulation with $\Delta\phi_i$: $-\Delta\phi$, 0, and $\Delta\phi$. When $\Delta\phi = \pi/2$, one measures three consecutive intensity values: $I(u,v)_{\Delta\phi=-\pi/2} = I_1$, $I(u,v)_{\Delta\phi=0} = I_2$, and $I(u,v)_{\Delta\phi=\pi/2} = I_3$, can calculate the phase distribution $\phi(u,v)$ and the module of the coherence function denoted by $\gamma(u,v)$, which is simplified as the fringe visibility or the *mutual* amplitude of two interfering waves. The four-step method is more accurate with the same step $\Delta\phi = \pi/2$. When four intensity distributions are collected, respectively, the phase shifts: $-\pi/2$, 0, $\pi/2$, π: $I(u,v)_{\Delta\phi=-\pi/2} = I_1$, $I(u,v)_{\Delta\phi=0} = I_2$, $I(u,v)_{\Delta\phi=\pi/2} = I_3$, and $I(u,v)_{\Delta\phi=\pi} = I_4$. The generalization of three- and four- is a more accurate N-step algorithm. Corresponding formulas to recover the $\phi(u,v)$ phase and the $\gamma(u,v)$ visibility distributions from the fringe patterns $I_1..I_l$ by the three-step-, the four-step, and the N-step PSI methods, are shown in Table 1.

Table 1. Three-, four-, and N-step PSI algorithms.

PSI Technique	Phase Distribution $\phi(u,v)$	Visibility Distribution $\gamma(u,v)$
Three-step: $\Delta\phi = -\pi/2, 0, \pi/2$	$tan^{-1}\left(\dfrac{I_1-I_3}{2I_2-I_1-I_3}\right)$	$\dfrac{\sqrt{\left((I_1-I_3)^2+(2I_2-I_1-I_3)^2\right)}}{I_1+I_3}$
Four-step: $\Delta\phi = -\pi/2, 0, \pi/2, \pi$	$tan^{-1}\left(\dfrac{I_4-I_2}{I_1-I_3}\right)$	$2\dfrac{\sqrt{\left((I_4-I_2)^2+(I_1-I_3)^2\right)}}{I_1+I_2+I_3+I_4}$
N-step: $\Delta\phi_i = 0...2\pi/(N-1)$	$tan^{-1}\left(\dfrac{\sum_{i=1}^{N} I_i \sin\Delta\psi_i}{\sum_{i=1}^{N} I_i \cos\Delta\phi_i}\right)$	$2\dfrac{\sqrt{\sum_{i=1}^{N}\left[I_i^2(\sin\Delta\psi_i)^2 + I_i^2(\sin\Delta\psi_i)^2\right]}}{\sum_{i=1}^{N} I_i}$

The implementation of the three-step PSI in the 180° rotational-shift interferometer is shown in Figure 3; here, the algorithm of the four-step method is complemented by the dashed blue color on the right. For illustration in the upper-right corner, four color panels simulated with *Proper* [27] show the mixture of semi-plane modulation with background phase perturbation pattern. We use the phase modulation of the half of pupil field and obtain intensities $I_{1..3}$ or $I_{1..4}$ according to the PSI algorithms.

Therefore, the wavefront with some initial phase distribution $\phi(u,v)$ undergoes an additional phase shift $\Delta\phi$, uniform in a half of the pupil plane. Wave propagation in the interfero-coronagraph arms follows the mutual image rotation to $\pm90°$ and results in a 180° rotational shift. An internal achromatic phase shift of π radian is implemented between the interferometer arms (schematically shown in Figure 3 in one interferometer arm). These processes determine the optical waves superposed by their interference with some fixed set of uniform phase modulations; see Figure 3. The corresponding phase shifts in the half-pupil make interference pattern intensities (as denoted by the superscripts indices (d)—lower half-field and (u)—upper half-field) suitable for PSI. These intensity maps can be easily constructed in a computer memory from field-halves to full fields; they are shown at the bottom of Figure 3 with the indices corresponding to three- and four-step PSI techniques as $I_{1..3,4}^{(u)}$, $I_{1..3,4}^{(d)}$. The implementation of the four-step method is shown in blue and dashed on the right. Finally, from intensities $I_{1..3,4}$, PSI technique recovers the phase $\phi(u,v)$ distribution and the fringe visibility $\gamma(u,v)$ distribution (following Table 1).

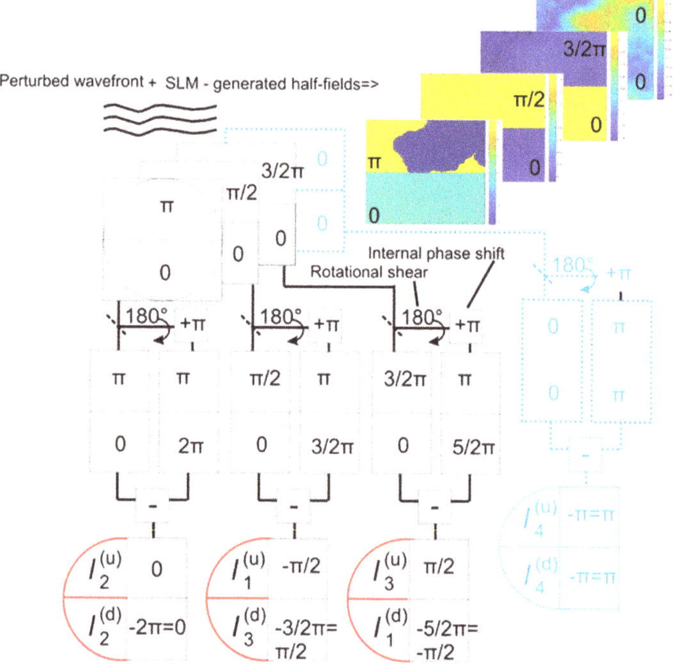

Figure 3. Phase modulation provided in the pupil semi-plane. Denoted are intensity values $I_{1..3}^{(u)}$, $I_{1..3}^{(d)}$ used by the three-step PSI; the subscripts (u) and (d) stand for the upper and lower halves of the pupil; an additional phase modulation with the corresponding intensities of $I_4^{(u)}$, $I_4^{(d)}$ for the four-step PSI algorithm is in blue.

Measured in this way, the phase distribution $\phi(u, v)$ in the pupil semi-plane is sufficient to compensate the observable wavefront aberrations by sending $-\phi(u, v)$ or $+\phi(u, v)$ distribution to AO LC SLM, depending on a chosen halfplane on SLM.

Interfero-coronagraph with an implemented 180° rotation-shift automatically suppresses the symmetric image components as originated from the initial even numbers (symmetric) wavefront aberration terms in an incident wave up-streaming to coronagraph. By the PSI modulation described above set by an LC SLM (as the main pixilated AO element), we measure distorted phase distribution $\phi_0(u, v)$ originally made from the asymmetric image components from the odd number of wavefront aberrations in an incident wave. Noting an off-axis planet PSF is an example of an odd number aberration. After the 180° interfero-coronagraph, odd number aberrations become added, with the aberration located geometrically symmetric respective to optical axis: $\phi_{\Sigma'}(u, v) = -\phi_{\Sigma'}(-u, -v)$. By this way, a *ghost* centro-symmetric copy of planet PSF is always observed. The subscript Σ' stands to define the succeeding pupil plane after (or downstream to) coronagraph, which is optically conjugated to the plane, denoted further as $\Sigma = \Sigma_0$, where the incident phase distribution $\phi_0(u, v)$ was considered.

These wavefront measurement and correction techniques by 180° interfero-coronagraph with an AO LC SLM before the 180° interfero-coronagraph are illustrated in Figure 4.

Here, we consider, in initial perturbed wavefront (phase distribution), two (red-marked) pixels which come later in the destructive interference process by their superposition. We denote by φ_1 and φ_2 the corresponding phases in these two pixels. The next row shows conjugated pixels in the plane after superposition the phases φ_1 and φ_2 are subtracted by interference. By PSI, one recovers phase pattern in (e.g., left) semi-plane as simplistic, and one considers phase difference $\varphi_1 - \varphi_2$ from the corresponding pixels. For following wavefront correction, one assumes a left semi-plane phase distribution and sends its negative value to AO LC SLM into the corresponding pixel in left semi-plane. To illustrate this correction, in Figure 4, we can see in the right column the left pixel becomes phase modulated, while the right pixel does not have any additional modulation. As the result of the superposition, we see the phase becomes further corrected in both pixels in both semi-planes.

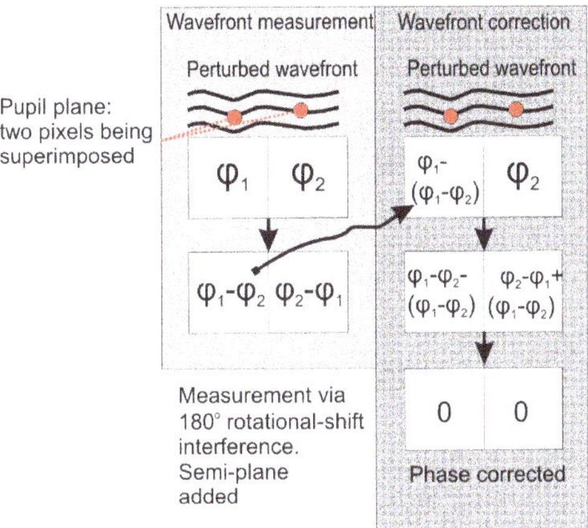

Figure 4. Illustration (in left column) for the measurement of wavefront by 180° interfero-coronagraph and (in right column) for the correction of wavefront by an AO LC SLM placed before the 180° interfero-coronagraph. (Note that here the corrected "left" side of SLM is shown. The same result could be achieved with "right" side just by altering the sign before $\varphi_1 - \varphi_2$).

3. Results

3.1. Numerical Simulation of Wavefront Control in an Optical Scheme with an Interfero-Coronagraph

To verify the applicability of the PSI technique to function in a 180° interfero-coronagraph, we have performed numerical simulations. In the *PROPER* [27] software package, we added several new blocks (i) to determine the phase and the amplitude distributions in optical pupil by PSI technique and (ii) to simulate a 180° interfero-coronagraph and to visualize its output. We considered the optical scheme shown in Figure 5, containing a telescope (shown by lens L_1) and optical elements (lenses $L_{2..6}$) in order to make several optical planes (with co-ordinates), $\Sigma' = (u', v')$, $\Sigma'' = (u'', v'')$, and $\Sigma''' = (u'', v'')$ being optically conjugated to the primary pupil $\Sigma_0 = (u, v)$. Aberrations of the telescope are resumed in Σ_0. AO element—LC SLM is installed in the secondary pupil Σ'. Interfero-coronagraph is placed in the space before the tertiary pupil Σ''. In the next pupil Σ'''', we place CCD_2, which is used to measure wavefront by recording a series of PSI images $I_{1..L}$. Another CCD_1 is the field camera to observe an image in the focal plane F'', which is illuminated (shown in blue) by some beam splitters, assumed to be dichroic or switchable.

We understand that a lot of optical elements and their optical surfaces are placed outside the pupil planes Σ_0, Σ', ... Σ'''. Optical surface defects can be optically characterized with power spectral density (PSD) for micro-roughness; these surface figures produce Fresnel-propagation-like aberrations which we can associate with non-common-path aberrations (NCPA). Strictly, the NCPAs have the difference from the aberrations detected by a wavefront sensor (WFS), important in that NCPAs degrade coronagraphic image quality. In our case, because wavefront measurement is organized after the coronagraph, NCPAs (Fresnel aberration screens) generate not only phase errors but also amplitude wavefront errors. In the model, NCPA can be introduced as a phase aberration screen in an intermediate Σ_{NCPA} plane, which is a non-conjugate optically to the pupil. By Fresnel propagation, NCPA results in a high spatial frequency amplitude (and phase) modulation of the pupil. Therefore, NCPAs' effect can be studied if one considers several aberration screens in mutually non-conjugated planes (Σ_{NCPA1}, …). In Figure 5, we show a single Σ_{NCPA} plane.

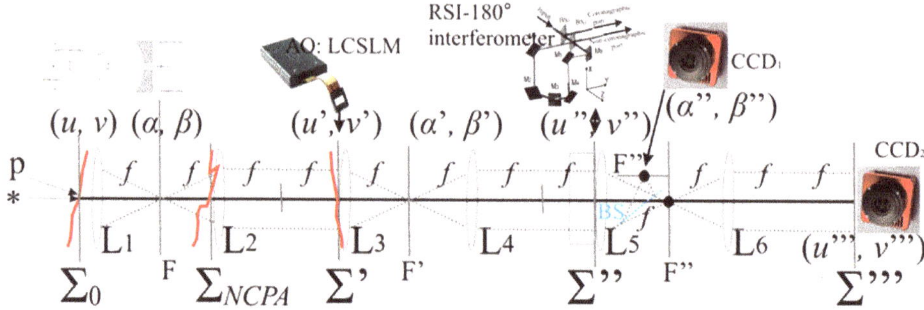

Figure 5. Optical scheme to assemble telescope by AO, interfero-coronagraph, field camera, and pupil camera for PSI technique.

To simplify the evaluation of aberrations' influence on a coronagraphic image and to study residuals after wavefront correction, Figure 5 shows a generalized optical system with unity magnification and with the equal focal lengths f of the optical lenses, while in a practical design, the magnifications and scales of the images will differ from the simulated parameters.

Figure 6 shows several results from the evaluation of the optical system (Figure 5). Examples of mutually non-correlated aberration phase screens in planes Σ_0 and Σ_{NCPA} are shown in Figures 6a and 6b, respectively. These aberrations do not include any classic geometric aberrations (coma, spherical, astigmatic, etc.) as represented by low-order Zernike polynomial decomposition. The shown aberrations consist only of micro-roughness errors, and they visually look like sky clouds. PSD was characterized by the local phase error $\delta\phi \sim \rho^{-2}$, where $\rho = \sqrt{u^2 + v^2}$ is the radial scale. The *RMS* error $\sigma = 10$ nm of the wavefront $\phi_{\Sigma_0}(u, v)$ in Σ_0, and the *RMS* $\sigma = 1$ nm error of the $\phi_{\Sigma_{NCPA}}(u, v)$ wavefront in Σ_{NCPA} were calculated at the characteristic length of $\rho = D/2$.

In Figure 6c–f, simulated coronagraphic images $I(x'', y'')$ are shown, as observed in the focal plane F'' by a CCD₁ field camera. A companion (planet) was computed with the PSFs peak-to-peak contrast $C = 10^{-9}$ between the point-like light sources planet and star. Contrast C was modeled as $C = \frac{I_{*,@\lambda}}{I_{p,@\lambda}}$ and the wavelength $\lambda = 500$ nm. The star (*) and planet (p) were separated at a 5 λ/D stellocentric distance. Panels (c), (d) were calculated with the wavefront distortions $\phi_{\Sigma_0}(u, v)$ by $\sigma = 10$ nm. In panel (c), the coronagraphic image $I(x'', y'')$ was computed without any wavefront correction. The image contains strong speckle effects, which completely masks the faint planet image. In panel (d), the coronagraphic image was computed as phase corrected, and we applied the phase screen of $-\phi_{\Sigma_0}(u', v')$ by AO LC SLM in the plane $\Sigma'(u', v')$. Now, both the planet PSF and its ghost symmetric copy were completely cleared of speckles. This is because the optically conjugated planes Σ_0 and Σ' are considered to contain the equal phase distributions, which were compensated for. In panels (e) and (f), besides the initial aberration $\phi_{\Sigma_0}(u, v)$, an additional $\phi_{\Sigma_{NCPA}}(u, v)$ aberration in different and not conjugated plane Σ_{NCPA} was computed. In panel (e), correction was not applied, and this image is nearly similar to that in panel (c). In panel (f), correction was applied by setting the phase screen $-\phi_{\Sigma'}(u', v')$ in the plane $\Sigma'(u', v')$; note that $-\phi_{\Sigma'}(u', v') \neq -\phi_{\Sigma_0}(u', v')$, which is non-equal to the correction used for result in (d) panel. Moreover, the amplitude distribution $P_{\Sigma'}(u', v')$ (not shown) becomes non-uniform across the pupil.

The panel (f) image as compared with the panel (d) image has residual speckles which are caused by non-corrected amplitude distortions in pupil Σ', as induced by phase aberration $\phi_{\Sigma_{\mathrm{NCPA}}}(u,v)$ in Σ_{NCPA}. Shown in panel (g), the radially averaged profile of coronagraphic image (f) robustly visualizes a companion in $5\,\lambda/D$ by the intensity maximum.

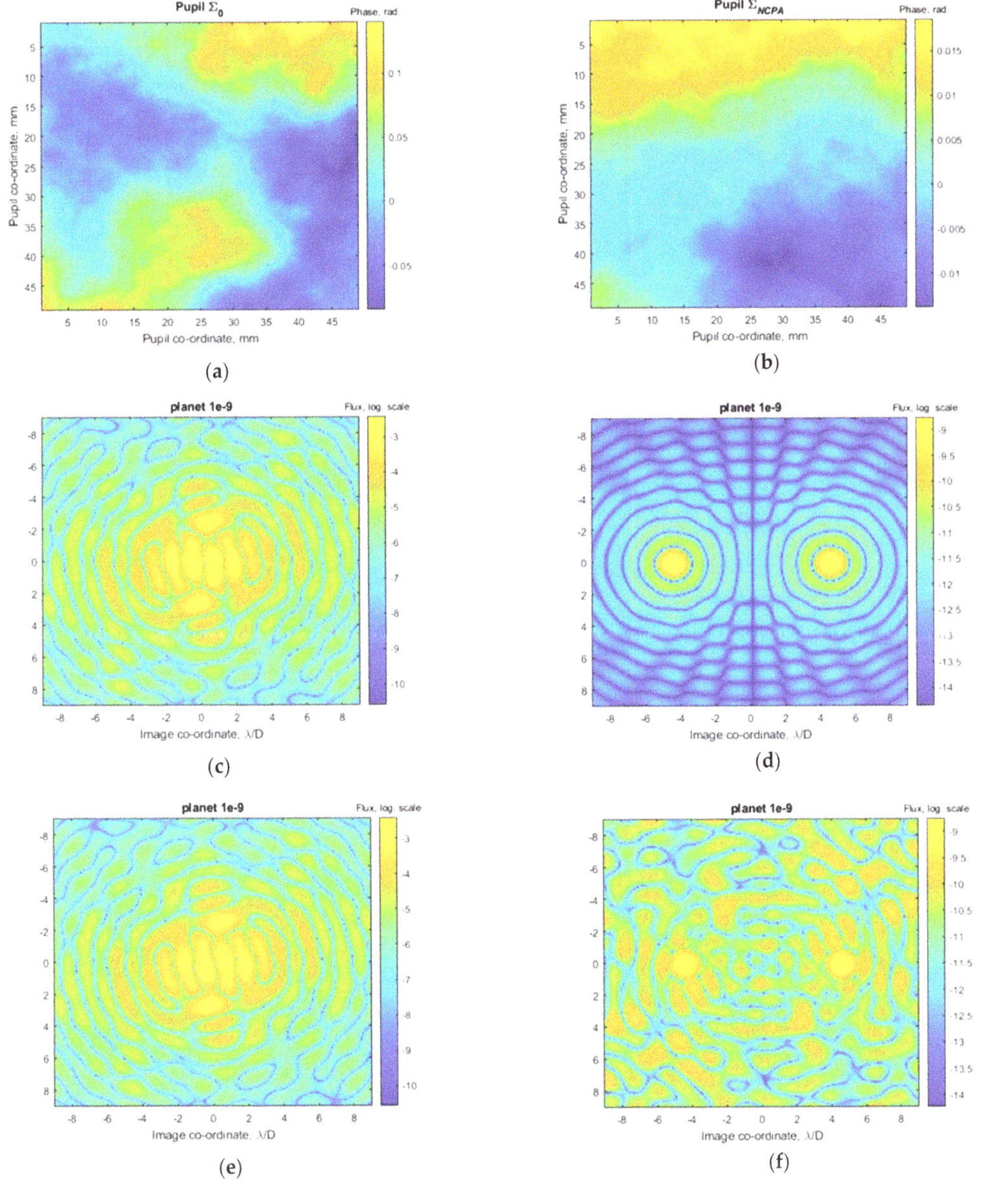

Figure 6. *Cont.*

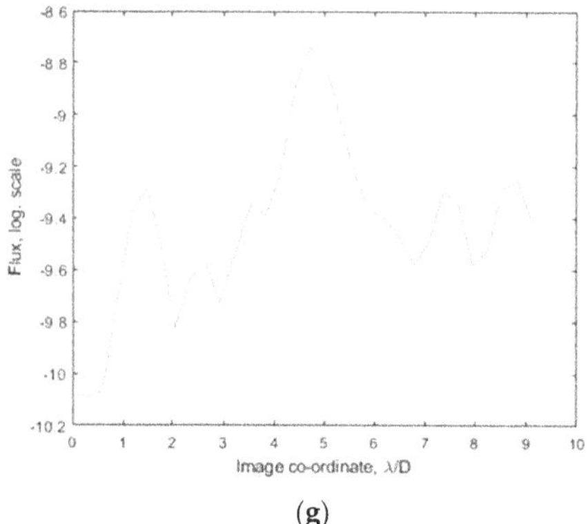

(g)

Figure 6. Simulated phase aberrations and corresponding coronagraphic images. Aberrations: (**a**)—phase distribution $\phi_{\Sigma_0}(u,v)$ with $\sigma = 10$ nm rms in Σ_0 plane and (**b**)—$\phi_{\Sigma_{\text{NCPA}}}(u,v)$ with $\sigma = 1$ nm rms in Σ_{NCPA}. (**c–f**)—evaluated coronagraphic images $I(x'', y'')$ at focus F'' (see optical scheme in Figure 5) on CCD_1 field camera, computed with the PSFs peak-to-peak contrast between point-like light sources: the planet and the star $C = 10^{-9}$. (**c,d**)—calculated with the wavefront distortion of $\phi_{\Sigma_0}(u,v)$. (**c**)—non-corrected wavefront. (**d**)—corrected by setting the phase screen $-\phi_{\Sigma'}(u',v') = -\phi_{\Sigma_0}(u,v)$ in the plane $\Sigma'(u',v')$; here, the planet PSF and its symmetric copy were completely cleared from the speckles. In panels (**e,f**), both aberrations $\phi_{\Sigma_0}(u,v)$ and $\phi_{\Sigma_{\text{NCPA}}}(u,v)$ in the different planes Σ_0 and Σ_{NCPA} were computed. (**e**)—any wavefront correction was not applied. (**f**)—wavefront correction was applied by setting phase screen $-\phi'_{\Sigma'}(u',v') \neq -\phi_{\Sigma_0}(u,v)$ in the plane $\Sigma'(u',v')$. (**g**)—radially averaged profile of coronagraphic image of panel (**f**).

Shown in Figure 6, the simulation results were obtained by the known phase distributions $\phi_{\Sigma'}(u',v')$, $\phi'_{\Sigma'}(u',v')$ in the plane of wavefront correction of $\Sigma' = (u', v')$. This is possible in the simulation; however, practically, one needs to measure phase distributions for its corrections. Therefore, next, we have modified our numerical experiment to check how to determine phase $\phi_{\Sigma_0}(u,v)$ by the PSI four-step phase-shifting technique downstream from the coronagraph. PSI was applied to half of the pupil plane $\Sigma'(u',v')$ for the wavefront correction in the AO LC SLM plane $\Sigma'(u',v')$. The images obtained by the simulation of the PSI technique have been found to coincide with coronagraphic images shown in Figure 6 in panels (c)–(f).

3.2. Lab Experiment to Verify Wavefront Correction

3.2.1. Schematics of Lab Experiment

Schematic and photo of the lab experiment, shown in Figure 7a,b, respectively.

In the lab experiment, light radiation from an *HeNe* laser passes through a spatial filter, a collimation lens L_2, a linear polarizer, and through a reflecting phase-only LC SLM propagating in the direction of the AIC-180° interfero-coronagraph. Lenses L_3–L_4 conjugate optically the pupil plane Σ' where the LC SLM is installed to the beam combiner plane of the AIC-180° interfero-coronagraph, where interfering waves are superposed. After the coronagraph, another beam splitter *BS* is located. In this direction after the beam splitter reflects, an image (focal) plane is formed and is observed by a CCD_1 field camera. CCD_1 stands to observe a coronagraphic image in the focal plane. In the other direction, the beam splitter transmits an image, and an additional lens L_6 stands to form an optical pupil plane image. This pupil image is observed by the second CCD_2 to measure the wavefront distribution according to the PSI algorithms; see Table 1.

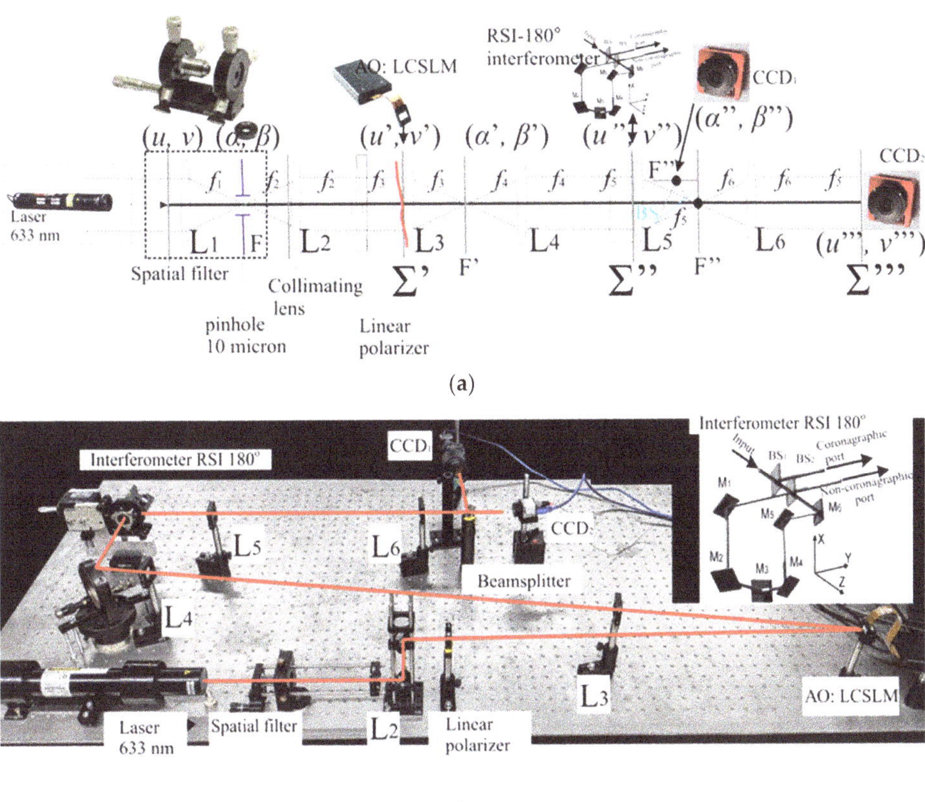

(a)

(b)

Figure 7. Optical scheme (**a**) and photo (**b**) of the lab experiment for wavefront correction with measurement of wavefront distortions after an interfero-coronagraph. Marked: *laser, spatial filter*: L_1 *lens*, L_2 *collimating lens, linear polarizer*, phase-only *LC SLM* (LC SLM shown in the in the transmitting mode in panel (**a**)), *interfero-coronagraph (AIC-180°)* (shown in enlarged size in tab (**b**)), *beam splitter, focusing lenses* L_3–L_5, CCD_1 (field camera in focal plane), focusing lens L_6 to form a pupil plane, CCD_2 (camera in pupil plane).

3.2.2. Wavefront Correction Maps in Pupil Plane

Some results of wavefront phase correction based on measurements in the pupil plane via CCD_2 applying PSI algorithms (modified as shown in Figure 3) are shown in Figure 8.

Figure 8 shows the correction of a nearly flat wavefront with the initial standard deviation $\sigma \approx \lambda/5$. The uncorrected wavefront is shown in (a) panel; then, the left and right parts were corrected alternately, as shown, respectively, in panels (b) and (c).

In the optical scheme, the spatial phase modulator LC SLM was installed before the interfero-coronagraph 180°; this schematic has allowed us to only symmetrically influence wavefront distortions simultaneously in two halves of the pupil plane. Because of this schematic feature, all asymmetric wavefront distortions are corrected alternately either in the left or the right halves of the pupil plane.

In Figure 8d, analyzing the radially averaged power spectrum density (PSD) of the pupil half-plane before correction (shown by the blue dashed line) and after (shown by the red solid line), one can see that heterogeneities with spatial frequencies (with characteristic size) less than 1/10 of the half-aperture can be effectively corrected (by more than an order of magnitude), while the higher spatial frequencies are not corrected. Mid- and high-frequency non-correction are explained by the presence of non-common path aberrations, which cause uncontrolled and, therefore, uncorrected amplitude errors in the mid and high frequencies.

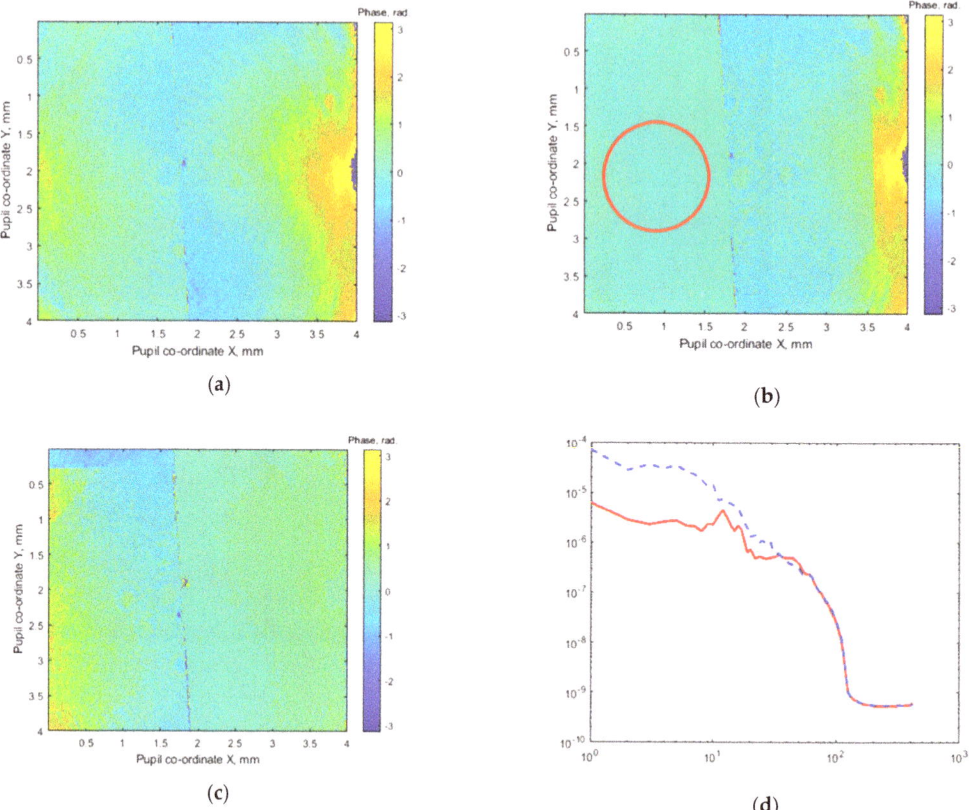

Figure 8. Correction of wavefront in left (right) semi-plane in pupil: (**a**)—uncorrected wavefront measured by interfero-coronagraph. (**b**,**c**)—consequently corrected wavefront in left and right semi-planes (the area within the shown red circle was used as corrected pupil to form focal images in following Figure 9). (**d**)—wavefront phase error power spectrum density (PSD) averaged over radial cross-section; before correction—blue (dashed) line, after correction—red (solid) line.

Used here is the adaptive optical element; the LC SLM was implemented in phase-only mode. However, it has a potential to correct both the amplitude- and phase wavefront distortions if the LC SLM is rotated respective to the polarization axis.

For this option, the PSI method shown in step-shift mode on the pupil semi-plane (for the 180° rotation shift plane interferometer) is well-suited to measure both the phase and amplitude wavefront distortions for wavefront control by LC SLM, assuming wavefront distortions originated in non-conjugated planes induce high frequency phase–amplitude aberration known as (non-common-path) aberrations (Σ_{NCPA}).

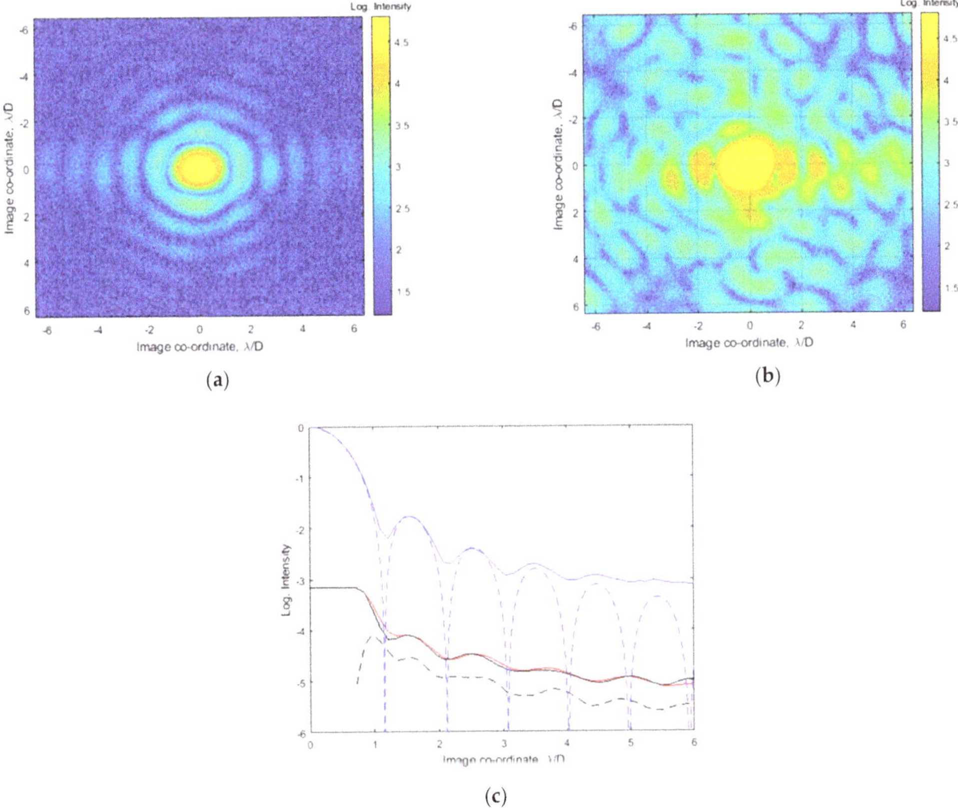

Figure 9. Experimental images: (**a**)—non-coronagraphic image (with exposure 0.02 ms). (**b**)—coronagraphic image by the phase correction down to with $\sigma \approx \lambda/40$ (with exposure $\approx$ 30 ms). (**c**)—averaged radial cross-sections: blue (solid) line for the non-coronagraphic image (**a**), red (solid) line for the coronagraphic image (**b**), blue (dashed) line for the radial cross-section of the theoretical PSF (non-coronagraphic), black (solid) line—cross section of another in time realization of the coronagraphic image, black (dashed) line—averaged cross section of the difference between two coronagraphic images.

3.2.3. Coronagraphic PSF Suppression in Focal Plane

Phase correction experiments (similar to those in the pupil plane shown in Section 3.2.2) were also observed in the focal (image) plane by CCD$_1$, where coronographic images were recorded. The non-coronagraphic and coronagraphic images shown in Figure 9a,b correspondingly were compared to measure a coronagraphic contrast. To obtain a non-coronographic image (after the coronagraph), the input wavefront contains a step-form phase modulation, and the phase gap between the left and right half-planes was about π (see Figure 4: $\varphi_1 - \varphi_2 = \pi$). By this method, the non-coronagraphic image looks similar to a theoretical PSF as the diffraction image of a point-like source with clearly observed airy rings. The non-coronagraphic frame was captured with 0.02 ms exposure.

The coronagraphic image in Figure 9b was obtained as a result of the applied wavefront correction algorithm. To register it, an exposure time of 30 ms (versus 0.02 ms for Figure 9a) was used to clearly visualize the speckle field pattern. Next, we resume the radially averaged intensity profiles (Figure 9a,b), taking into account their different exposures to plot them in Figure 9c (blue (solid) and red (solid) lines corresponding to the non-coronagraphic PSF and coronagraphic PSF). The PSF profiles were normalized to the value of the maximum intensity of the non-coronographic image (the blue curve value at stellocentric equals unity). The same figure shows the scaled radially averaged intensity profiles of the coronographic images captured in two consequent frames (black and red curves in Figure 9c). In the region of the central maximum of the PSF (in region of 0 along the stellocentric horizontal axis) of the coronographic images, a "flat top" is observed. This is due to the saturation effect of the CCD_1 matrix (see CCD_1 in the optical scheme in Figure 7) by 30 ms exposure of the coronographic image. Therefore, the represented non-coronagraphic(Figure 9a) and coronagraphic images (Figure 9b) are shown with their scale respective of their exposure ratio $30/0.02 = 1500$.

Figure 9b corresponds to Figure 8b, which is taken simultaneously but in the pupil plane. In Figure 8b, the red circle defines the approximate area of the pupil; this was taken by an additional diaphragm (which is not shown in Figure 7a), the diaphragm was used for the images in Figure 9a,b. The phase map in pupil was measured having the standard deviation of $\sigma \approx \lambda/40$ over the corrected pupil area, marked with a red circle.

Figure 9c enabled the estimation of the coronagraphic contrast (as the ratio of the non-coronagraphic PSF maximum to the scattered light coronagraphic intensity at a given stellocentric distance), which gives $\sim 10^5$ at $>4\ \lambda/\mathrm{D}$.

In Figure 9c are plotted: the radial cross-section of the theoretical PSF $(2J_1(r)/r)^2$—the blue dashed line, the averaged radial cross-section of non-coronagraphic PSF (a)—blue solid line, two averaged radial cross-sections of coronagraphic realizations at different times—solid black and red lines (b), the black dashed line made by the difference between the two coronagraphic images obtained at different time. The difference in coronagraphic images implies post-processing, in which so-called static speckles are allowed to subtract and show the possibility of an additional gain in coronagraphic contrast. Assuming possible post-processing, in the stellocentric region, wider than the second airy ring ($>2\ \lambda/\mathrm{D}$), we experimentally obtained a coronagraphic contrast of 10^5 above the PSF value at the maximum, (see black dashed line). Post-processing can improve this value by an order of magnitude, which has not yet been shown in the presented laboratory experiment and will require a greater number of processed images.

3.3. Constraints and Their Overcome

3.3.1. Using an LC SLM for Extreme Wavefront Correction

In the laboratory experiment, we used an LC SLM commercially available (in the Russian Federation) instead of a precision deformable mirror (DM). The LC SLM choice has some advantages and some disadvantages, which are worth noting in the context of the liquid crystal spatial light modulator-based adaptive optics to function for the stellar coronagraph.

Among the *advantages* of the LC SLM [28] is (i) the significantly higher number of addressable pixels over $1500 \times 1000 = 1.5 \times 10^6$ compared to the lower number of DM actuators $\sim\!10^3...10^4$. (ii) An amplitude-phase modulation and a general polarization modulation are both possible when using polarization devices. Amplitude-phase modulation occurs, e.g., when two linear polarizers are placed, one before and one after the LC SLM, and polarizers are oriented at their transmission axis at different angles from the direction of the main working polarization axis of the LC SLM. (iii) It is possible to realize an arbitrary wavefront surface, including wavefront discontinuities, which is practically impossible with the DM.

The main *drawback* of AO LC SLM is the significantly lower accuracy of phase modulation, nominally $\lambda/250 \approx 2\pi/2^8$ (i.e., $\lambda/256$), at which the corrected wavefront does not have a sufficient quality for observing exoplanets. However, AO LC SLM phase modulation accuracy can be enhanced if an initial phase distortion is within, e.g., $\lambda/10$ P-V and has to be corrected. Such a calibrated AO LC SLM performs $\lambda/2500 \approx 0.1\ \pi/2^8$ radian accuracy.

There are also pixel boundary effects different from the DM due to the physical boundaries of the pixels; here, the relative sizes of the modulated and non-modulated zones are described by the fill factor. Different to LC SLM, a DM has no visible boundaries (on the side of the deformable mirror) between the actuators, but the customized function (named as formfactor) is used to accurately approximate the phase distribution over a discrete number of actuators, taking into account the features of the interconnection between the actuators in deformable mirror to crosstalk the neighboring actuators. Finally, the LC SLM chromaticity, due to the dispersion properties of the liquid crystal material, should be taken into account compared to the DM with the effective free-space dispersion d to modify an optical path length $\frac{1}{\lambda}$.

Let us discuss the main drawback in the insufficient accuracy of the liquid crystal spatial modulator and sketch possible ways to function the LC SLM as an adaptive optics to correct the wavefront for stellar coronagraph. A way to effectively increase the phase modulation accuracy (by more than an order of magnitude) is possible if one applies the principle of an (extremely) unbalanced interferometer [29]. Thus, in a two-beam unbalanced interferometer, two coherent waves are superposed with unequal amplitudes; for example, differing by an order of magnitude (or, in the general case, by a factor of k). Inside the unbalanced interferometer in one arm, the phase of the wave having a smaller amplitude is corrected within the limited accuracy of LC SLM in phase modulation. In the other arm of the interferometer, the phase of the more intensive wave remains unmodulated by an LC SLM. These two waves interfere after a beam combiner.

If one measures the depth of the phase modulation of the resulting wave after the interference process, phase modulation depth will be less than the depth of the phase modulation of the weak amplitude wave. This principle can effectively reduce the modulation error. A detailed description of this method can be found in [29], noting that under the atmosphere, the wavefront is distorted by turbulence and contains a significant wavefront error exceeding several wavelengths in optical range at ~1000 nm. Unbalanced in amplitude, the interferometer can increase the accuracy of the wavefront correction if the wavefront error is substantially smaller than the wavelength λ (at least by an order of ~$\lambda/10$...~$\lambda/100$). It is, therefore, possible to use two-pixel actuators with accuracy on the order of $\lambda/100$, one mounted at the input before the unbalanced interferometer and the other inside the interferometer in the smaller amplitude arm, for accurate wavefront correction in ground-based telescope mode. In an orbital space telescope, a diffraction-limited image is obtained so that the distortion of the wavefront is usually several times smaller than the central wavelength λ. The LC SLM can also be used here in an unbalanced interferometer in a quasi-static mode.

3.3.2. Using an LC SLM for Wavefront Correction in Phase–Amplitude Mode

For the phase modulation mode, the linear polarization at the input has to be set parallel to the working (active) axis of the LC SLM. Here, the corresponding Jones vectors are:

$$\begin{bmatrix} E_x \\ E_y \end{bmatrix} = \begin{bmatrix} e^{i\varphi} & 0 \\ 0 & 1 \end{bmatrix} \begin{bmatrix} 1 \\ 0 \end{bmatrix} = e^{i\varphi} \begin{bmatrix} 1 \\ 0 \end{bmatrix}. \tag{3}$$

Suppose the AO LC SLM is set by its main working axis rotated in respect to the polarization axis of the incoming wave by a non-zero angle. Instead of phase modulation, a modulation of polarization in general form is carried out.

To organize a phase–amplitude modulation mode, we consider a linear polarization at the input that is set at a certain angle—for example, 45 degrees—to the direction of the working (active) axis of the LC SLM. Additionally, to prevent a polarization modulation, we consider using a polarizer (parallel to the initial linear polarization):

$$\begin{bmatrix} E_x \\ E_y \end{bmatrix} = \mathbf{R}_{-45} \begin{bmatrix} 1 & 0 \\ 0 & 0 \end{bmatrix} \mathbf{R}_{45} \begin{bmatrix} e^{i\varphi} & 0 \\ 0 & 1 \end{bmatrix} \frac{1}{\sqrt{2}} \begin{bmatrix} 1 \\ 1 \end{bmatrix} = \frac{1}{2\sqrt{2}} \left(e^{i\varphi} + 1 \right) \begin{bmatrix} 1 \\ 1 \end{bmatrix}, \tag{4}$$

where $\mathbf{R}_\beta = \begin{bmatrix} \cos\beta & -\sin\beta \\ \sin\beta & \cos\beta \end{bmatrix}$ is the rotation matrix by angle β.

Further, we consider the fact that the interfero-coronagraph is a shear interferometer. Physically, it causes the coherent superposition (summation) of two waves. Two different pixels of LC SLM—let us denote them as "1" and "2"—become superimposed by an anti-phase to process the destructive interference. For a rotational-shear interferometer RSI-180° (see Figure 2), "1" and "2" are centro-symmetric relative to the rotation axis. When considering the modulation given by (4) set in two pixels "1" and "2" with independently controlled phases φ_1 and φ_2, it becomes possible to compensate for the initial phase and amplitude modulation in these pixels $A_{01}e^{i\varphi_{01}}$, $A_{02}e^{i\varphi_{02}}$. By φ_1 and φ_2 control, we can equalize the optical fields in pixels in the form of the equal complex amplitudes and phases. When solving the following equation:

$$A_{01}e^{i\varphi_{01}}\left(e^{i\varphi_1}+1\right) = -A_{02}e^{i\varphi_{02}}\left(e^{i\varphi_2}+1\right) \tag{5}$$

we search the roots of φ_1 and φ_2, reducing the number of variables from four to two: $\Delta\varphi = \varphi_{01} - \varphi_{02}$ means the difference of initial phases and $a = A_{01}/A_{02}$ means the ratio of amplitudes:

$$ae^{i\Delta\varphi}\left(e^{i\varphi_1}+1\right) = -\left(e^{i\varphi_2}+1\right). \tag{6}$$

We checked the solution existence for (6). The study of its practical applicability is scheduled for a future work. We hope to use this approach to organize phase–amplitude wavefront corrections aiming to work out a non-common path aberration effect, an amplitude imbalance from some kind of Fresnel type wavefront distortion that originated in the non-conjugated plane to the pupil.

4. Discussion

We have investigated, both in a theoretical model and in a laboratory experiment, a new technique to correct the wavefront required to observe exoplanets in an astronomical diffraction-limited image in the vicinity of a star at stellocentric distance of about several diffraction radii from the parent star. The proposed methodology is workable in terms of measuring and correcting the wave front. In particular, the imaging of the Earth in the vicinity of the Sun, with contrast 10^9, requires a correction accuracy better than $\lambda/500$ on controllable 500×500 pixels.

This is achieved by a combination of extremely precise active adaptive optics (ExAO) systems and the coronagraph. In the process of correcting the wavefront, a nontrivial task is solved to measure the wavefront after the coronagraph, where aberrations of the non-common path are taken into account. Therefore, widely used wavefront sensors (e.g., Shack–Hartmann principle based sensors) appear inaccurate and contain additional sources of errors.

The new, currently proposed method for measuring and correcting the wavefront demonstrates in a simple experiment the wavefront quality is better than $\lambda/40$, which is still about an order of magnitude worse than the required value ($\lambda/500$–$\lambda/1000$). An achievement of the target accuracy in wavefront correction will be improved in future work with the possible involvement of additional techniques. With the correction accuracy achieved, the experimental coronagraphic contrast is shown to be better than 10^5 at a stellocentric distance of more than 2 diffraction radii ($\sim$2 λ/D) at a wavelength of 633 nm.

To sum up, the obtained results agree with the theoretical estimates. For the standard deviation σ of the phase of wavefront, a primitive estimation of starlight suppression via the destructive interference process is:

$$\frac{I_0}{I} = 1 - \cos(\sigma) \approx \frac{\sigma^2}{2}. \tag{7}$$

which, for $\sigma = 2\pi/40$, gives 10^2. Therefore, for a coronagraphic PSF at a distance of 5 λ/D, (7) shows the coronagraphic contrast estimate of 10^5, which was experimentally demonstrated.

Author Contributions: Conceptualization, methodology, A.T. and A.Y.; software, I.S.; experiment, A.T., A.Y., A.K. and A.L.; writing—original draft preparation, A.Y.; supervision, A.T. and O.K. All authors have read and agreed to the published version of the manuscript.

Funding: This research received no external funding.

Institutional Review Board Statement: Not applicable.

Informed Consent Statement: Not applicable.

Data Availability Statement: The data presented in this study are available on request from the corresponding author.

Acknowledgments: The authors are grateful to the Government of the Russian Federation and the Ministry of Higher Education and Science of the Russian Federation for support under Grant No. 075-15-2020-780 (N13.1902.21.0039).

Conflicts of Interest: The authors declare no conflict of interest.

References

1. Traub, W.; Oppenheimer, B. Direct imaging of exoplanets. In *Exoplanets*; Seager, S., Ed.; University of Arizona Press: Tucson, AZ, USA, 2011; pp. 111–156. Available online: https://www.amnh.org/content/download/53052/796511/file/DirectImagingChapter.pdf (accessed on 19 January 2023).
2. Guyon, O. Extreme Adaptive Optics. *Annu. Rev. Astron. Astrophys.* **2018**, *56*, 315–355. [CrossRef]
3. Proposed Missions—Terrestrial Planet Finder. Available online: https://www2.jpl.nasa.gov/missions/proposed/tpf.html (accessed on 19 January 2023).
4. Lindler, D.J. TPF-O design reference mission. In *UV/Optical/IR Space Telescopes: Innovative Technologies and Concepts III*; SPIE: Bellingham, WA, USA, 2007; Volume 6687, pp. 406–414. [CrossRef]
5. Large Binocular Telescope Interferometer. Available online: https://nexsci.caltech.edu/missions/LBTI/ (accessed on 19 January 2023).
6. Noecker, M.C.; Zhao, F.; Demers, R.; Trauger, J.; Guyon, O.; Kasdin, N.J. Coronagraph instrument for WFIRST-AFTA. *J. Astron. Telesc. Instrum. Syst.* **2016**, *2*, 011001. [CrossRef]
7. The Nancy Grace Roman Space Telescope. Available online: https://www.jpl.nasa.gov/missions/the-nancy-grace-roman-space-telescope (accessed on 19 January 2023).
8. Guyon, O.; Pluzhnik, E.A.; Kuchner, M.J.; Collins, B.; Ridgway, S.T. Theoretical Limits on Extrasolar Terrestrial Planet Detection with Coronagraphs. *Astrophys. J. Suppl. Ser.* **2006**, *167*, 81–99. [CrossRef]
9. Baudoz, P.; Rabbia, Y.; Gay, J. Achromatic interfero coronagraphy. *Astron. Astrophys. Suppl. Ser.* **2000**, *141*, 319–329. [CrossRef]
10. Tavrov, A.V.; Kobayashi, Y.; Tanaka, Y.; Shioda, T.; Otani, Y.; Kurokawa, T.; Takeda, M. Common-path achromatic interferometer–coronagraph: Nulling of polychromatic light. *Opt. Lett.* **2005**, *30*, 2224–2226. [CrossRef] [PubMed]
11. Aimé, C.; Ricort, G.; Carlotti, A.; Rabbia, Y.; Gay, J. ARC: An Achromatic Rotation-shearing Coronagraph. *Astron. Astrophys.* **2010**, *517*, A55. [CrossRef]
12. Tavrov, A.; Korablev, O.; Ksanfomaliti, L.; Rodin, A.; Frolov, P.; Nishikwa, J.; Tamura, M.; Kurokawa, T.; Takeda, M. Common-path achromatic rotational-shearing coronagraph. *Opt. Lett.* **2011**, *36*, 1972–1974. [CrossRef] [PubMed]
13. Frolov, P.; Shashkova, I.; Bezymyannikova, Y.; Kiselev, A.; Tavrov, A. Achromatic interfero-coronagraph with variable rotational shear: Reducing of star leakage effect, white light nulling with lab prototype. *J. Astron. Telesc. Instrum. Syst.* **2015**, *2*, 011002. [CrossRef]
14. Pueyo, L.; Kay, J.; Kasdin, N.J.; Groff, T.; McElwain, M.; Give'On, A.; Belikov, R. Optimal dark hole generation via two deformable mirrors with stroke minimization. *Appl. Opt.* **2009**, *48*, 6296–6312. [CrossRef] [PubMed]
15. Give'On, A. A unified formailsm for high contrast imaging correction algorithms. In *Techniques and Instrumentation for Detection of Exoplanets IV*; SPIE: Bellingham, WA, USA, 2009; Volume 7440, pp. 112–117. [CrossRef]
16. Riggs, A.E.; Ruane, G.; Coker, C.T.; Shaklan, S.B.; Kern, B.D.; Sidick, E. Fast linearized coronagraph optimizer (FALCO) I: A software toolbox for rapid coronagraphic design and wavefront correction. In *Space Telescopes and Instrumentation 2018: Optical, Infrared, and Millimeter Wave*; SPIE: Bellingham, WA, USA, 2018; Volume 10698, pp. 878–888. [CrossRef]
17. Griewank, A.; Walther, A. *Evaluating Derivatives: Principles and Techniques of Algorithmic Differentiation*, 2nd ed.; Society for Industrial and Applied Mathematics: Philadelphia, PA, USA, 2008; ISBN 978-0-89871-659-7. Available online: https://epubs.siam.org/doi/pdf/10.1137/1.9780898717761.fm (accessed on 19 January 2023).
18. Paul, B.; Mugnier, L.; Sauvage, J.-F.; Ferrari, M.; Dohlen, K. High-order myopic coronagraphic phase diversity (COFFEE) for wave-front control in high-contrast imaging systems. *Opt. Express* **2013**, *21*, 31751–31768. [CrossRef] [PubMed]
19. Riggs, A.J.E.; Kasdin, N.; Groff, T.D. Recursive starlight and bias estimation for high-contrast imaging with an extended Kalman filter. *J. Astron. Telesc. Instrum. Syst.* **2016**, *2*, 011017. [CrossRef]
20. Baudoz, P.; Boccaletti, A.; Baudrand, J.; Rouan, D. The Self-Coherent Camera: A new tool for planet detection. *Proc. Int. Astron. Union* **2005**, *1*, 553–558. [CrossRef]

21. Groff, T.D.; Riggs, A.J.E.; Kern, B.; Kasdin, N.J. Methods and limitations of focal plane sensing, estimation, and control in high-contrast imaging. *J. Astron. Telesc. Instrum. Syst.* **2015**, *2*, 011009. [CrossRef]
22. Nishikawa, J.; Murakami, N. Unbalanced nulling interferometer and precise wavefront control. *Opt. Rev.* **2013**, *20*, 453–462. [CrossRef]
23. Malacara, D. *Optical Shop Testing*, 3rd ed.; John Wiley & Sons, Inc.: Hoboken, NJ, USA, 2006; ISBN 9780471484042. [CrossRef]
24. Huang, Y.; Liao, E.; Chen, R.; Wu, S.-T. Liquid-Crystal-on-Silicon for Augmented Reality Displays. *Appl. Sci.* **2018**, *8*, 2366. [CrossRef]
25. Hénault, F.B.; Carlotti, A.; Verinaud, C. Phase-shifting coronagraph. In *Techniques and Instrumentation for Detection of Exoplanets VIII*; SPIE: Bellingham, WA, USA, 2017; Volume 10400, pp. 418–439. [CrossRef]
26. Hénault, F. Phase-shifting technique for improving the imaging capacity of sparse-aperture optical interferometers. *Appl. Opt.* **2011**, *50*, 4207–4220. [CrossRef] [PubMed]
27. PROPER Optical Propagation Library. Available online: https://proper-library.sourceforge.net (accessed on 19 January 2023).
28. HOLOYEY Spatial Light Modulators. Available online: https://holoeye.com/spatial-light-modulators (accessed on 19 January 2023).
29. Shashkova, I.; Shkursky, B.; Frolov, P.; Bezymyannikova, Y.; Kiselev, A.; Nishikawa, J.; Tavrov, A.V. Extremely unbalanced interferometer for precise wavefront control in stellar coronagraphy. *J. Astron. Telesc. Instrum. Syst.* **2015**, *2*, 011011. [CrossRef]

Article

Statistical Tool Size Study for Computer-Controlled Optical Surfacing

Weslin C. Pullen [1], Tianyi Wang [2], Heejoo Choi [1,3], Xiaolong Ke [4], Vipender S. Negi [5,6], Lei Huang [2], Mourad Idir [2] and Daewook Kim [1,3,7,*]

[1] James C. Wyant College of Optical Sciences, The University of Arizona, 1630 E. University Blvd., P.O. Box 210094, Tucson, AZ 85721, USA; wpullen@email.arizona.edu (W.C.P.); hchoi@optics.arizona.edu (H.C.)

[2] National Synchrotron Light Source II (NSLS-II), Brookhaven National Laboratory, P.O. Box 5000, Upton, NY 11973, USA; tianyi@bnl.gov (T.W.); lhuang@bnl.gov (L.H.); midir@bnl.gov (M.I.)

[3] Large Binocular Telescope Observatory, University of Arizona, Tucson, AZ 85721, USA

[4] School of Mechanical and Automotive Engineering, Xiamen University of Technology, Xiamen 361024, China; kexiaolong@xmut.edu.cn

[5] Council of Scientific and Industrial Research-Central Scientific Instruments Organisation (CSIR-CSIO), Chandigarh 160030, India; vipender@live.com

[6] Academy of Scientific and Innovative Research (AcSIR), Ghaziabad 201002, India

[7] Department of Astronomy and Steward Observatory, University of Arizona, 933 N. Cherry Ave., Tucson, AZ 85721, USA

* Correspondence: dkim@optics.arizona.edu

Abstract: Over the past few decades, computer-controlled optical surfacing (CCOS) systems have become more deterministic. A target surface profile can be predictably achieved with a combination of tools of different sizes. However, deciding the optimal set of tool sizes that will achieve the target residual error in the shortest run time is difficult, and no general guidance has been proposed in the literature. In this paper, we present a computer-assisted study on choosing the proper tool size for a given surface error map. First, we propose that the characteristic frequency ratio (CFR) can be used as a general measure of the correction capability of a tool over a surface map. Second, the performance of different CFRs is quantitatively studied with a computer simulation by applying them to guide the tool size selection for polishing a large number of randomly generated surface maps with similar initial spatial frequencies and root mean square errors. Finally, we find that CFR = 0.75 achieves the most stable trade-off between the total run time and the number of iterations and thus can be used as a general criterion in tool size selection for CCOS processes. To the best of our knowledge, the CFR is the first criterion that ties tool size selection to overall efficiency.

Keywords: computer-controlled optical surfacing; fabrication; tool size; tool influence function; manufacturing

Citation: Pullen, W.C.; Wang, T.; Choi, H.; Ke, X.; Negi, V.S.; Huang, L.; Idir, M.; Kim, D. Statistical Tool Size Study for Computer-Controlled Optical Surfacing. *Photonics* **2023**, *10*, 286. https://doi.org/10.3390/photonics10030286

Received: 10 February 2023
Revised: 4 March 2023
Accepted: 6 March 2023
Published: 9 March 2023

1. Introduction

Computer-controlled optical surfacing (CCOS) [1,2] systems have been successfully used to fabricate high-precision optics in various cutting-edge applications, such as telescopes for space exploration [3–5], X-ray mirrors for synchrotron radiation and free-electron laser facilities [6–10], and optics in EUV lithography [11,12]. Different CCOS systems use different tools, which can be adopted based on the requirements for the precision and shape of the desired optical surface.

CCOS uses tools that are much smaller (i.e., sub-aperture tools) than the optical surface to correct the local errors. All CCOS techniques are mathematically modeled and have become much more deterministic [13,14], which enables a desired optical surface to be predictably achieved with a combination of tools with different sizes. In the CCOS process, a tool is simulated by its material removal footprint, known as its tool influence function (TIF). It is well known that certain TIF sizes have limits on the feature sizes within the optical surface that they can correct [2,15–17]. Basically, larger TIFs with higher peak removal rates (PRRs) are preferred because they remove material faster. However, while any given TIF can correct features larger than the size of the TIF completely, they cannot correct features that are smaller than the TIF footprint well. On the other hand, if the TIF is too small, then the surfacing efficiency will be low, and unexpected mid-to-high-frequency errors may be left on the optical surface. Therefore, choosing the optimal set of TIF sizes that achieves the target residual surface error with the shortest run time has been a difficult problem in practical CCOS processes. This problem is especially important in the fabrication of large optics, where a small improvement of efficiency leads to a great reduction in manpower and financial resources.

Reducing the fabrication time through more efficient and deterministic computer-controlled grinding, polishing, and metrology will allow larger-aperture space telescopes to be launched since the available budget will extend further [18]. As an example, saving a part of the overall fabrication time and iterative fabrication loops for each of the 798 hexagonal segments of the European Extremely Large Telescope (E-ELT) [19] will save significant time and uncertainties in the overall schedule.

Conventionally, a set of tools was empirically determined by a fabrication artisan's experience. However, multiple iterations of the trial-and-error cycle are usually required to approach the target residual surface error, which is inefficient. Additionally, this method highly relies on the expertise of the artisan and thus cannot be formalized as a general guidance.

Quantitative characterization of the correction capability of a certain TIF has been attempted by examining the power spectral density (PSD) of the TIF [15–17] in the literature. The PSD uses the Fourier transform to decompose a TIF and a surface error map into different spatial frequencies with their respective amplitudes. The amplitudes quantify the contributions of certain spatial frequencies to the entire surface error map. With the help of the PSD, Zhou et al. theoretically analyzed the removal characteristics in the CCOS process using a sinusoidal surface error map that only contained a single spatial frequency. A material removal availability (MRA) value equal to the ratio between the target material removal volume and the actual (or predicted) material volume removed was proposed as an indicator of the correction capability of a particular TIF [15]. The MRA can be used to determine how well a TIF can correct certain spatial frequencies. However, the concept was only verified for individual frequencies, and the relationship between the MRA and the total run time was unclear. Wang et al. presented a procedure for using the PSD to calibrate a specific TIF to determine its capability of correcting features that are smaller than the TIF [16,17]. However, this procedure requires multiple real fabrication runs and metrology to feed back to the result, specifically focusing on the smoothing efficiency of a TIF. Therefore, the method is not generally applicable to other TIFs without running the same procedure.

In this study, we present a computer-assisted analysis of the general guidance to choosing proper tool sizes for a given surface error map. First, the concept of the characteristic frequency ratio (CFR) is developed from Fourier theory and calibrated with a reference single-frequency sinusoidal surface as a proper measure of the correction capability of a certain TIF. Second, the relationship between the CFR and the run time is quantitatively studied via massive computer simulations, where different combinations of CFRs are applied as a reference in choosing the tool sizes for a large number of randomly generated surface error maps with similar initial spatial frequencies and root mean square (RMS) values. The statistics for each CFR combination, including the average of the run time, the standard deviation of the run time, and the number of iterations, are summarized and compared. Finally, the simulation results demonstrate that the CFR of 0.75 achieved the most stable trade-off between the total run time and the number of iterations and thus could be selected as a general efficiency criterion in choosing the tool sizes in CCOS. To the best of our knowledge, this is the first statistical study of tool size selection, and the CFR is the first general criterion that ties the TIF correction capability to the total run time.

The rest of this paper is organized as follows. The necessary background of Fourier theory as applied to analyze surface errors is briefly reviewed in Section 2, followed by a detailed explanation of the novel characteristic frequency (CF) of a TIF and its calibration in Section 3. Section 4 provides examples of determining the CFs for three standard TIF shapes, and then Section 5 describes the proposed CFR criterion. The computer-assisted study of different CFRs is discussed in Section 6, which includes a discussion on the practical applicability and limitations of the study. Section 7 concludes the paper.

2. Fourier Analysis of Surface Error

2.1. The Fourier Transform

In optical fabrication, surface errors are usually described as a 2D matrix of the difference between the measured and target surface shapes. This surface error map consists of error features which vary in lateral size and magnitude. The discrete Fourier transform (DFT) of the surface error is defined as follows:

$$\mathcal{Z}(u,v) = l_x l_y \sum_{x=0}^{N_x-1} \sum_{y=0}^{N_y-1} Z(x,y) e^{-i2\pi\left(\frac{u}{N_x}x + \frac{v}{N_y}y\right)},\tag{1}$$

where $\mathcal{Z}(u,v)$ is the 2D spatial frequency spectrum of the surface error $Z(x,y)$ and $l_x \times l_y$ is the pixel size. Here, $l_x = N_x/L_x$ and $l_y = N_y/L_y$, where N_x and N_y are the numbers of sample points in the x and y directions, respectively, and L_x and L_y are the periodicity in the x and y directions, respectively. The DFT decomposes the surface error into all the spatial frequencies which are present in the measurement, represented by sinusoidal features and their respective amplitudes.

2.2. Power Spectral Density

The power spectral density (PSD) of a surface is a statistical tool that decomposes a surface into contributions from different spatial frequencies. Figure 1 illustrates the relationship between the measured surface profile and its one-dimensional PSD curve, namely that the PSD is the Fourier transform of the auto-correlation function of the surface error map, which contains the power across a range of frequencies. An important realization is that the PSD of the mean-removed surface error gives the surface variance of each spatial frequency present in the measurement.

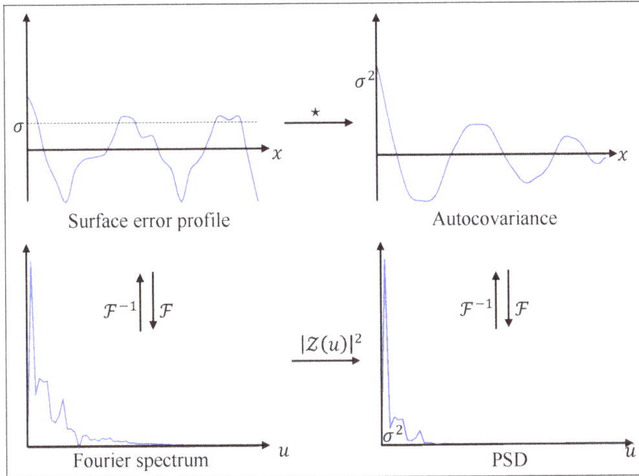

Figure 1. Schematic of the relationship between relevant surface parameters, where "$\star$" is the auto-correlation operator, $\mathcal{F}$ and $\mathcal{F}^{-1}$ represent the forward and inverse Fourier transforms, respectively, and σ^2 is the variance of the surface profile.

It is worth mentioning that in the real implementation of a DFT defined in Equation (1), it is assumed that $l_x = l_y = 1$ so that $1/L_x L_y = 1/N_x N_y$ [20]. Therefore, according to Figure 1, the PSD based on Equation (1) can be calculated as follows:

$$\mathcal{P}(u,v) = \frac{1}{N_x N_y} |\mathcal{Z}(u,v)|^2. \tag{2}$$

2.3. Encircled Error

In optical metrology, the encircled energy has been used to measure the concentration of energy of a point spread function (PSF) at the image plane. Analogous to the encircled energy of a PSF, we define the encircled error (EE) of a PSD as follows:

$$E(r) = \frac{\sum\limits_{\theta=0}^{2\pi} \sum\limits_{\rho=0}^{r} \mathcal{P}(\rho,\theta)}{\sum\limits_{\theta=0}^{2\pi} \sum\limits_{\rho=0}^{R} \mathcal{P}(\rho,\theta)}, \tag{3}$$

where $\mathcal{P}(\rho,\theta)$ is $\mathcal{P}(u,v)$ transformed to the polar coordinate system, ρ is the sampled frequency measured radially from the central frequency bin, θ is the azimuthal angle covering the PSD, and R is the maximum spatial frequency within the PSD. As an example, Figure 2a shows a randomly generated surface with a total RMS error of 98.8 nm. Figure 2b is its two-dimensional (2D) PSD map, and Figure 2c gives the corresponding EE. Recalling that the PSD of the mean-removed surface error is the surface variance contribution of each spatial frequency contained in the measurement, the EE then reveals what percentage of the error is due to spatial frequencies lower than a given frequency.

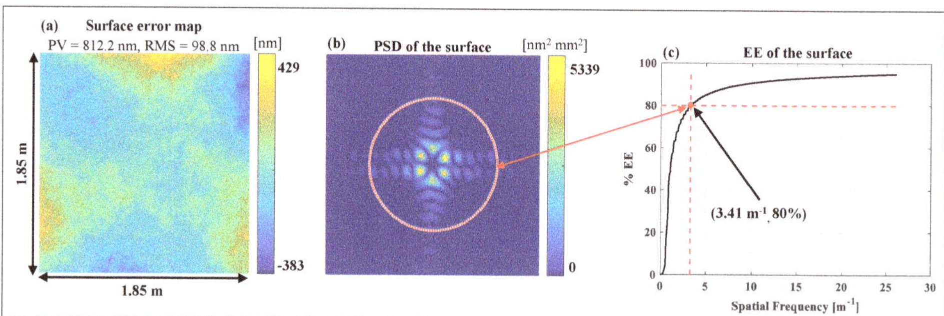

Figure 2. An example surface error map (**a**) and its 2D PSD map (**b**). The EE of the PSD (**c**) demonstrates that 80% of the RMS error was due to spatial frequencies lower than 3.41 m^{-1}.

2.4. Characteristic Frequency of a Surface Error Map

For an optical image, a typical criterion for the encircled energy is the radius of the PSF at which 50% or 80% of the energy is encircled. As an analogy for this study, we define the characteristic frequency (CF) of a given surface error map to be located at EE = 80% as

$$f^c_{SURF} = \arg_r[E(r) = 80\%], \tag{4}$$

where f^c_{SURF} refers to the CF of the surface error map. For example, as shown in Figure 2b,c, the CF of this surface error map is $f^c_{SURF} = 3.41$ m^{-1}.

3. The Reference TIF

Zhou, et al. discussed how the amplitude frequency spectrum, given by the Fourier transform of a tool influence function (TIF), is a measure of the correctability of that TIF for localized errors of the same spatial frequencies [15]. We used a reference surface and a reference TIF to calibrate the CF of the TIF.

3.1. The Reference Surface

Since the Fourier series decomposes surface errors into sinusoidal patterns, we define a reference surface error map containing a single Fourier mode in the x direction as

$$Z_{REF}(x,y) = A \cdot \cos(2\pi f x) + A, \tag{5}$$

where A is the amplitude and f is the single spatial frequency of the surface error. The reference surface error map and its profile along the x direction are shown in Figure 3a,b, respectively, where the size of the map is 100 mm $\times$ 100 mm, and A = 10 nm and $f = 0.05$ mm^{-1} are used in this study. It is worth mentioning that A and f could be set arbitrarily and not affect the outcome of the analysis. We also note that the surface error was piston-adjusted to have no negative values, since CCOS processes are only capable of removing material.

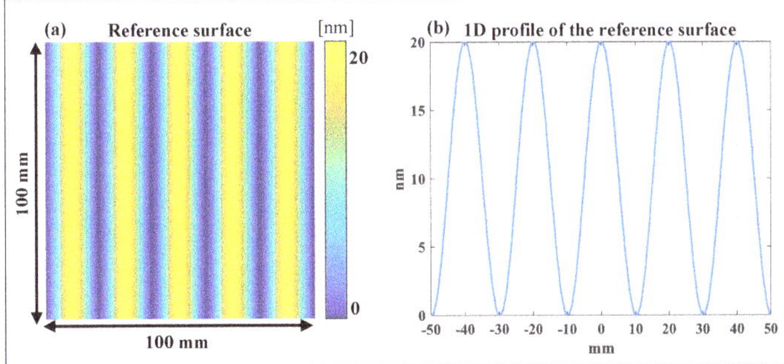

Figure 3. (**a**) The reference surface error map and (**b**) its 1D profile along the x direction.

3.2. Preston's Equation and the Line TIF

The material removal process is classically defined by Preston's equation [2] as follows:

$$\frac{\partial Z(x,y)}{\partial t} = \kappa \cdot P(x,y) \cdot V(x,y) \tag{6}$$

where the material removal rate per unit time $\partial Z(x,y)/\partial t$ is proportional to the contact pressure $P(x,y)$ and the relative velocity $V(x,y)$ between the tool and the workpiece. Preston's constant, κ, is used to consider additional factors that contribute to friction between the tool and the workpiece, such as slurry and polishing interface material. This equation is used to theoretically define a static TIF, which gives the material removal rate of a polishing process if the tool is parked in one spot and allowed to run for one unit of time. According to Equation (6), a TIF which matches the shape of the error feature will perfectly correct that error in the shortest amount of time (i.e., the most efficient TIF for that feature).

The static TIF considers only the motion of the tool without traveling across the workpiece. Many CCOS processes use a spiral tool path, and therefore the Ring TIF was conceived [21]. The Ring TIF considers the motion of tools traveling across the workpiece as the workpiece rotates under the tool and is therefore a function of the radial position from the center of the workpiece. To simplify this technique, we define the Line TIF, which is defined in the same way as the Ring TIF, only a Cartesian raster tool path is assumed rather than a spiral one. Converting the 2D static TIF to the Line TIF is as simple as summing down the columns (or equivalently the rows, depending on the major direction of the raster path) of the static TIF matrix.

3.3. The Reference TIF

Therefore, the ideal (i.e., reference) Line TIF $I_{REF}(x,y)$ for the single-frequency surface error map is simply one period of the sinusoid of the same frequency, defined by

$$I_{REF}(x,y) = \frac{PRR}{2} \cdot \cos(2\pi f x) + \frac{PRR}{2}, \quad f = 0.05 \, \text{mm}^{-1}, \tag{7}$$

where PRR is the peak removal rate of the TIF and the TIF is piston-adjusted by $PRR/2$ so that there are no negative removal rates. It is obvious from Equations (5) and (7) that the ideal TIF to correct the reference surface in Figure 3 is the sinusoidal reference TIF with a frequency $f = 0.05 \, \text{mm}^{-1}$. Therefore, the CF of the reference TIF is defined to be $f^c_{TIF} = 0.05 \, \text{mm}^{-1}$. Figure 4a shows the 1D profile of the reference TIF in the x direction with PRR = 1 nm/s.

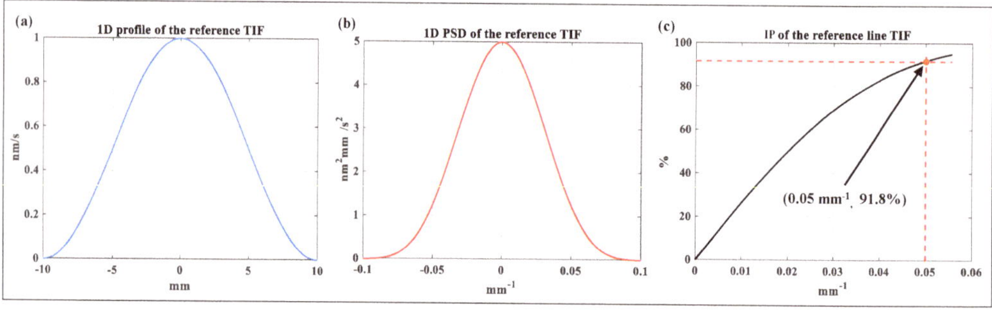

Figure 4. (**a**) Reference Line TIF with characteristic frequency $f^c_{TIF} = 0.05 \, \text{mm}^{-1}$, (**b**) 1D PSD of the reference Line TIF, and (**c**) IP of the reference TIF.

3.4. Integrated PSD and the Characteristic Frequency of a TIF

To generalize the CF to any TIF, we take the analysis from Zhou et al. further by defining the Integrated PSD (IP) of a TIF to calibrate a TIF's CF. As shown in Figure 4b,c, the IP is simply a one-dimensional (1D) equivalent of an EE calculation (see Equation (3)) performed on the 1D PSD of the Line TIF. For the reference TIF with $f^c_{TIF} = 0.05 \, \text{mm}^{-1}$, as demonstrated in Figure 4c, the corresponding IP occurs at 91.8%. Therefore, we define the CF for any TIF as the spatial frequency where IP = 91.8%; in other words, we have

$$f^c_{TIF} = \arg_{r}[I(r) = 91.8\%]. \tag{8}$$

4. Extending the Calibration to Other TIF Shapes

When utilizing the calibration of the f^c_{TIF} (see Equation (8)) by means of the reference TIF illustrated in Section 3.1, we can determine the size of any kind of TIF by simply applying a scale factor, depending on the given TIF shape. In the following subsections, we provide examples that employ Equation (8) to determine the appropriate size of three kinds of typical TIFs in CCOS to correct the reference surface in Figure 3, namely Gaussian TIFs [10], Spin TIFs [22], and Orbital TIFs [22], with the objective that the CF of these TIFs matches the only error frequency present in the reference surface in Section 3.1.

4.1. The Gaussian TIF

The zero-mean, rotationally symmetric Gaussian TIF can be defined as follows:

$$I_G(x,y) = \text{PRR} \cdot \exp\left(\frac{x^2+y^2}{2\sigma^2}\right), \tag{9}$$

where σ is the standard deviation of the Gaussian distribution that defines the size of the Gaussian TIF. The size of a Gaussian TIF can be defined by the full width at half maximum (FWHM), defined as $\text{FWHM} = 2\sigma\sqrt{2\ln2}$. Then, the scale factor for the Gaussian TIF is FWHM/f^c_{TIF}.

4.2. The Spin TIF

The Spin TIF is derived from Equation (6) as follows:

$$I_S(x,y) = \begin{cases} \kappa \cdot P(x,y) \cdot \omega \cdot \rho(x,y), & \rho \leq R_T, \\ 0, & \rho > R_T \end{cases} \tag{10}$$

where ω is the angular velocity of the spinning motion of the machine tool, R_T is the radius of the actual tool, and $\rho(x,y)$ is the radial distance from the center of the tool. The size of the Spin TIF is defined only by the size of the tool such that $R_{TIF} = R_T$. The scale factor for the Spin TIF is then R_{TIF}/f^c_{TIF}.

4.3. The Orbital TIF

The Orbital TIF is also derived from Equation (6) but is much more complicated than the previous two TIFs. Dong et al. provided a well-organized derivation of the Orbital TIF [23], which is summarized as follows:

$$I_O(x,y) = \begin{cases} \dfrac{\pi}{30} \cdot \kappa \cdot P(x,y), & \rho \leq R_T - R_O, \\[2mm] \dfrac{1}{30} \cdot \kappa \cdot P(x,y) \cdot \omega \cdot R_O \cdot \arccos\left(\dfrac{\rho^2(x,y) + R_O^2 - R_T^2}{2 \cdot \rho(x,y) \cdot R_O}\right), & R_T - R_O < \rho \leq R_T + R_O \cdot \\[2mm] 0, & \rho > R_T + R_O \end{cases} \tag{11}$$

where R_O is the radius of the orbital stroke and all the other variables are the same as those presented in Equation (10). Unlike the previous two TIFs, the Orbital TIF is defined by two parameters: R_T and R_O, where $R_{TIF} = R_O + R_T$. Similar to the Spin TIF, the scale factor for the Orbital TIF is R_{TIF}/f^c_{TIF}.

4.4. Determining the TIF Size to Match a Desired CF

In order to determine the desired TIF size for a given CF, we simply generate an arbitrarily sized TIF, calculate the scale factor, and multiply by the desired CF (i.e., $R_{TIF,desired} = R_{TIF,arbitrary}/f^c_{TIF,arbitrary} * f^c_{TIF,desired}$). This procedure is demonstrated as follows. First, the 2D TIF is generated based on its governing equation (described above) at an arbitrary size. Next, the 1D Line TIF is obtained by summing down the columns of the 2D TIF, from which the 1D PSD is calculated. The IP is then calculated, where we can determine the f^C_{TIF} for the respective size and shape of the TIF in use. Finally, the scale factor is calculated, and the 2D TIF is scaled accordingly. Figure 5 shows the Gaussian, Spin, and Orbital TIFs scaled to have $f^c_{TIF} = 0.05\,\text{mm}^{-1}$.

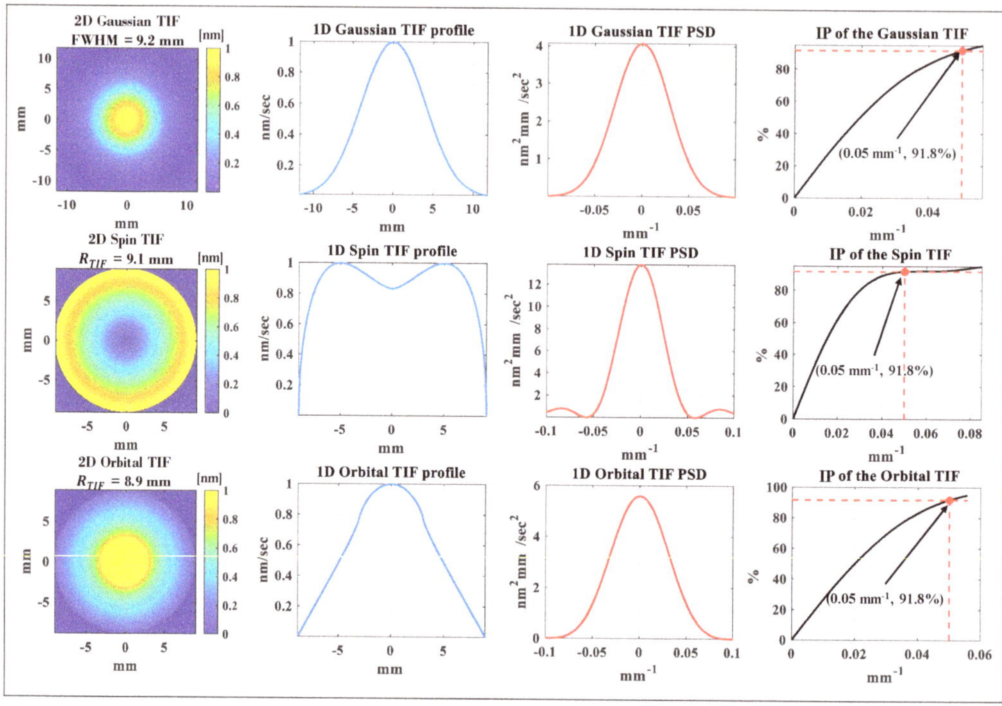

Figure 5. Typical TIFs scaled to have a characteristic frequency of $f_c = 0.05$ mm^{-1}

5. The Characteristic Frequency Ratio

We defined f^c_{SURF} for a surface error map and calibrated f^c_{TIF} for a TIF, and we illustrated the method of determining the size of a TIF using f^c_{TIF}. Now, we need a new criterion that can combine f^c_{SURF} with f^c_{TIF} to guide tool size selection based on a particular surface error map. To accomplish this, we define the characteristic frequency ratio (CFR), which is simply the ratio between the CF of a TIF and the CF of a surface error map such that

$$\mathrm{CFR} = \frac{f^c_{TIF}}{f^c_{SURF}}. \tag{12}$$

This simple ratio allows us to quantitatively set the TIF size based on the spatial frequency distribution of the surface error. A CFR of one, based on the definition given in Section 2.4, theoretically implies that the given TIF will correct 80% of the surface error. However, this setting may cost too much processing time, negatively influencing the processing efficiency. In addition, because this TIF only corrects the lower 80% frequency modes, the remaining 20% of the error is now due to higher spatial frequencies not suited for the initial TIF, and thus a new, smaller TIF is now required. Although the frequency content of the new residual error map has changed, the CFR should be able to be used once again to set this TIF size appropriately. Therefore, a well-selected CFR for each TIF is critical for the overall accuracy and efficiency of the fabrication process.

6. Computer-Assisted Study of the Optimal CFR for Tool Size Selection

An optimal CFR should consider the following aspects. First, it should balance the accuracy and efficiency. In other words, we expect that the target residual RMS error can be achieved in the shortest available total run time of all the tools. Secondly, the number of iterations (i.e., the number of tools employed to achieve the target) should be as small as possible, since frequent changing of tools also influences the overall efficiency. Finally, the selected CFR should be stable so that the same CFR can be applied to select the tool size in each iteration. Based on these philosophies, a computer simulation was designed and conducted to statistically study the optimal CFR.

6.1. Simulation Specifications

To study the optimal CFR, different CFR values were applied to select tool sizes for many initial surface error maps that were randomly generated with the same RMS value.

In detail, similar to the surface error map shown in Figure 3a, for a single test case, we first generated a random surface error map with a size of 1.85 m × 1.85 m by using the measured PSD trend of the aspheric DKIST primary mirror [24], adding random amounts of low-to-medium spatial frequency errors and scaling to the same total RMS.

Next, a CFR value was selected to investigate, with CFR ∈ {0.5, 0.6, 0.7, 0.8, 0.9}. The CFR, along with the CF of the given surface error map, was used to define the TIF size as detailed in Section 4. The robust iterative Fourier transform-based dwell time algorithm (RIFTA) [25] was used to determine the residual error and the total run time for each iteration, since the RIFTA is fast and minimizes the residual and run time simultaneously. Afterward, on the residual error map, another TIF was chosen with the same method. This procedure was repeated for this test case until the surface was within the target RMS bounds, which were set to be 15% ± 3% of the initial RMS. Each test case (i.e., a certain CFR value) was run on 30 separate randomly generated surface error maps, and the average total run time, run time standard deviation, and average number of iterations over the 30 trials were recorded.

Figure 6 depicts one set of this process on a single initial surface error map. The selected CFR was 0.7, and dwell time optimization with the RIFTA was performed on the residual surface errors iteratively until the target RMS was reached.

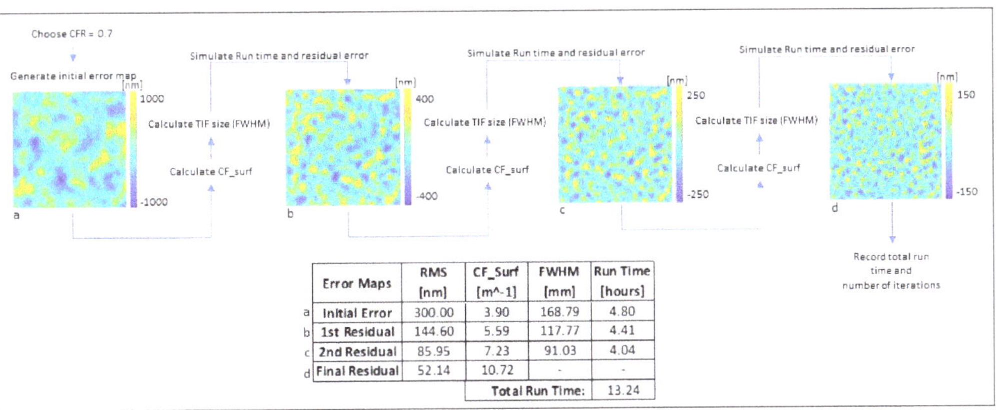

Error Maps		RMS [nm]	CF_Surf [m^-1]	FWHM [mm]	Run Time [hours]
a	Initial Error	300.00	3.90	168.79	4.80
b	1st Residual	144.60	5.59	117.77	4.41
c	2nd Residual	85.95	7.23	91.03	4.04
d	Final Residual	52.14	10.72	-	-
				Total Run Time:	13.24

Figure 6. Simulation example for one iteration of test case with CFR = 0.7 showing (**a**) initial error map, (**b**) first residual error map, (**c**) second residual error map, and (**d**) final residual error map, with a table keeping track of the RMS value and CF of each map, the FWHM of the Gaussian TIF used on each map, and the calculated run time required for each run.

6.2. Simulation Results

The results of the simulation are shown in Figure 7. To compare the results of each case, we defined a figure of merit (FoM) for one case to be the root sum square (RSS) of the average run time, standard deviation of the run time, and average number of iterations, each normalized by the maximum case value for its respective category. A smaller FoM value thus corresponds to a more efficient CFR. As shown in Figure 7, the small FoM values appeared between CFR = 0.7 and CFR = 0.8.

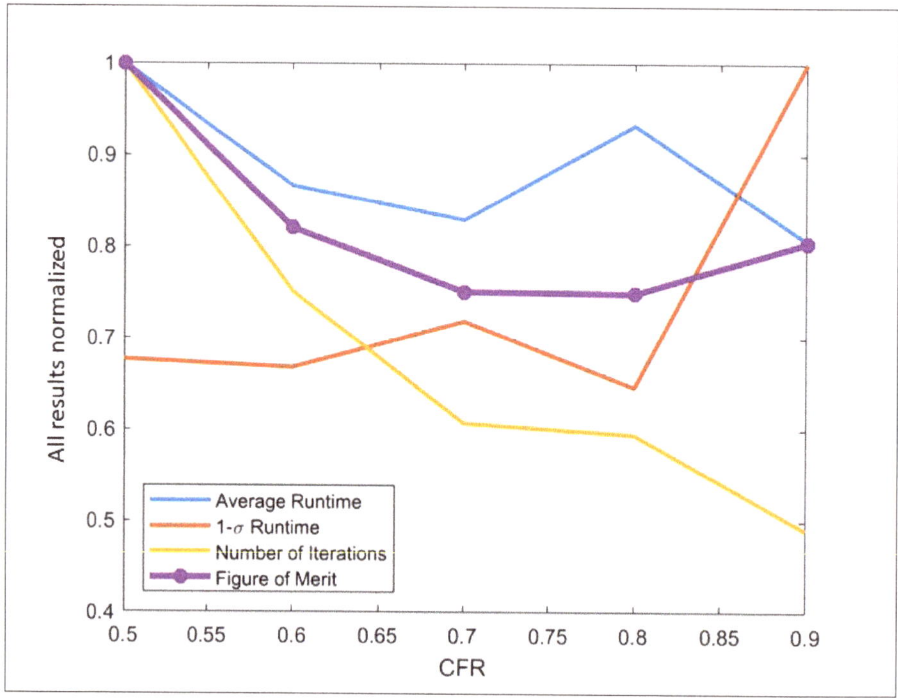

Figure 7. Simulation results for the average total run time (blue), the standard deviation of the total run time (red), the average number of iterations (yellow), and the figure of merit (purple).

The average run time of each case is shown by in the blue line in Figure 7, which was calculated as the mean run time of the 30 runs of the same CFR. It is obvious that CFR = 0.5 and CFR = 0.8 do not appear to be a good choice because of their longer run times. The standard deviation of the total run time of each case is shown in red in Figure 7, which represents the stability of the selected CFR. It was found that CFR = 0.9 was not very stable. Finally, the average number of iterations of each case that was spent to achieve the target residual error is given in yellow in Figure 7. A small number of iterations indicates a higher overall efficiency for the selected CFR value, primarily due to the down time caused by tool changes. Therefore, larger CFR values tend to be more efficient under this definition than smaller CFR values. It is clear by interpolation that the consistent use of a CFR within 0.75 yielded the most stable balance among the accuracy, total run time, and number of required iterations. Thus, it was selected as a general criterion in choosing tool sizes in CCOS.

7. Conclusions

In this paper, we proposed a straightforward characteristic frequency ratio (CFR) to guide tool size selection in computer-controlled optical surfacing (CCOS) processes. The proposed CFR was statistically studied via computer simulation, and it was the first general criterion that considered both the residual errors and total run time.

The CFR is defined as the ratio between the characteristic frequencies (CFs) of a surface error map and a tool influence function (TIF). While the CF of a surface error map is defined according to the proposed encircled error metric, the CF of a TIF is calibrated based on a reference surface error map containing a single sinusoidal frequency and a reference TIF derived from Preston's equation. Based on the novel integrated PSD of a TIF metric, the method for generalizing the calibrated CF to determine the sizes of different kinds of TIFs in typical CCOS processes was then presented, verifying the applicability of the method. Finally, the proposed idea was statistically studied with a well-designed computer simulation on many initial surface error maps and different CFRs. The simulation results demonstrate that a CFR of 0.75 consistently achieves the most stable balance among the residual error, total run time, and total number of iterations and thus can be chosen as a simple criterion to guide tool size selection.

Author Contributions: Conceptualization, methodology, data curation, and writing—original draft preparation, W.C.P.; validation and writing—review and editing, T.W.; validation and writing— review and editing, H.C.; validation and writing—review and editing, X.K.; validation and writing— review and editing, V.S.N.; validation and writing—review and editing, L.H.; validation and writing— review and editing, M.I.; supervision, project administration, and writing—review and editing, D.K. All authors have read and agreed to the published version of the manuscript.

Funding: This work was partially supported by the Basic Energy Science Office of the US Department of Energy (DOE) through FWP PS032. This research was partially carried out at the Optical Metrology Laboratory at National Synchrotron Light Source II, a US DOE Office of Science User Facility operated for the DOE Office of Science by Brookhaven National Laboratory (BNL) under Contract No. DE-SC0012704. This work was partially performed under the BNL LDRD 17-016 'Diffraction limited and wavefront preserving reflective optics development'.

Institutional Review Board Statement: Not applicable.

Data Availability Statement: The data that support the findings of this study are available from the leading author, [W.P.], upon reasonable request.

Acknowledgments: Recognition is given to Jim Burge for inadvertently inspiring the thoughts that gave rise to this work.

Conflicts of Interest: The authors declare no conflict of interest.

References

1. Jones, R.A. Optimization of computer controlled polishing. *Appl. Opt.* **1977**, *16*, 218–224. [CrossRef]
2. Cheng, H. *Independent Variables for Optical Surfacing Systems*; Springer: Berlin/Heidelberg, Germany, 2016.
3. Fanson, J.; Bernstein, R.; Angeli, G.; Ashby, D.; Bigelow, B.; Brossus, G.; Bouchez, A.; Burgett, W.; Contos, A.; Demers, R.; et al. Overview and status of the Giant Magellan Telescope project. In Proceedings of the Ground-based and Airborne Telescopes VIII. International Society for Optics and Photonics, Online, 14–22 December 2020; Volume 11445, p. 114451F.
4. Ghigo, M.; Vecchi, G.; Basso, S.; Citterio, O.; Civitani, M.; Mattaini, E.; Pareschi, G.; Sironi, G. Ion figuring of large prototype mirror segments for the E-ELT. *Adv. Opt. Mech. Technol. Telesc. Instrum.* **2014**, *9151*, 225–236.
5. Kim, D.; Choi, H.; Brendel, T.; Quach, H.; Esparza, M.; Kang, H.; Feng, Y.T.; Ashcraft, J.N.; Ke, X.; Wang, T.; et al. Advances in Optical Engineering for Future Telescopes. *Opto-Electron. Adv.* **2021**, *4*, 210040-1. [CrossRef]
6. Schindler, A.; Haensel, T.; Nickel, A.; Thomas, H.J.; Lammert, H.; Siewert, F. Finishing procedure for high-performance synchrotron optics. *Opt. Manuf. Test. V* **2003**, *5180*, 64–72.
7. Beaucamp, A.; Namba, Y. Super-smooth finishing of diamond turned hard X-ray molding dies by combined fluid jet and bonnet polishing. *CIRP Ann.* **2013**, *62*, 315–318. [CrossRef]
8. Thiess, H.; Lasser, H.; Siewert, F. Fabrication of X-ray mirrors for synchrotron applications. *Nucl. Instrum. Methods Phys. Res. Sect. A Accel. Spectrometers Detect. Assoc. Equip.* **2010**, *616*, 157–161. [CrossRef]
9. Wang, T.; Huang, L.; Choi, H.; Vescovi, M.; Kuhne, D.; Zhu, Y.; Pullen, W.C.; Ke, X.; Kim, D.W.; Kemao, Q.; et al. RISE: robust iterative surface extension for sub-nanometer X-ray mirror fabrication. *Opt. Express* **2021**, *29*, 15114–15132. [CrossRef]
10. Wang, T.; Huang, L.; Vescovi, M.; Kuhne, D.; Zhu, Y.; Negi, V.S.; Zhang, Z.; Wang, C.; Ke, X.; Choi, H.; et al. Universal dwell time optimization for deterministic optics fabrication. *Opt. Express* **2021**, *29*, 38737–38757. [CrossRef]
11. Weiser, M. Ion beam figuring for lithography optics. *Nucl. Instrum. Methods Phys. Res. Sect. B Beam Interact. Mater. Atoms.* **2009**, *267*, 1390–1393. [CrossRef]
12. Wischmeier, L.; Graeupner, P.; Kuerz, P.; Kaiser, W.; van Schoot, J.; Mallmann, J.; de Pee, J.; Stoeldraijer, J. High-NA EUV lithography optics becomes reality. *Extrem. Ultrav. (EUV) Lithogr. XI* **2020**, *11323*, 1132308.
13. Han, Y.; Duan, F.; Zhu, W.; Zhang, L.; Beaucamp, A. Analytical and stochastic modeling of surface topography in time-dependent sub-aperture processing. *Int. J. Mech. Sci.* **2020**, *175*, 105575. [CrossRef]
14. Chaves-Jacob, J.; Beaucamp, A.; Zhu, W.; Kono, D.; Linares, J.M. Towards an understanding of surface finishing with compliant tools using a fast and accurate simulation method. *Int. J. Mach. Tools Manuf.* **2021**, *163*, 103704. [CrossRef]
15. Zhou, L.; Dai, Y.; Xie, X.; Li, S. Frequency-domain analysis of computer-controlled optical surfacing processes. *Sci. China Ser. E Technol. Sci.* **2009**, *52*, 2061–2068. [CrossRef]

16. Wang, J.; Fan, B.; Wan, Y.; Shi, C.; Zhuo, B. Method to calculate the error correction ability of tool influence function in certain polishing conditions. *Opt. Eng.* **2014**, *53*, 075106. [CrossRef]
17. Wang, J.; Hou, X.; Wan, Y.; Shi, C. An optimized method to calculate error correction capability of tool influence function in frequency domain. *Optifab 2017* **2017**, *10448*, 104481Z.
18. Trumper, I.; Hallibert, P.; Arenberg, J.W.; Kunieda, H.; Guyon, O.; Stahl, H.P.; Kim, D.W. Optics technology for large-aperture space telescopes: from fabrication to final acceptance tests. *Adv. Opt. Photonics* **2018**, *10*, 644–702. [CrossRef]
19. Graves, L.R.; Smith, G.A.; Apai, D.; Kim, D.W. Precision Optics Manufacturing and Control for Next-Generation Large Telescopes. *J. Int. Soc. Nanomanuf.* **2019**, *2*, 65–90. [CrossRef]
20. Jacobs, T.D.; Junge, T.; Pastewka, L. Quantitative characterization of surface topography using spectral analysis. *Surf. Topogr. Metrol. Prop.* **2017**, *5*, 013001. [CrossRef]
21. Kim, D.W.; Kim, S.W.; Burge, J.H. Non-sequential optimization technique for a computer controlled optical surfacing process using multiple tool influence functions. *Opt. Express* **2009**, *17*, 21850–21866. [CrossRef]
22. Kim, D.W.; Park, W.H.; Kim, S.W.; Burge, J.H. Parametric modeling of edge effects for polishing tool influence functions. *Opt. Express* **2009**, *17*, 5656–5665.
23. Dong, Z.; Cheng, H.; Tam, H.Y. Modified subaperture tool influence functions of a flat-pitch polisher with reverse-calculated material removal rate. *Appl. Opt.* **2014**, *53*, 2455–2464. [CrossRef] [PubMed]
24. Kim, D.W.; Oh, C.J.; Lowman, A.; Smith, G.A.; Aftab, M.; Burge, J.H. Manufacturing of super-polished large aspheric/freeform optics. *Adv. Opt. Mech. Technol. Telesc. Instrum. II* **2016**, *9912*, 99120F.
25. Wang, T.; Huang, L.; Kang, H.; Choi, H.; Kim, D.W.; Tayabaly, K.; Idir, M. RIFTA: A Robust Iterative Fourier Transform-based dwell time Algorithm for ultra-precision ion beam figuring of synchrotron mirrors. *Sci. Rep.* **2020**, *10*, 8135. [CrossRef] [PubMed]

Article

EXPLANATION: Exoplanet and Transient Event Investigation Project—Optical Facilities and Solutions

Gennady Valyavin [1,*], Grigory Beskin [1,2], Azamat Valeev [1,3,4], Gazinur Galazutdinov [1,4], Sergei Fabrika [1], Iosif Romanyuk [1] Vitaly Aitov [1], Oleg Yakovlev [1,5], Anastasia Ivanova [5], Roman Baluev [3], Valery Vlasyuk [1], Inwoo Han [6], Sergei Karpov [1,2,7], Vyacheslav Sasyuk [2], Alexei Perkov [8], Sergei Bondar [8,†], Faig Musaev [1,†], Eduard Emelianov [1], Timur Fatkhullin [1], Sergei Drabek [1], Vladimir Shergin [1], Byeong-Cheol Lee [6], Guram Mitiani [1], Tatiana Burlakova [1,4], Maksim Yushkin [1], Eugene Sendzikas [1], Damir Gadelshin [1], Lisa Chmyreva [1], Anatoly Beskakotov [1], Vladimir Dyachenko [1], Denis Rastegaev [1], Arina Mitrofanova [1], Ilia Yakunin [1,3], Kirill Antonyuk [1,4], Vladimir Plokhotnichenko [1], Alexei Gutaev [1,2], Nadezhda Lyapsina [1], Vladimir Chernenkov [1], Anton Biryukov [2,9], Evgenij Ivanov [1,8], Elena Katkova [8], Alexander Belinski [9], Eugene Sokov [3,10], Alexander Tavrov [5], Oleg Korablev [5], Myeong-Gu Park [11], Vladislav Stolyarov [1,12], Victor Bychkov [1], Stanislav Gorda [13], A. A. Popov [13] and A. M. Sobolev [13]

1 Special Astrophysical Observatory, Russian Academy of Sciences, Nizhnij Arkhyz 369167, Russia
2 Engelhardt Observatory, Kazan Federal University, Kazan 420008, Russia
3 The Faculty of Mathematics and Mechanics, Department of Astronomy, Saint Petersburg State University, 7-9 Universitetskaya Emb., Saint Petersburg 199034, Russia
4 Federal State Budget Scientific Institution Crimean Astrophysical Observatory of RAS, Nauchny, 298409 Crimea
5 Space Research Institute, Russian Academy of Sciences, 84/32 Profsoyuznaya Str., Moscow 117997, Russia
6 Korea Astronomy and Space Science Institute, 776 Daedeokdae-ro, Yuseong-gu, Daejeon 34055, Republic of Korea
7 CEICO, Institute of Physics, Czech Academy of Sciences, 18200 Prague, Czech Republic
8 Research and Production Corporation "Precision Systems and Instruments", Moscow 111024, Russia
9 Sternberg Astronomical Institute, Moscow M.V. Lomonosov State University, Universitetskij Pr. 13, Moscow 119992, Russia
10 Central (Pulkovo) Observatory, Pulkovskoe Shosse 65, Saint Petersburg 196140, Russia
11 Department of Astronomy and Atmospheric Sciences, Kyungpook National University, Daegu 41566, Republic of Korea
12 Cavendish Laboratory, University of Cambridge, Cambridge CB3 0HE, UK
13 Astronomical Observatory, Institute for Natural Sciences and Mathematics, Ural Federal University, 19 Mira Street, Ekaterinburg 620002, Russia
* Correspondence: gvalyavin@sao.ru
† Passed away.

Citation: Valyavin, G.; Beskin, G.; Valeev, A.; Galazutdinov, G.; Fabrika, S.; Romanyuk, I.; Aitov, V.; Yakovlev, O.; Ivanova, A.; Baluev, R.; et al EXPLANATION: Exoplanet and Transient Event Investigation Project—Optical Facilities and Solutions. *Photonics* **2022**, *9*, 950. https://doi.org/10.3390/photonics9120950

Received: 29 September 2022
Accepted: 21 November 2022
Published: 8 December 2022

Publisher's Note: MDPI stays neutral with regard to jurisdictional claims in published maps and institutional affiliations.

Abstract: Over the past decades, the achievements in astronomical instrumentation have given rise to a number of novel advanced studies related to the analysis of large arrays of observational data. One of the most famous of these studies is a study of transient events in the near and far space and a search for exoplanets. The main requirements for such kinds of projects are a simultaneous coverage of the largest possible field of view with the highest possible detection limits and temporal resolution. In this study, we present a similar project aimed at creating an extensive, continuously updated survey of transient events and exoplanets. To date, the core of the project incorporates several 0.07–2.5 m optical telescopes and the 6-m BTA telescope of the Special Astrophysical Observatory of RAS (Russia), a number of other Russian observatories and the Bonhyunsan observatory of the Korea Astronomy and Space Science Institute (South Korea). Our attention is mainly focused on the description of two groups of small, wide-angle optical telescopes for primary detection. All the telescopes are originally designed for the goals of the project and may be of interest to the scientific community. A description is also given for a new, high-precision optical spectrograph for the Doppler studies of transient and exoplanet events detected within the project. We present here the philosophy, expectations and first results obtained during the first year of running the project.

Keywords: astronomical telescopes; photomerty; spectroscopy; transient events; exoplanets

1. Introduction

The achievements of recent decades in the development of wide-angle telescopes and large-format image sensors (CCD, CMOC, etc.), including mosaic ones, have made a true revolution in astronomy and astrophysics, giving the green light to extensive studies of billions of events in outer space available for observations from Earth. This produced new fundamental discoveries and gave fresh knowledge about the Universe. Among these are the discovery of planets orbiting the stars other than the Sun (exoplanets) and massive studies of high-energy gamma-bursts from relativistic objects in the Universe. Let us briefly discuss some of these advanced studies.

Thanks to the efforts of the *Kepler* [1] and *Corot* [2] space missions, as well as the successful operation of various ground-based photometric surveys of exoplanets, such as the famous SuperWASP [3] or a lesser known *Kourovka search for planets* [4], we now have a list of several thousand extrasolar planets and candidates. Based on these studies, the basic physical properties of a large proportion of the confirmed planets are known. Exoplanets cover a wide range of masses, chemical compositions and distances from their host stars. Despite this progress, statistics of the new exoplanet and exoplanet candidates requires more space and ground-based projects aimed at studying the new, as well as the already discovered extrasolar planets and candidates.

The problem of studying the rapidly changing cosmic events from the optical sources (transient events, or transients) was first formulated by H. Bondi in 1970 [5]. To detect and study such sources, wide-angle instruments with high-performance detectors having a temporal resolution of at least a fraction of a second are needed. The latter requirement is due to the short duration (up to 0.01 s) of the considered phenomena. For instance, subsecond highly polarized synchrotron spikes with front durations up to 0.1 s have been first detected in the optical observations of UV Ceti with the 6-m telescope of the SAO RAS [6]. Subsequently, synchrotron flares from red dwarfs were recorded in the millimeter range [7]. To search for and study such events, sufficiently short exposures are required, at least at the level of hundredths of a second, and/or high speeds, up to tens of degrees per second (satellites, space debris, meteors and fireballs).

Nowadays, studies of transients require very wide-angle telescopes in combination with the temporal resolution down to milliseconds. This is due to the advent of a fundamentally new task of searching for optical transients accompanying fast radio bursts [8] and impulses of gravitational waves [9], the duration of which lies in the millisecond range. A number of traditional tasks in the field of optical transient research do also justify the need of bringing the temporal resolution to 0.1 s or below.

In this paper, we present a joint Russian-Korean ground-based astronomical project abbreviated EXPLANATION (EXoPLANet And Transient events InvestigatiON). The project is aimed at a massive photometric, speckle-interferometric, spectral, and radio bolometric search for non-stationary events in the Universe, as well as the study of exoplanets. The core of the project incorporates several 0.07–2.5 m Russian and Korean optical telescopes, as well as the giant 6-m BTA telescope.

In the next section, titled 'What is the EXPLANATION?', we briefly discuss the philosophy of the project and its shareable instrumentation. Then, the 'Design of the main optical facilities' Section presents the details on two main groups of small, wide-angle optical telescopes for the primary detection, a new, high-precision optical spectrograph for the Doppler studies of transient events and exoplanets, and some other classical facilities of the project. The section called as 'Expectations' presents the results we expect from the project. The Results and Discussion sections present the first results obtained in the course of the project and general discussion, respectively.

2. What Is the EXPLANATION?

The Explanation project is aimed at both conducting its own ground-based surveys of exoplanets/transient events, and supporting the space missions aimed at the same goals. Its principal difference from the other research is that it integrates a three-level scheme for the diagostics and investigation of non-stationary events, employing a complete range of the instrumentation needed. The scheme consists of a wide-angle optical survey of high temporal resolution (the primary detection level), the second level of examination of the initially detected objects with high angular resolution, and the third level of expert examination of the most interesting objects that have been selected from the first and second levels.

The first level consists of two groups of fast-response wide-angle telescopes with the aperture diameters from 0.07 m to 0.5 m. A detailed description of these instruments is the main goal of this study.

The second level is represented by a group of classic 0.6–1.3 m telescopes. The task for these telescopes is to perform additional studies of the primarily selected targets with higher angular resolution in order to measure their coordinates and multicolor broad-band photometric characteristics with the necessary accuracy.

Finally, the core of the third, expert level of the project involves large Russian and Korean optical telescopes with the apertures ranging from 1.8 to 6 m. The workload of these telescopes is to perform expert photometric, polarization, spectral and speckle-interferometric studies of objects selected during the previous stages. This level also assumes the study of targets from other missions, including the space-based, as already mentioned above.

3. Optical Design of Project Facilities

In this section, we describe two groups of wide-angle robotic telescopes of the first level of research—a nine-channel Mini-MegaTORTORA (MMT-9) survey telescope, consisting of nine 0.07-m telescopes (channels), and a group of three+ 0.5-m 2 square degree telescopes.

3.1. MMT-9

The Mini-MegaTORTORA (MMT-9) is a wide-field optical monitoring system with high temporal resolution [10] built for and owned by the Kazan Federal University. It has been operated since 2014 under an agreement between the Kazan Federal University and Special Astrophysical Observatory, Russia.

MMT-9 is a successor to the simpler single-channel FAVOR and TORTORA cameras that were operated between 2004 and 2014. Their primary task was an untriggered search for the optical components of gamma-ray bursts, and it proved successful with the discovery of a bright and rapidly variable optical emission from GRB 080319B [11] by the TORTORA camera. The design of MMT-9 builds on this experience, and is aimed towards both increasing the simultaneously observable sky area, maintaining the high temporal resolution, and allowing for a multicolor mode of observations.

All these requirements are solved by the current Mini-MegaTORTORA implementation as a set of nine individual channels installed in pairs on five equatorial mounts. Every channel has a celostate mirror mounted before the Canon EF85/1.2 objective for rapid (faster than 1 s) adjustment of the objective direction within a limited range (approximately 10 degrees in any direction). This allows for either mosaicing a larger field of view, or for pointing all the channels in one direction. In the latter case, a set of color (Johnson's B, V or R) and polarimetric (three different directions) filters may be inserted before the objective to maximize the information acquired for the observed region of the sky (performing both the three-color photometry and polarimetry). High temporal resolution is ensured by using Andor Neo sCMOS detectors with 2560×2160 pixels 6.4 μm each, providing the 9×11-degree field of view for every channel, and allowing it to operate with the exposure times as short as 0.1 s.

The implementation of this original operating mode of the MMT-9 system is provided by a two-level scheme of its positioning and maintenance. The first level uses a traditional clock mechanism based on the EQ-6 equatorial mount with a two-coordinate retargeting speed of about $1°/s$ and a tracking accuracy of about $3''$ on a scale of 20 min. On the second level, a quick change of the field of view of each MMT-9 channel is carried out in the range of ± 8 degrees clockwise, and ± 16 degrees in declination by moving the aforementioned celostate mirror.

It is mounted on a gimbal suspension, and is capable of rotating about two axes with the help of two TGY-MG958 servos. During the observations, the mirror is clamped by brake magnetic shoes and the drive is de-energized. When changing the observed field, power is supplied to the brake coils and the steering machine. The brake coil has two poles and the brake shoe has two mating poles. As a result, at the time of power supply, the shoe is thrown away by the magnetic field, and the mirror is moved to the desired angle by the servo drive. After this operation is completed, power is removed from the machine and the brake, the brake shoe is pulled back, the mirror is braked, and a new observational process begins. A feature of the celeste design is the manufacture of the articulation axes of the gimbal frames on the steel tapes. To attenuate the current surge and save the moment when transferring the mirror in the power circuit drives, the afterburner tanks of a large denomination are provided. For more details, see [12].

Figure 1 shows a general view of two channels with open covers, allowing to see the mirrors of the integral units (the assembly elements of their frames are shown in the top plot of Figure 2). The EQ-6 equatorial mount is in the center between the channels. On the continuation of its hour axis, a switching and ventilation unit is placed, containing power supplies and switching boards. An air duct is located in the mount support, purifying and dehumidifying (the camera) by means of a climate control system, the air that ventilates the housing optical channels (Figure 3). The MMT-9 system, together with the shelter is shown in Figure 4.

The instrument as a whole allows for simultaneous observations of up to 30×30 degrees of the sky. Typically, the exposures of 0.1 s (10 frames per second, with an effective detection limit of about $V = 11$ mag) are used during the monitoring. A dedicated fast differential imaging pipeline allows for a real-time analysis of the data in this mode in order to detect and characterize rapidly variable objects on sub-second time scales and optionally initiate the follow-up observations.

Apart from the high temporal resolution monitoring, the MMT-9 also performs a routine photometric sky survey with a low temporal resolution, only with a deeper limiting magnitude (down to about $V = 13.5$ mag). More than 1.7 million images, covering every point of the northern sky 5000–20,000 times with the exposures of 20 to 60 s were acquired in this mode.

All these images are catalogued and also analyzed by a dedicated photometric pipeline that processes and calibrates them to the Johnson V band by deriving the photometric equation for every individual image, and then statistically regressing for individual object colors, assuming that their brightness variations are uncorrelated with the changes in the atmospheric conditions. About 30 billion photometric measurements are stored in the database and may be queried by position to extract the light curves of any object detectable by the MMT-9 (The database of images and photometric measurements of the MMT-9 Sky Survey is publicly available at http://survey.favor2.info/).

The accuracy of photometric measurements reachable on the MMT-9 is limited due to very broad point spread function (PSF), and strong pixel-to-pixel inhomogeneity of the CMOS chips of the detectors. On average, the light curve scatter of brighter non-variable objects is about 0.02 magnitudes.

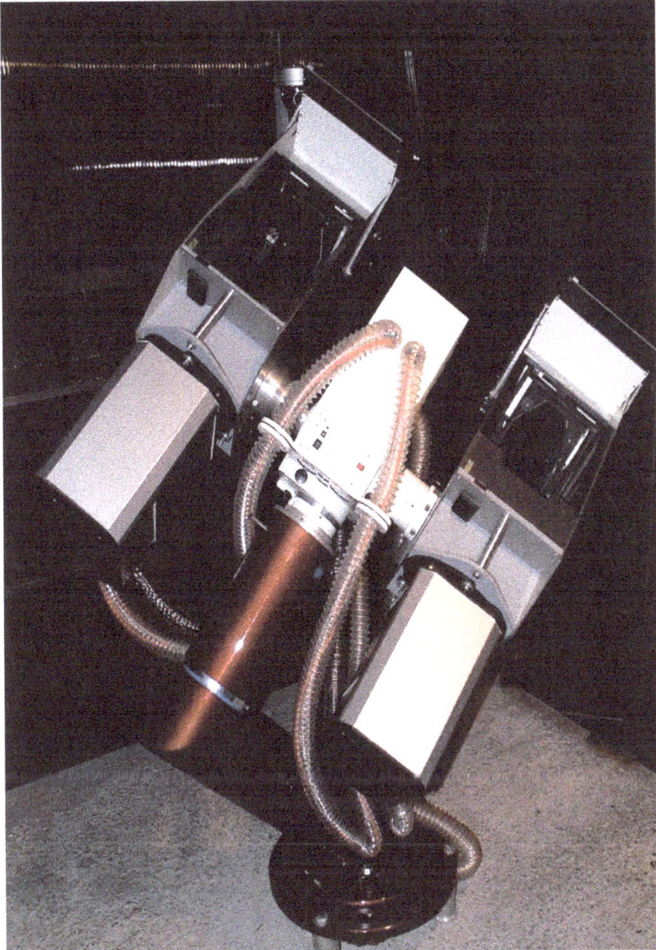

Figure 1. General view of the mount with two channels.

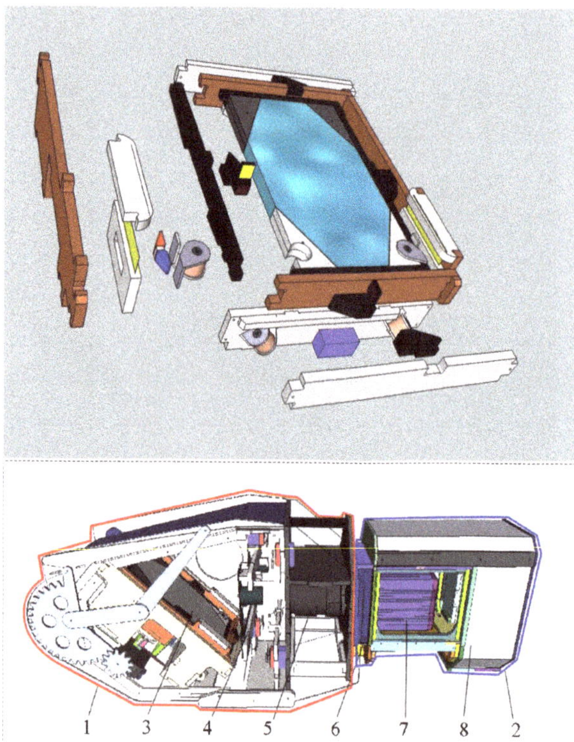

Figure 2. Top panel—celostate mirror frame components. **Bottom panel**—a scheme of the optical channel of MMT-9: 1. Unit of celostate mirror and filters. 2. Light receiver. 3. Mirror of the celostate block. 4. Block of filters. 5. Lens. 6. Flange for fastening the chamber block. 7. Light detector. 8. Commutation compartment.

Figure 3. General view of the assembled entire Mini-MegaTORTORA system.

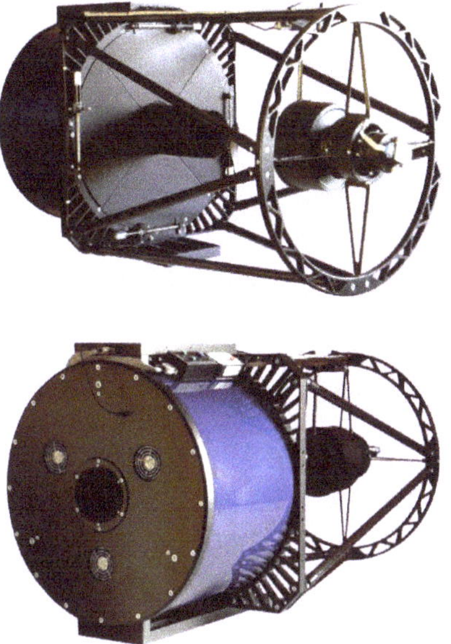

Figure 4. General view of the 0.5-m telescope of the array.

3.2. Array of 0.5-m Telescopes

In addition to the MMT-9, the survey level of the project is performed by an array of several 0.5 m robotic telescopes. The operation of the array is organized in a way similar to the MMT-9 and was started at the SAO RAS in 2019 by a consecutive development of the first AS500 telescope followed by the second and third ones (fall 2021). Additionally, we plan to install still more (three or more) telescopes with the same apertures over the next few years. Three completely automated telescopes are already operational.

The 0.5-m telescopes we use are designed and manufactured by ASTROSIB (Novosibirsk, Russia). These are classic Ritchey-Chrétien telescopes with the main mirror 0.5 m in diameter. This element is still the same for all the telescopes in the array. Slight differences begin from this point onwards: telescopes No. 1 and 2 are equipped with robotic mounts (10 Micron Mount GM 4000 HPS) made in Italy. Telescope No. 3 has a mount manufactured by Astrosib. The general view of the telescope is presented in Figure 4.

There are also some differences in the design of the 'all-sky' telescope domes. Namely, telescope No. 1 is equipped with a dome manufactured by Baader (Germany), while Nos. 2 and 3 have domes manufactured by Astrosib.

Originally, we have chosen an optical design for the telescopes that provides a maximum field of view of about 1.5 degrees in diameter, which implies using a lens corrector and installing a light detector in the primary focus, where the aperture ratio F/2.7 is implemented. In the future, for a number of telescopes, detectors will be installed in the position commonly used in the Ritchey-Chrétien systems, the Cassegrain F/8 focus. The purpose of this replacement is to improve the scale of the image for precise tasks.

The telescopes are focused by means of standard telescope focusing devices. The large format cameras are manufactured by FingerLake Instrumentation (USA) and based on the CCD devices featuring a 4096 × 4096 KAF16803 front-illuminated array with pixels sized 9 micron. The photometric system is formed by 50 mm turret filters with 5 positions for Johnson's broadband filters. Observational data are recorded in standard 4152 × 4128 16-bit FITS files.

Currently, the array operates independently of the MMT-9 photometric complex, mainly focused on the search for new exoplanet candidates (the first results are outlined below). However, in the near future, the 0.5-m telescopes along with the MMT-9 will become part of a unified robotic photometric complex controlled by a single software package. The general view of the array is presented in Figure 5.

Figure 5. General view of the array. **Top plot**: a project. **Bottom plot**: the three operating telescopes.

3.3. A Fiber-Fed High Resolution Planetary Spectrograph

In this chapter, we present a scheme of a high-precision fiber-optic spectrograph operating in combination with the 6-m BTA telescope of the SAO RAS. The need to create such an instrument is directly related to the main goals of the project: the study of exoplanets and the diagnosis of bright transient sources selected in the course of the photometric surveys presented above and other studies.

A description of the general concept and optical layout of the spectrograph can be found in Valyavin et al. [13–16]. In our solution, we followed the classical scheme known as the "white pupil" [17]. In contrast to the traditional schemes, the white pupil design uses two off-axis collimators. One of the collimators operates with the echelle grating in the quasi-Littrow configuration, and the other collimator forms the pupil plane at its focus by constructing an undispersed image of the echelle grating there. This is where the cross-dispersion unit and then the focusing optics with the CCD are accommodated. The main advantage of the selected spectrograph layout is its compactness. In this version, the cross-dispersion unit is located exactly in the pupil plane, thereby making it possible to minimize the size of this unit and the focusing camera, and substantially reduce the cost of the entire instrument.

The spectrograph's optical design is presented in Figure 6. The off-axis mirror collimators ((2) and (5) in Figure 6) have parabolic-shaped surfaces with a focal distance of 2175 mm. Echelle grating (3) consists of a mosaic of two standard echelle gratings, each with a blaze angle of 76° (an an R4 grating). The cross disperser (6) consists of a prism made of the OHARA PBMy glass, and a diffraction 300 lines/mm structure at the exit end. The CCD detector (8) is a 4K × 4K camera with a pixel size of 15 microns (http://www.e2v.com). We use an F/2 lens combination consisting of six spherical optical elements with an effective focus of 470 mm as a focusing camera (7).

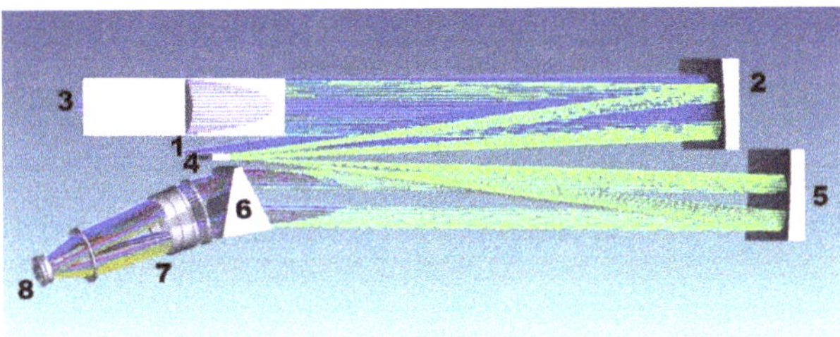

Figure 6. Optical scheme of the spectrograph.

At present, the focusing camera is still under construction. However, the availability of simpler standard lenses that can still be used in combination with our spectrograph, resulting in the loss of about 1 stellar magnitude allowed us to start regular scientific observations at the spectrograph in combination with the 6-m telescope. In these conditions, the configuration of the instrument at R = 50,000–65,000 provides observations of a 10–11 magnitude star under the normal weather conditions (the seeing of 1."5 at the SAO RAS) with hour-long exposures and the signal-to-noise ratio S/N = 100.

By using a 63-m fiber assembly, which feeds the light from the star accumulated in the prime focus of the 6-m telescope to a special thermally and mechanically isolated room with the spectrograph (see Valyavin et al. [16] for details), the presently reached accuracy of the radial-velocity measurements for cool stars is up to 3 m/s. In the final configuration, the spectrograph is planned to be equipped with an interferometric control system based on a vacuumized and stabilized FabryPerot interferometer for yet higher, up to 50 cm/s accuracies. The first scientific results with this instrument have already been presented in [18,19].

3.4. Other Facilities

The second and third levels in the project's hierarchy are covered by a group of optical telescopes with the apertures ranging from 0.6 and higher. The most significant of them are listed below.

The Zeiss-1000 telescope of the SAO RAS with a primary mirror diameter of 1 m. This is a classical optical telescope with a focal ratio of F/13. Its unvignetted field is 45 arcminutes wide, and the typical angular resolution under the North Caucasus weather conditions is about 1.5 arcseconds. The instrument is equipped with a complete range of photometric/polarimetric equipment for the expert diagnostics of transient events and exoplanet candidates detected in the course of the wide-field search described above.

The AZT-11 Astronomical Reflecting Telescope of the Crimean Astrophysical Observatory (CrAO). This is a 1.25-m Ritchey-Chretien reflector with a focal length of 16 m, two focuses and all the necessary equipment: a 5-channel photopolarimeter developed by V. Piirola and installed in the main focus and a CCD-photometer at the auxiliary focus. The telescope is basically used for the photometric and polarimetric observations of various space objects: variable stars of different types, active galactic nuclei, exoplanets, asteroids, comets, etc.

The RC600 0.6 m telescope of the Caucasian Mountain Observatory (CMO) of Moscow State University (SAI MSU) is a Ritchey Chretien telescope with a 600 mm main mirror diameter and a focal length of 4200 mm, manufactured by ASA (Austria). The instrument is installed on a German parallactic mount ASA DDM160 with direct drive engines with absolute encoders; the ScopeDome 55M slit dome is used as a covering. The RC600 is used for precise photometric observations of exoplanet transits. The photometric unit is equipped with an FLI CenterLine double filter wheel for installing eight 50 × 50 mm filters. The following sets of photometric filters, made using the interference technology are available: U, B, V, Rc, Ic, g′, r′, i′, and Clear glass. The receiver is an Andor iKon-L BV CCD camera, 2048 × 2048 pixels, with a pixel size of 13.5 microns.

Additionally, several 1-meter class telescopes of the Kourovka Observatory (Russia) and the Korea Astronomy and Space Sci. Institute (KASI, Korea) are involved in the photometric part of the project. All these instruments have all the necessary tools for conducting high-precision photometric studies of identified transient events and transits of exoplanets, including large super-Earths—see, for example, the observation of the super-Earth HD219134-b carried out recently with the SAO RAS 1-m telescope [20].

Finally, the top, expert level of the project consists of the following optical telescopes: the 1.8-m telescope of the Bonhuynsan astronomical observatory of KASI (Korea), the CMO 2.5-m telescope, and the SAO RAS 6-m telescope. Instrumentation of these telescopes used in the context of EXPLANATION will be soon described in detail in an another paper.

4. Operating Modes

Table 1 provides a brief overview on all the telescopes described above and those already functioning in the project. The first column of the table indicates the name of the telescope, the second gives the size of its aperture, the third presents the field of view, the fourth lists the angular resolution, while the fifth column gives the number of working channels of the telescope (in fact, the number of telescopes in the MMT-9 and the 0.5-m array). We can observe that the telescopes cover completely different fields of view and angular resolutions. However, all of them are aimed at solving one problem—the search and a comprehensive study of transient events and exoplanets.

Table 1. Explanation telescopes.

Telescope	Aperture Size	Field of View	Angular Resolution	Number of Channels
	m	arcmin	arcsec	
BTA	6	1–6	0.06–0.6	1
2.5-m CMO	2.5	10–40	0.15	1
1.8-m Bohuynsan	1.8	10–40	0.1	1
AZT-11	1.25	12	0.6	1
1.2-m Kourovka	1.2	90	0.15	1
ZEISS-1000	1.2	8	0.8	1
RC-600	0.6	22	0.5	1
0.5-m Array	0.5	90	0.5–1.4	3
MMT-9	0.07	540 × 660	10	9

To date, the entire complex is organized for operation in two main modes: the traditional monitoring of the celestial sphere in order to search for the transiting planets and other variable events near stars and galaxies (Basic mode), and an express analysis of a bright flash from one or another transient event registered in any wide-angle systems (Alert mode). Let us now describe both of these modes.

In the Basic mode, everything goes according to the traditional scheme of searching for transiting planets or other periodic variable events. At the first level of the search for such events with the wide-angle photometric systems, the search is carried out for several months in the selected areas of the sky (with the 0.5-m array) and with a continuous scanning of the Northern Sky (with the MMT-9). The accumulated observational material is constantly analyzed in order to search for regular variability in stars and other galactic or extragalactic objects. As soon as a regular variability owing to a periodic transit of an exoplanet across the disk of a star, or for other reasons, is detected, the coordinates of the star and the ephemeris of the event are broadcast from the first level to the higher levels for the subsequent analysis of the event with the expert telescopes. The operation in the Basic mode is schematically represented in Figure 7 by black arrows. The main feature of the mode is that it does not require any urgent study of detected events. However, things are different in the Alert mode.

If a sudden flare is detected during the MMT-9 photometric survey, or upon receipt of a message from any space mission (usually X-ray), the MMT-9 generates two independent alerts to the array of 0.5-m telescopes and to the second-level telescopes, which embark on the photometric monitoring examination of the event. Unfortunately, the telescopes of the second and third levels (except for the BTA, see below) are not always able to connect to the study immediately after receiving an alert. Therefore, their operation in this mode is considered in a long-term context. In the transient event studies, however, expert observations of the first seconds and hours after the onset of a gamma burst are of particular scientific importance. For this purpose, we have provided a special scheme (see the red arrows in Figure 7) with the participation of 0.5-m telescopes and the BTA.

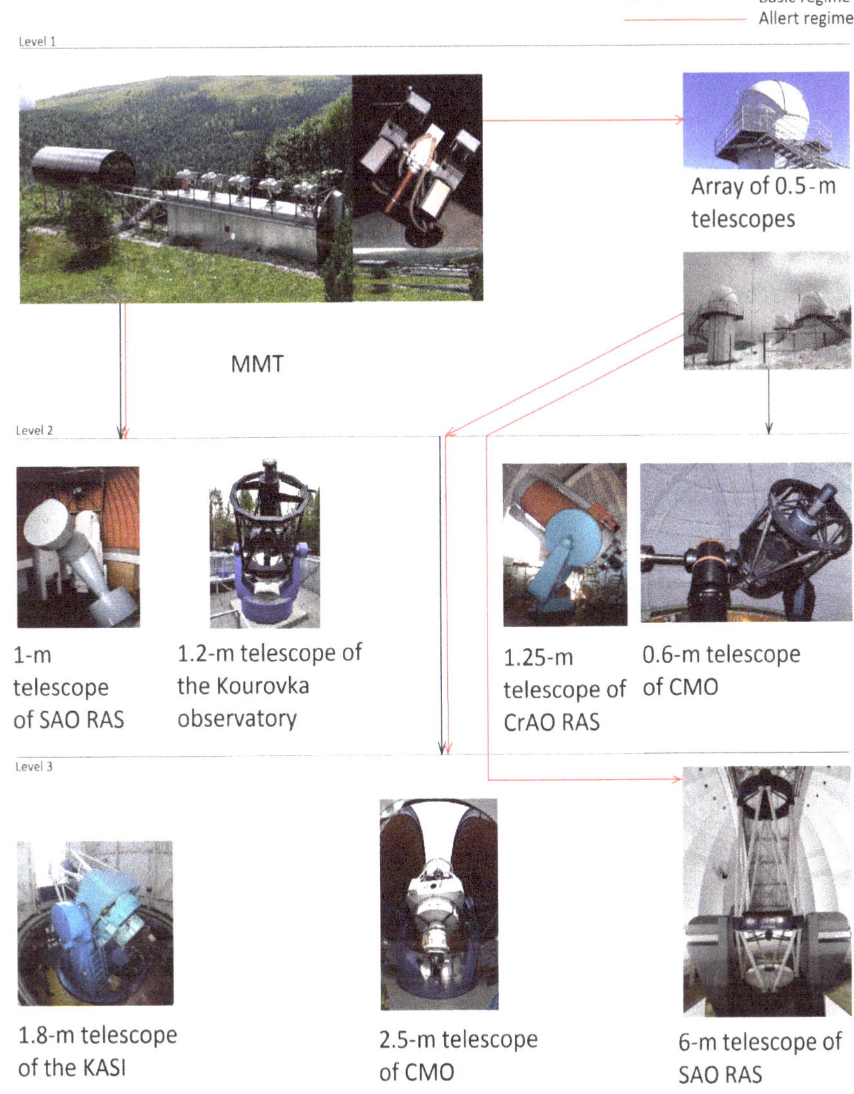

Figure 7. Explanation operating modes.

After receiving an alert from the MMT-9, within less than one minute, the 0.5-m telescopes interrupt their operation in the basic mode and embark on targeting at the event, using the roughly generated MMT-9 coordinates in order to refine the coordinates and transmit them to the BTA. The whole process takes 2–5 min, after which the updated coordinates are transmitted to the BTA. There, by agreement with the observer (1 min), the 6-m telescope is pointed at these coordinates and an expert observation of the event starts within 5 to 15 min. The goal of such an elaborate scheme is to start the spectral monitoring of the evolving transient with a large telescope as soon as possible. In our case, we count on the initiation of such observations at times from 8 to 21 min. Employing such a workflow in the astronomical practice is fundamentally new and innovative. We believe this to be a novelty of the Explanation project, in contrast to other projects, where large telescopes within the traditional schemes are usually employed in such studies only with a long delay.

5. Expectations

From the scientific perspective, our expectations from the project are mainly based on our experience of several years of MMT-9 service. During the 8 years of MMT-9 operation, it successfully detected and catalogued more than 300,000 meteor events (The database of meteors detected by MMT-9 is publicly available at http://mmt.favor2.info/meteors/) [21], and more than 300,000 passes of 10,000 different artificial satellites (The database of photometric measurements of satellites detected by MMT-9 is publicly available at http://mmt.favor2.info/satellites/) [22], as well as observing the bright prompt optical emission of GRB 160625B [23] and detecting a large number of other transient events, including rapid optical flashes from the glinting satellites [24]. About 9000 flares with a sub-second duration, all belonging to the satellites, have been automatically followed up.

Figure 8 demonstrates the sky coverage (the number of frames covering any given position) of the Mini-MegaTORTORA Sky Survey, as well as the typical light curve scatter for individual stars as a function of their mean magnitude. These figures clearly illustrate the sky coverage efficiency, which we believe to be quite high, 100% of the Northern Sky, and about 10% of the Southern Sky. The bottom plot gives evidence of the effective search for new bright transient events up to 13^m.

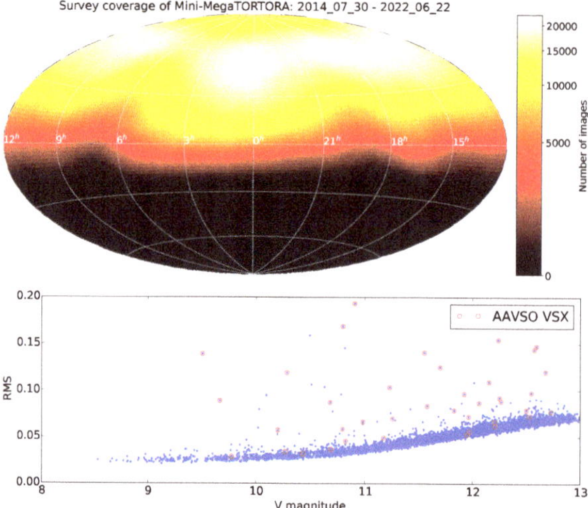

Figure 8. Top panel: the sky coverage of the Mini-MegaTORTORA Sky Survey, i.e., the number of survey frames covering a given position in the sky. **Bottom panel**: a typical RMS along the light curves in the survey data as a function of mean magnitude. The red circles mark the known variable stars from the AAVSO VSX database.

The array of the 0.5-m robotic telescopes makes the search for transient events four stellar magnitudes deeper, though, unfortunately, with a 100 times smaller coverage efficiency. Nevertheless, increasing the number of telescopes in the array (only three telescopes are presently in operation) and effective interaction of the array with the MMT-9 within a joint programme planner of the observations shall minimize this drawback.

6. Results

In this section, we briefly discuss a couple of examples. Namely, the latest transient and an exoplanet obtained with the observational facilities of the Explanation survey. Please note: Due to the fact that the project has just been started, we cannot yet present in this paper the results exhaustively illustrating the work of the project as a whole in all the operating modes (in particular, the Alert mode). However, the examples below demonstrate the efficiency of the wide-angle telescopic systems and present a new result, discovery of an exoplanet candidate, completely derived within the Basic mode.

6.1. Detecting an Optical Flare Accompanying the Extremely Bright Gamma-ray Burst GRB 210619B, and Investigating the Nature of This Event

An optical flare accompanying the gamma-ray burst GRB 210619B [25–28] was detected during the automated monitoring of the celestial sphere in June 2021 by a group of telescopes including the MMT-9. The optical source was registered by the MMT-9 55 s after several space telescopes had simultaneously observed a powerful gamma radiation. The system realigned itself to point towards the region where the gamma-ray source was located after receiving a GCN alert with its coordinates. Observations were carried out synchronously in four channels with the temporal resolutions of 1, 5, 10 and 30 s in the B,V bands and in white light. The scientific details on this discovery are presented in our special study [29].

6.2. Searching for Exoplanet Candidates

As part of the search and study of non-stationary events, the project also obtained the first detections of new exoplanet candidate transits. At the time of writing, there are already about a dozen such candidates, detected within the project by the 0.5-m telescope array. A part of these results has been published in [30]. All results will be published in detail after their validation with the 1–2 m and 6-m telescopes. In this paper, we shall limit ourselves to demonstrating one of them, detected by the 0.5-m telescope array, and refined by the 1-m telescope of the second level, Zeiss-1000.

In search for transiting exoplanets, we are looking for and modeling short-term brightness dips in their host stars. Details about this procedure can be found in one of our previous papers (see, for example, [20] and references therein). Briefly, the modeling process is as follows.

The model computes the transit shape in a given spectral band as a function of the relative exoplanet radius expressed as a fraction of the host star radius, equilibrium temperature (Teq) of the exoplanet, the physical characteristics of the star itself, and the impact parameter. The equilibrium temperature of an exoplanet determines the additional intrinsic luminosity when computing the transit depths based on black body relations. Such an approximation is sufficient for the vast majority of the practically realizable cases. The physical properties of the host star (its effective temperature Teff, surface gravity log g, and chemical composition) are used [20] in computations of the linear and squared limb darkening coefficients for a given spectral band.

Figure 9 shows the phase curve of a periodic transit event for a 17th magnitude orange/red dwarf star. The effective temperature of the star is 4957 K and $R\star = 0.7$ solar radii, the corresponding brightness decrease in white light during the transit suggests its exoplanetary nature with a planet of 1.64_{MJup}. This result has been obtained with the telescopes of different project levels. Once the first positive registration of the transit event was received from the array of the 0.5-m telescopes, other telescopes joined the study as well. In Figure 9, the results of transit observations obtained with different telescopes are marked by different colors. In particular, black dots illustrate transit observations with the SAO RAS 1-m Zeiss-1000 telescope. The green line is the model of the event.

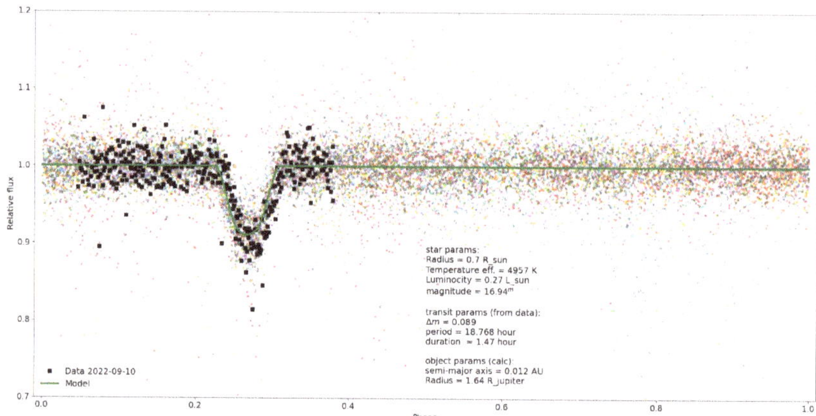

Figure 9. An exoplanet candidate detected within the Explanation project.

7. Discussion

We have briefly described the new joint Russian-Korean EXPLANATION project, its aims, philosophy, instrumental base, and presented the first results. Generally, the project consists of two large blocks: photometric surveys, and a group of observational facilities and methods for extended studies of objects detected in the course of the surveys. In contrast to the second block, the instrumentation for which has already been elaborated and we have no plans of altering it over the next decade, but the photometric surveys deserve some more discussion.

The configuration of the photometric instruments presented here will be further developed into a more complete network of telescopes with small to medium apertures. In this regard, we also present here (Beskin et al., this issue) one of the ideas for the upgrade. The idea has already been introduced by some of the authors of this paper in [31] within the framework of the SAINT photometric project. We are sure that the upgrade of the wide-field observing facilities as proposed by the SAINT will fundamentally magnify the scientific output of the EXPLANATION project.

Author Contributions: Supervision, G.V.; writing—original draft preparation, G.V. and G.B.; conceptualization, G.V., G.B., A.V., G.G., and S.F.; methodology, all authors. All authors have read and agreed to the published version of the manuscript.

Funding: The study was funded by the Ministry of Science and Higher Education of Russian Federation, project 075-15-2020-780.

Institutional Review Board Statement: Not applicable.

Informed Consent Statement: Not applicable.

Data Availability Statement: Not applicable.

Acknowledgments: The study is supported by the Ministry of Science and Higher Education of Russian Federation, project 075-15-2020-780. MGP was supported by the National Research Foundation of Korea(NRF No-2018R1A6A1A06024970) funded by the Ministry of Education, Korea.

Conflicts of Interest: The authors declare no conflict of interest.

References

1. Dupree, A. NASA's Kepler Mission: A Search for Habitable Planets In Proceedings of the Joint Fall 2009 Meeting of the New England Section of the APS and AAPT, Durham, NH, USA, 16–17 October 2009; abstract F1.002.
2. Baglin, A.; Auvergne, M.; Barge, P.; Michel, E.; Catala, C.; Deleuil, M.; Weiss, W. The CoRoT mission and its scientific objectives. *AIP Conf. Proc.* **2007**, *895*, 201–209.
3. Street, R.A.; Pollaco, D.L.; Fitzsimmons, A.; Keenan, F.P.; Horne, K.; Kane, S.; Collier Cameron, A.; Lister, T.A.; Haswell, C.; Norton, A.J.; et al. SuperWASP: Wide Angle Search for Planets. *ASP Conf. Ser.* **2003**, *294*, 405–408.
4. Burdanov, A.Y.; Benni, P.; Krushinsky, V.V.; Popov, A.A.; Sokov, E.N.; Sokova, I.A.; Rusov, S.A.; Lyashenko, A.Y.; Ivanov, K.I.; Moiseev, A.V.; et al. First results of the Kourovka Planet Search: Discovery of transiting exoplanet candidates in the first three target fields. *MNRAS* **2016**, *461*, 3854–3863. [CrossRef]
5. Bondi, H. Astronomy of the Future. *Q. J. R. Astron. Soc.* **1970**, *11*, 443–450.
6. Beskin, G.; Karpov, S.; Plokhotnichenko, V.; Stepanov, A.; Tsap, Y. Discovery of the Sub-second Linearly Polarized Spikes of Synchrotron Origin in the UV Ceti Giant Optical Flare. *Publ. Astron. Soc. Aust.* **2017**, *34*, e010. [CrossRef]
7. MacGregor, M.A.; Weinberger, A.J.; Wilner, D.J.; Kowalski, A.F.; Cranmer, S.R. Detection of a Millimeter Flare from Proxima Centauri. *ApJLett* **2018**, *855*, L2. [CrossRef]
8. Thornton, D.; Stappers, B.; Bailes, M.; Barsdell, B.; Bates, S.; Bhat, N.D.R.; Burgay, M.; Burke-Spolaor, S.; Champion, D.J.; Coster, P.; et al. A Population of Fast Radio Bursts at Cosmological Distances. *Science* **2013**, *341*, 53–56. [CrossRef] [PubMed]
9. Abbott, B.P.; Abbott, R.; Abbott, T.D.; Abernathy, M.R.; Acernese, F.; Ackley, K.; Adams, C.; Adams, T.; Addesso, P.; Adhikari, R.X.; et al. GW151226: Observation of Gravitational Waves from a 22-Solar-Mass Binary Black Hole Coalescence. *Phys. Rev. Lett.* **2016**, *116*, 241103. [PubMed]
10. Beskin, G.M.; Karpov, S.V.; Biryukov, A.V.; Bondar, S.F.; Ivanov, E.A.; Katkova, E.V.; Orekhova, N.V.; Perkov, A.V.; Sasyuk, V.V. Wide-field optical monitoring with Mini-MegaTORTORA (MMT-9) multichannel high temporal resolution telescope. *Astrophys. Bull.* **2017**, *72*, 81–92. [CrossRef]
11. Beskin, G.; Karpov, S.; Bondar, S.; Greco, G.; Guarnieri, A.; Bartolini, C.; Piccioni, A. Fast Optical Variability of a Naked-eye Burst Manifestation of the Periodic Activity of an Internal Engine. *ApJ* **2010**, *719*, L10–L14. [CrossRef]
12. Karpov, S.; Beskin, G.; Bondar, S.; Perkov, A.; Ivanov, E.; Guarnieri, A.; Bartolini, C.; Greco, G.; Shearer, A.; Sasyuk, V. *Gamma-ray Bursts: 15 Years of GRB Afterglows*; Castro-Tirado, A.J., Gorosabel, J., Park, I.H., Eds.; EAS Publications Series; EDP Sciences: Les Ulis, France, 2013; Volume 61, pp. 465–469.
13. Valyavin, G.; Bychkov, V.; Yushkin, M.; Galazutdinov, G.A.; Drabek, S.V.; Shergin, V.S.; Sarkisyan, A.N.; Semenko, E.A.; Burlakova, T.E.; Kravchenko, V.M.; et al. High-resolution fiber-fed echelle spectrograph for the 6-m telescope. I. Optical scheme, arrangement, and control system. *Astrophys. Bull.* **2014**, *69*, 224–239. [CrossRef]
14. Valyavin, G.; Bychkov, V.; Yushkin, M.; Galazutdinov, G.A.; Drabek, S.V.; Shergin, V.S.; Sarkisyan, A.N.; Semenko, E.A.; Perkov, A.V.; Sazonenko, D.A.; et al. High-Resolution Fiber-Fed Echelle Spectrograph for the SAO 6-m Telescope. *ASP Conf. Ser.* **2015**, *494*, 305–307.
15. Valyavin, G.; Musaev, F.; Perkov, A.; Bychkov, V.D.; Yushkin, M.V.; Galazutdinov, G.A.; Drabek, S.V.; Shergin, V.S.; Sazonenko, D.A.; Kukushkin, D.E.; et al. The High-Resolution Fiber-Fed Echelle Spectrograph for the SAO RAS 6-m Telescope: First Spectra. *ASP Conf. Ser.* **2019**, *518*, 242–246.
16. Valyavin, G.; Musaev, F.; Perkov, A.; Aitov, V.N.; Bychkov, V.D.; Drabek, S.V.; Shergin, V.S.; Sazonenko, D.A.; Kukushkin, D.E.; Galazutdinov, G.A.; et al. High-Resolution Fiber-Fed Spectrograph for the 6-m Telescope of the Special Astrophysical Observatory of the Russian Academy of Sciences: Assessment of Efficiency. *Astrophys. Bull.* **2020**, *75*, 191–197. [CrossRef]
17. Dekker, H.; D'Odorico, S.; Kaufer, A.; Delabre, B.; Kotzlowski, H. Design, construction, and performance of UVES, the echelle spectrograph for the UT2 Kueyen Telescope at the ESO Paranal Observatory. *SPIE Conf. Proc.* **2000**, *4008*, 534–545.
18. Burlakova, T.E.; Valyavin, G.G.; Aitov, V.N.; Galazutdinov, G.A.; Valeev, A.F.; Yakunin, I.A.; Gadelshin, D.R.; Bychkov, V.D.; Tavrov, A.V.; Ivanova, A.E.; et al. Observations of Radial Velocity Variability in Stars from the Spectra Obtained with the BTA Fiber-Fed Spectrograph in High Spectral Resolution Mode. *Astrophys. Bull.* **2020**, *75*, 482–485. [CrossRef]
19. Gadelshin, D.R.; Valyavin, G.G.; Lee, B.-C.; Jeong, G.; Inwoo, H.; Galazutdinov, G.A.; Aitov, V.N.; Yakunin, I.A.; Burlakova, T.E.; Valeev, A.F. Mass Constraint of Several Transiting Planets. *Astrophys. Bull.* **2020**, *75*, 437–439. [CrossRef]
20. Valyavin, G.G.; Gadelshin, D.R.; Valeev, A.F.; Burlakova, T.E.; Antonyuk, K.A.; Galazutdinov, G.A.; Pit, N.V.; Moskvitin, A.S.; Sokov, E.N.; Sokova, I.A.; et al. Exoplanet Studies. Photometric Analysis of the Transmission Spectra of Selected Exoplanets. *Astrophys. Bull.* **2018**, *73*, 225–234. [CrossRef]
21. Karpov, S.; Orekhova, N.; Beskin, G.; Biryukov, A.; Bondar, S.; Ivanov, E.; Katkova, E.; Perkov, A.; Sasyuk, V. Meteor observations with Mini-Mega-TORTORA wide-field monitoring system. In *IV Workshop on Robotic Autonomous Observatories*; Caballero-García, M.D., Pandey, S.B., Hiriart, D., Castro-Tirado, A.J., Eds.; Revista Mexicana de Astronomía y Astrofísica (Serie de Conferencias); Instituto de Astronomía: Mexico City, Mexico, 2016; Volume 48, pp. 97–98.
22. Karpov, S.; Orekhova, N.; Beskin, G.; Biryukov, A.; Bondar, S.; Ivanov, E.; Katkova, E.; Perkov, A.; Sasyuk, V. Massive photometry of low-altitude artificial satellites on Mini-Mega-TORTORA. In *IV Workshop on Robotic Autonomous Observatories*; Caballero-García, M.D., Pandey, S.B., Hiriart, D., Castro-Tirado, A.J., Eds.; Revista Mexicana de Astronomía y Astrofísica (Serie de Conferencias); Instituto de Astronomía: Mexico City, Mexico, 2016; Volume 48, pp. 112–113.

23. Zhang, B.-B.; Zhang, B.; Castro-Tirado, A.J.; Dai, Z.G.; Tam, P.-H.T.; Wang, X.-Y.; Hu, Y.-D.; Karpov, S.; Pozanenko, A.; Zhang, F.-W.; et al. Transition from fireball to Poynting-flux-dominated outflow in the three-episode GRB 160625B. *Nat. Astron.* **2018**, *2*, 69–75. [CrossRef]
24. Karpov, S.; Beskin, G.; Biryukov, A.; Bondar, S.; Ivanov, E.; Katkova, E.; Orekhova, N.; Perkov, A.; Plokhotnichenko, V.; Sasyuk, V.; et al. Observations of Transient Events with Mini-MegaTORTORA Wide-Field Monitoring System with Sub-Second Temporal Resolution. In *V Workshop on Robotic Autonomous Observatories*; Caballero-García, M.D., Pandey, S.B., Hiriart, D., Castro-Tirado, A.J., Eds.; Revista Mexicana de Astronomía y Astrofísica (Serie de Conferencias); Instituto de Astronomía: Mexico City, Mexico, 2019; Volume 48, pp. 30–38.
25. D'Avanzo, P.; Bernardini, M.G.; Lien, A.Y.; Melandri, A.; Page, K.L.; Palmer, D.M.; Sbarrato, T.; Neil Gehrels Swift Observatory Team. GRB 210619B: Swift detection of a bright burst and optical counterpart. In *GRB Coordinates Network, Circular Service*; NASA: Washington, DC, USA, 2021; No. 30261.
26. Zhao, Y.; Xiong, S.L.; Huang, Y.; Xie, S.L.; Xiao, S.; Cai, C.; Liu, J.C.; Li, C.Y.; Zhang, Y.Q.; Xue, W. C.; et al. GECAM detection of GRB 210619B. In *GRB Coordinates Network, Circular Service*; NASA: Washington, DC, USA, 2021; No. 30264.
27. Svinkin, D.; Golenetskii, S.; Frederiks, D.; Ulanov, M.; Tsvetkova, A.; Lysenko, A.; Ridnaia, A.; Cline, T.; Konus-Wind Team. Konus-Wind detection of GRB 210619B. In *GRB Coordinates Network, Circular Service*; NASA: Washington, DC, USA, 2021; No. 30276.
28. Poolakkil, S.; Meegan, C.; Fermi GBM Team. GRB 210619B: Fermi GBM detection. In *GRB Coordinates Network, Circular Service*; NASA: Washington, DC, USA, 2021; No. 30279.
29. Oganesyan, G.; Karpov, S.; Jelínek, M.; Beskin, G.; Ronchini, S.; Banerjee, B.; Branchesi, M.; Strobl, J.; Polasek, C.; Hudec, R.; et al. Exceptionally bright optical emission from a rare and distant gamma-ray burst. *arXiv* **2021**, arXiv:2109.00010.
30. Yakovlev, O.Y.; Valeev, A.F.; Valyavin, G.G.; Tavrov, A.V.; Aitov, V.N.; Mitiani, G.S.; Korablev, O.I.; Galazutdinov, G.A.; Beskin, G.M.; Emelianov, E.V.; et al. Exoplanet Two-Square Degree Survey with SAO RAS Robotic Facilities. *Front. Astron. Space Sci.* **2022**, *9*, 903429. [CrossRef]
31. Beskin, G.M.; Karpov, S.V.; Plokhotnichenko, V.L.; Bondar, S.F.; Perkov, A.V.; Ivanov, E.A.; Katkova, E.V.; Sasyuk, V.V.; Shearer, A. Wide-field subsecond temporal resolution optical monitoring systems for the detection and study of cosmic hazards. *Phys. Uspekhi* **2013**, *56*, 836–842. [CrossRef]

Article

Image Degradation Model for Dynamic Star Maps in Multiple Scenarios

Haima Yang [1,2], Yan Jin [1], Yinan Hu [3,*], Dawei Zhang [1], Yong Yu [4], Jin Liu [5], Jun Li [1], Xiaohui Jiang [6] and Xiaojun Yu [7]

[1] School of Optical-Electrical and Computer Engineering, University of Shanghai for Science and Technology, Shanghai 200093, China
[2] Key Laboratory of Space Active Opto-Electronics Technology, Chinese Academy of Sciences, Shanghai 200083, China
[3] Shanghai Institute of Technical Physics, Chinese Academy of Sciences, Shanghai 200083, China
[4] Shanghai Astronomical Observatory, Chinese Academy of Sciences, Shanghai 200030, China
[5] School of Electronic and Electrical Engineering, Shanghai University of Engineering Science, Shanghai 201620, China
[6] School of Mechanical Engineering, University of Shanghai for Science and Technology, Shanghai 200093, China
[7] Shanghai East China Railway Electrification Engineering Co., Ltd., Shanghai 200041, China
* Correspondence: 12120082@bjtu.edu.cn

Abstract: To meet the ground test requirements of star sensors, we establish the star map simulation algorithm and the interactive interface in multiple scenarios. The combination of the degradation model of star points, the imaging noise model, and the attitude disturbance model is introduced to solve the problem of different patterns of noise existing in the actual measurement, improving the traditional simulation model. In addition, a user-friendly interface design makes it easier for both scholars and average individuals to understand the parameters and then generate static single-frame star maps—or a series of dynamic sequence star maps—under various conditions. The results of the proposed star map simulation method are highly comparable to the actual captured star images, and this method can be applied for the tests and calibrations of star sensors.

Keywords: star map; image degradation; dynamic environment; aerospace; noise model

Citation: Yang, H.; Jin, Y.; Hu, Y.; Zhang, D.; Yu, Y.; Liu, J.; Li, J.; Jiang, X.; Yu, X. Image Degradation Model for Dynamic Star Maps in Multiple Scenarios. *Photonics* **2022**, *9*, 673. https://doi.org/10.3390/photonics9100673

Received: 16 August 2022
Accepted: 15 September 2022
Published: 20 September 2022

Publisher's Note: MDPI stays neutral with regard to jurisdictional claims in published maps and institutional affiliations.

1. Introduction

Star sensors play an essential role in aerospace attitude detection. Star sensors need to undergo rigorous performance tests before the engineering of applications. Generally, there are three ways to calibrate star sensors with high accuracy: numerical simulation, hardware simulation, and outfield observation experiments. Outfield observation experiments are conducted by placing a star sensor in an open area with low atmospheric turbulence and relatively weak stray light. Such an approach is costly and time-consuming and is usually only used as the last step in a star sensor validation experiment. Hardware simulation refers to utilizing a starfield simulator in a laboratory environment to simulate an infinity star point. The star sensor carries out identification and attitude calculation by observing the simulated star images. The disadvantages of this method are that the starfield simulator is expensive, and the parameters of the starfield simulator, such as field of view and aperture—which are difficult to match precisely with the star sensor—are not adjustable. Therefore, the versatility of the hardware simulation is limited. Compared with these two methods, numerical simulation is a highly flexible and the least resource-intensive approach. In routine performance tuning or laboratory experiments, simulated star maps are widely used because of their low cost and ease of use [1–4].

Many scholars have invested in studying the use of star sensors from digital simulation, i.e., the study of star map simulation algorithms. In a star map simulation, the establishment of the noise model is essential. To solve this problem, some scholars simulated the Gaussian grayscale distribution of star point images by considering the stellar image shift caused by the satellite motion [5]. Some scholars have introduced new theories, such as the effect of rotation around the optical axis on star point imaging, the star color index, and the idea of black-body radiation, to evaluate the quality of star points in a simulation result [6,7]. To enlarge the simulated scenarios in the simulation system, nebulae, moonlight, and Earth-obscured background are added into the field of view in the star sensors [8,9]. The literature [10] proposed a fast pixel-discretization algorithm based on the convolutional surface model to simulate dynamic real-time star maps and established a complete description of the star point motion trajectory model. However, the parameters in this algorithm are difficult to understand, limiting its usage. Researchers [11] designed a star map simulation algorithm for arbitrary exposure time length in a high-fidelity manner. However, this method is performed in an ideal state, without incorporating various error factors into the model.

Here, we design a set of algorithms and interfaces for simulating star maps in multiple scenarios from a numerical simulation approach. The algorithm includes the simulation of static single-frame star maps and dynamic sequence star maps, taking into account various noise sources. The noise may come from the electronic noise caused by the imaging system, the overall noise amplitude variation due to atmospheric condition or stray light, and the attitude disturbance caused by the different motion states between the imaging platform and the tracking star target. With various parameter inputs, noise levels and dynamic trajectories are visualized, and star maps close to the actual images are generated, which can provide a source of data for the parametric testing of the star sensor.

2. Motion Blur Degradation Model

2.1. The Process of Image Degradation

For a ground-based star sensor, during the process of image acquisition, there will be many challenges, such as the shift and superposition of light beams. All these difficulties can result in the degradation of image quality [12]. Figure 1 shows the general model of image degradation.

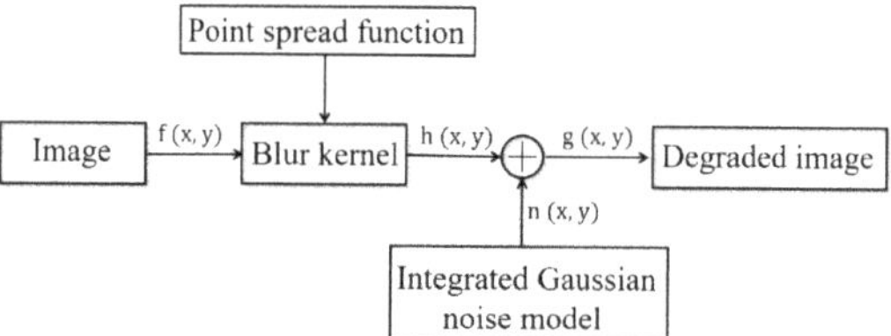

Figure 1. The general model of image degradation.

In Figure 1, if h (x, y) is a linear spatially invariant process, then the degraded image in the spatial domain is represented as:

$$g (x, y) = h (x, y) \times f (x, y) + n (x, y) \tag{1}$$

in which f (x, y) represents a static, two-dimensional image, which degenerates to g (x, y) under the interference of additive noise n (x, y) through blur kernel h (x, y). * denotes a general convolution operator.

2.2. Simulations of Defocus Factor

In general, h (x, y) in the degradation model includes defocus blur and motion blur. To realize the out-of-focus impact more accurately, we use a point spread function model to simulate the image degradation with the following formula [13,14]:

$$g_{ij} = \frac{A}{2\pi\sigma^2} \int_{i-0.5}^{i+0.5} \int_{j-0.5}^{j+0.5} exp[-\frac{\left(x_{i,j} - x_m\right)^2 + \left(y_{i,j} - y_m\right)^2}{2\sigma^2}]dxdy \tag{2}$$

where A denotes the energy grayscale coefficient, which is related to the total illumination, i.e., magnitude, of the imaging point on the photosensitive surface of the star sensor. σ denotes the Gaussian dispersion radius of the energy distribution in the star point region, indicating the degree of defocusing. $(x_{i,j}, y_{i,j})$ is any pixel within the range of the diffuse pixel point. (x_m, y_m) denotes the projected position of the star on the imaging plane of the star sensor. We can evaluate a static star point in a simulation in terms of two key parameters: the gray energy factor A and the Gaussian dispersion radius σ [15]. Table 1 lists the parameters of the simulated image in Figure 2, which shows the combination of different values of A and σ, representing various results of image degradation that may occur under real circumstances.

Table 1. Parameters of the simulated image.

Position	A	σ_x	σ_y
(15, 15)	613	1.0	1.0
(30, 30)	2113	2.0	1.5
(40, 10)	4113	2.5	3.0

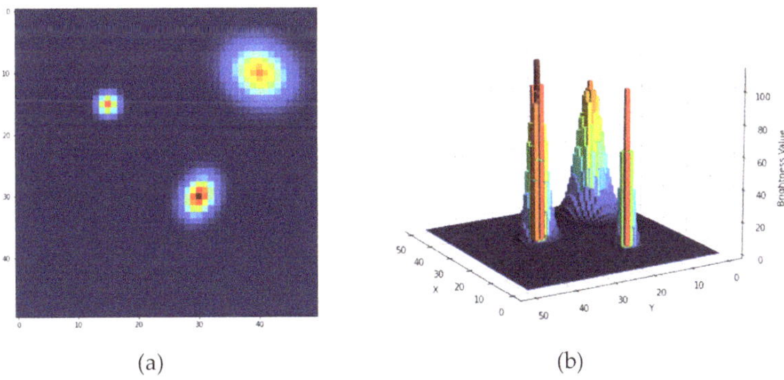

(a)   (b)

Figure 2. The combination of different A and σ: (a) 2D; (b) 3D.

2.3. Integrated Gaussian Noise Model

In addition to the simulation for the background star point and target star point, the star sensor simulation also needs to simulate the effect of noise to improve the simulation's authenticity. The random noise contained in the star map is mainly ambient noise and detector noise, such as dark current noise and output noise. These types of noise can be defined as Gaussian noise, due to their randomness [16]. The literature [2,7] states that the simulation of ground-based imaging systems must consider the effects of atmospheric influences and uneven background illumination, such stray light, moonlight, and sunlight, on the captured images. These effects will lead to a change in the overall brightness value of the image from weak to strong. Here, we use Equation (3) to represent the integrated Gaussian noise model:

$$N_{img} = f(x, y) \times \psi \times \eta(0, std), \eta\sim Gaussian(0, std) \tag{3}$$

where $f(x, y)$ is the original image. ψ is designed to control the shape of the noise area, and it can be expressed by functions. Here is an example of a linear expression. If we set $\psi_y(w, k) = w \times y + k$, a mathematical model of monotonic increase or monotonic decrease, the image result will have a gradient effect on the y-axis. That is, the magnitude of the image will change from dark to light or from light to dark. ψ can also be expressed in terms of other functions, such as polar coordinates. With the use of polar coordinates, a circular, radial noise pattern can be clearly expressed. $\eta(0, std)$ is the amount that controls the magnitude of the noise, which follows a Gaussian distribution. The default mean of η is 0, and std represents the standard deviation. The larger the value of std, the larger the range of noise fluctuation.

Figure 3 shows the simulations of the integrated Gaussian noise model with different inputs. The simulated image is 8 bits, and the image array is 512×512, where the Gaussian dispersion radius of the star point is between 1 and 4. The star point circled in the figures is the same star point (with the same A, σ, and location) in the three images.

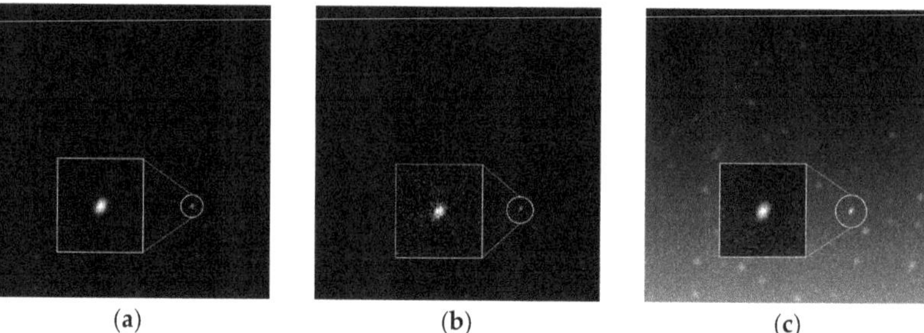

(a)   (b)   (c)

Figure 3. The simulations of the integrated Gaussian noise model with different inputs: (a) the simulation result without the noise model; (b) the simulation result with $N_{img} = f(x, y) \times 1 \times \eta(20, 5)$; (c) the simulation result with $N_{img} = f(x, y) \times \psi_y \cdot \eta(20, 5)$, $\psi_y = y + 5$.

3. Simulations for Dynamic Sequence Star Maps

3.1. The Energy Distribution Model of Trajectory

Because of the relative motion between the ground-based imaging system and the celestial body, this study will complete a star map simulation in a dynamic environment. In the dynamic environment, the target star point will be dragged to form a trajectory on the imaging plane during the camera exposure time, which is called a "smearing star point." The length and direction of this trailing path can reflect the motion information of the target star point [17]. To summarize, if there is a relative motion between the star point and the ground-based imaging system, the target is in the form of a trajectory line segment. Particularly if the imaging system has tracked and located a target object, the target star point appears relatively stationary in the image.

To achieve a star streak quickly, the exposure time T is divided into N segments at equal intervals. Accordingly, the trailing path of the star point is also divided into N segments. When interval $\delta = T/N$ is short enough, each part of the trailing path can be approximated as a point. Through such an approximation process, the dynamic trajectory of the star point can be approximated as a superposition of N static points. We set the total energy during the exposure time of the star point to $E_0 = \sum_{n=1}^{n=N} E_{i,i+1}$, and the energy distribution of each static star point simulating the dynamic trajectory is shown in Equation (4). In the interval from t = i to t = i + 1, the simulation formula for the energy of a star point is:

$$E_i(x,\ y) = \frac{E_{i,i+1}}{2\pi\sigma^2} \int \int \exp[-\frac{(x-x_i)^2+(y-y_i)^2}{2\sigma^2}]dxdy \tag{4}$$

Assuming that the dispersed spots of the above Gaussian distribution are equal in radius, the energy distribution function of the star image in each small exposure time is superimposed, and the total energy distribution function of the star image is as follows:

$$E(x,\ y) = \sum_{n=1}^{n=N} E_i(x,\ y) \tag{5}$$

3.2. Attitude Disturbance Model

When the ground-based imaging system tracks a moving target star point, there are attitude disturbances between the measurement. The attitude disturbance mainly refers to the jitter of the ground-based platform brought about by its working condition and the motion lag of the camera system when it does not accurately track the motion of the target object.

Figure 4 shows a schematic diagram of the projection trajectory of the projection of the star point on the imaging plane under the influence of attitude disturbance. When rotating in any direction within $O_sx_sy_sz_s$, the coordinate system measured around the star sensor, the angular velocity can be decomposed onto the 3 axes of $O_sx_sy_sz_s$. By setting the star image on the imaging plane Oy on the fixed axis of the uniform angular velocity motion along the clockwise direction, the angular velocity is recorded as ω_z. By setting the ith instant in time, the star point position vector $R_i = (x_i, y_i)$, $|R_i| = \sqrt{x_i^2+y_i^2}$. The angle between the R_i and the Oy axis is denoted θ_i, so the velocity of the star at the ith instant in time can be expressed as:

$$\begin{cases} v_{ix} = \omega_z|R_i|\cos\theta_i + v_{0x} = \omega_z y_i + v_{0x} \\ v_{iy} = -\omega_z|R_i|\sin\theta_i + v_{0y} = -\omega_z x_i + v_{0y} \end{cases} \tag{6}$$

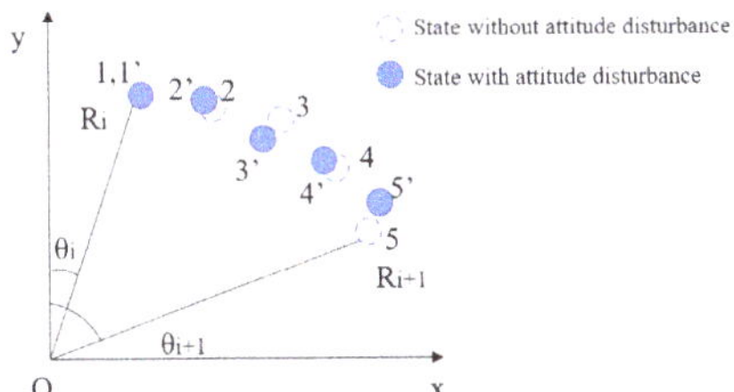

Figure 4. The schematic of attitude disturbance.

Given the time elapsed from the ith instant in time to the i + 1st instant in time, the amount of star point displacement [18] can be expressed as:

$$\begin{cases} \Delta x_i = x_{i+1} - x_i = v_{ix} \cdot \Delta t = (\omega_z y_i + v_{0x}) \cdot \Delta t \\ \Delta y_i i = y_{i+1} - y_i = v_{iy} \cdot \Delta t = (-\omega_z x_i + v_{0y}) \cdot \Delta t \end{cases} \tag{7}$$

To achieve the effect of motion blur, we propose that $N_{\Delta x}$ be added as a blur factor to the expression of the appropriate amount of the star position:

$$\begin{cases} x_{i+1} = x_i + \Delta x_i + N_{\Delta x}, N_{\Delta x} \sim \text{Gaussian}(0, n) \\ y_{i+1} = y_i + \Delta y_i + N_{\Delta y}, N_{\Delta y} \sim \text{Gaussian}(0, m) \end{cases} \tag{8}$$

where the default mean of the Gaussian distributed random number is 0. n and m represent the standard deviation of the set Gaussian random number. The values of n and m can be same. The larger the n and m, the greater the deviation of the simulated star point from the preset position.

In conclusion, the description of the star trajectory can be expressed by Equation (9):

$$\begin{cases} x_i(t) = x_{i0} + \int_{t0}^{t0+T} v_{ix}(\tau)d\tau \\ y_i(t) = y_{i0} + \int_{t0}^{t0+T} v_{iy}(\tau)d\tau \end{cases} \tag{9}$$

Therefore, we use the proposed methods to simulate the dynamic trajectory of a star point. Figures 5–7 show the superposition of the different states when the simulated star trajectory is a linear equation according to $\begin{cases} x_{i+1} = x_i + 1 + N_{\Delta x} \\ y_{i+1} = y_i + 1 + N_{\Delta y} \end{cases}$. Figure 8 shows the superposition of the different states when the simulated star trajectory is a curvilinear equation according to $\begin{cases} x_{i+1} = x_i + 1 + N_{\Delta x} \\ y_{i+1} = -0.006x_i^2 + 2.7x_i - 27 + N_{\Delta y} \end{cases}$.

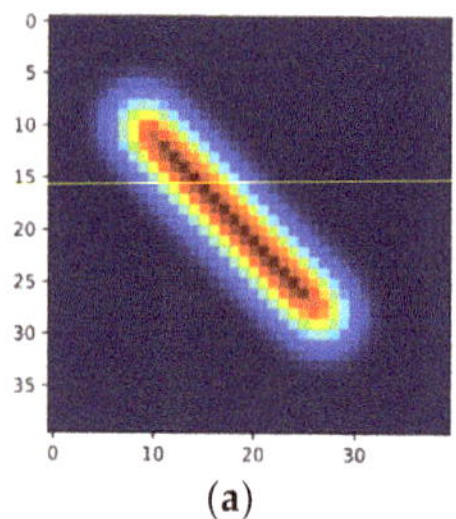 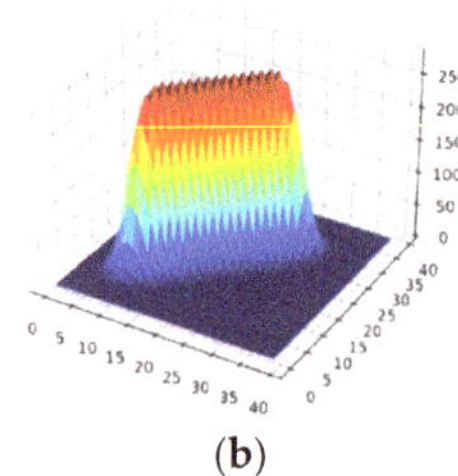

(a)　　　　　　　　　　　(b)

Figure 5. Simulation results with $A = 2000$, $\sigma = 2$, $N_{img} = f(x, y) \times 1 \times \eta(0,0)$, $N_{\Delta x}, N_{\Delta y} \sim \text{Gaussian}(0,0)$: (**a**) 2D; (**b**) 3D.

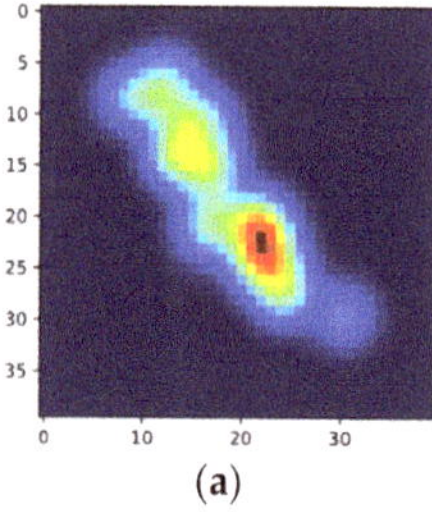 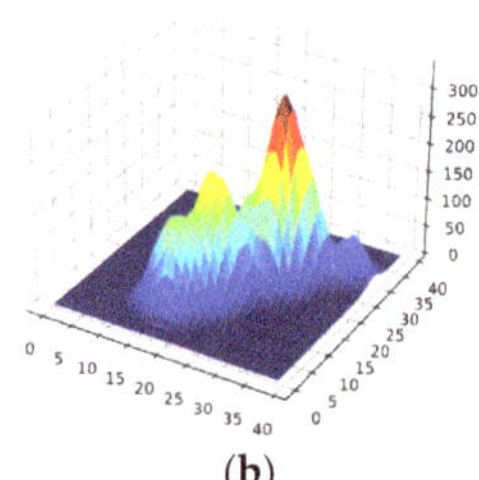

(a)　　　　　　　　　　　(b)

Figure 6. Simulation results with $A = 2000$, $\sigma = 2$, $N_{img} = f(x, y) \times 1 \times \eta(0,0)$, $N_{\Delta x}, N_{\Delta y} \sim \text{Gaussian}(0,2)$: (**a**) 2D; (**b**) 3D.

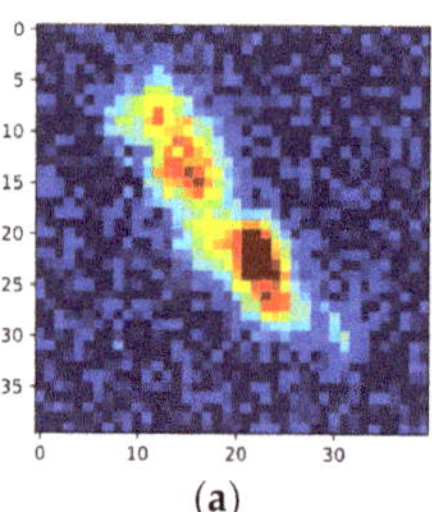 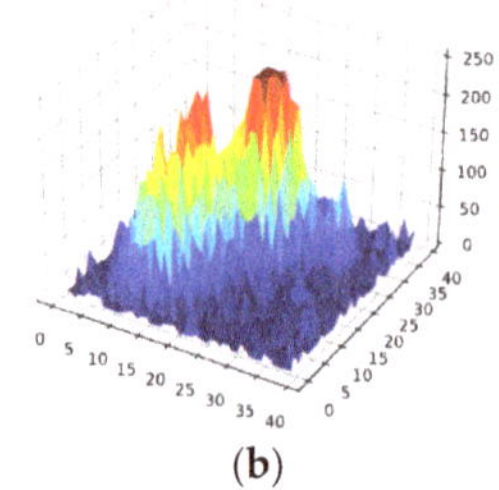

(a)　　　　　　　　　　　(b)

Figure 7. Simulation results with $A = 2000$, $\sigma = 2$, $N_{img} = f(x, y) \times 1 \times \eta(0,20)$, $N_{\Delta x}, N_{\Delta y} \sim \text{Gaussian}(0,2)$: (**a**) 2D; (**b**) 3D.

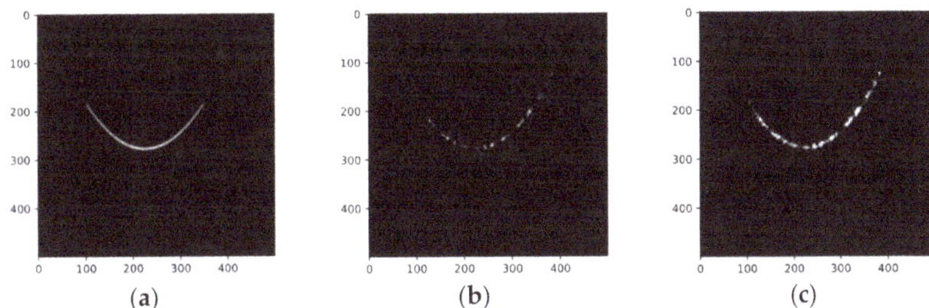

Figure 8. Simulation results of simulating star point trajectories as curvilinear equations: (a) results with $A = 2000$, $\sigma = 2$, $N_{img} = f(x, y) \times 1 \times \eta(0,0)$, $N_{\Delta x}, N_{\Delta y} \sim$ Gaussian$(0,0)$; (b) results with $A = 2000$, $\sigma = 2$, $N_{img} = f(x, y) \times 1 \times \eta(0,0)$, $N_{\Delta x}, N_{\Delta y} \sim$ Gaussian$(0,2)$; (c) results with $A = 2000$, $\sigma = 2$, $N_{img} = f(x, y) \times 1 \times \eta(0,40)$, $N_{\Delta x}, N_{\Delta y} \sim$ Gaussian$(0,2)$.

4. Interface Function Description

The field of view of the simulated detector in the proposed method is $12° \times 12°$, and the size of its plane array is 512 pixels $\times$ 512 pixels. The magnitude sensitivity is 8.0 apparent magnitude, the simulated imaging system has a frame rate of 50 Hz, and the exposure time for each frame is 20 ms. The interface uses the Tkinter module based on Python 3.9.

There are two functions in the designed interface. First, it is the static single-frame star map simulation. Figure 9 shows the initial interface of this function. In this state, the default number of the target star point is 1. The adjustable parameters are the Gaussian noise level of the image, the number of background stars, the maximum value of target brightness, and the minimum value of target brightness. Users can generate single-frame simulated star maps, with specified parameters, in this mode. Additionally, a function to display images is added, which can also be used separately as a picture viewer. After generating the simulated star maps, the primary information of the target star point, such as brightness information, position coordinates, and Gaussian dispersion radius, can be viewed in "Target Star Information."

Figure 9. The interface of static single-frame star map simulation.

The second function is the dynamic sequence star map simulation, and its user interface is shown in Figure 10. In this mode, the simulation can provide two motion states for a target star point. One of the states is that the ground-based camera system has followed the target point to be measured. Under this state, by default, the target in the first frame is in the center of the image. In addition to the size and brightness of a star point, it is also possible to simulate errors in the actual situation by inputting the magnitude of the attitude disturbance of a target point and the average and standard deviation of Gaussian noise. The default exposure time of each frame of the system is 20 ms. By adjusting the size of the "Dynamic Frame Count," users can input the imaging time of the simulated ground-based camera system. The other state is that the ground-based camera system has not followed the target point to be measured. Under this state, both the target and background star points have different movement direction speed sizes, compared with the ground-based camera system. The user must input the x-direction and y-direction offset to complete the exact movement simulation of the target star point. Since it would take time to generate a large number of background star points in real-time to meet the requirements, the way to handle the background star points is by loading a pre-generated binary file. This file is the sum of a specified number of star trajectories laid out according to a preset angle. The user can also change this, according to specific requirements.

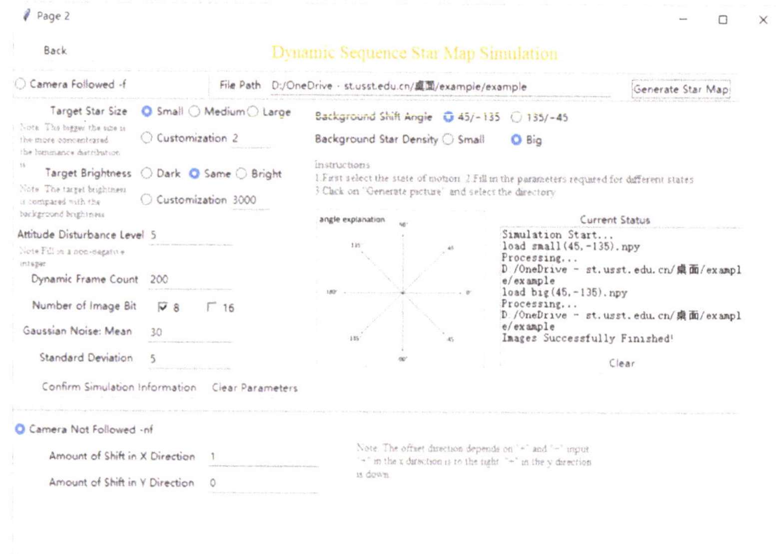

Figure 10. The interface of the dynamic sequence star map simulation.

Figure 11 shows the first, the 90th, and the 150th frames of the 200-frame sequence of star maps simulated under the conditions of the selected states and parameters depicted in Figure 10. The target to be measured is marked with a red circle, and the coordinates of the target star point in different frames are listed in the white textboxes.

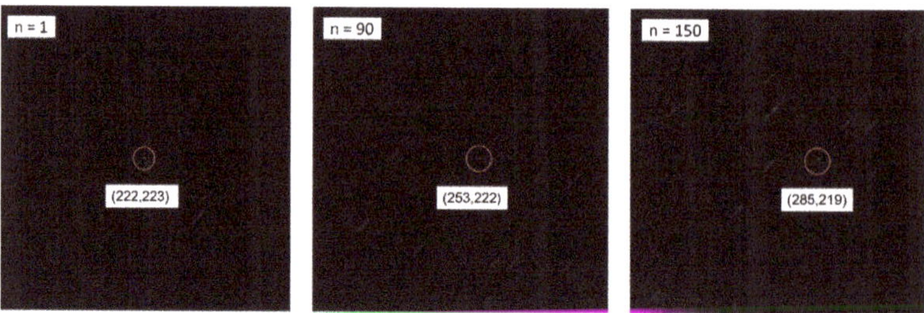

Figure 11. Dynamic sequence star maps.

5. Results

5.1. Image Quality Assessment

Figure 12 shows simulation results using the method proposed in this paper. Figure 12a,b shows the static single-frame star map simulation results, and Figure 12c shows the dynamic sequence star map. Figure 13 displays actual photos from the Shanghai Sheshan Observatory and from astronomer Michael A. Earl. From the gray histograms of the two types of images, we can determine that the actual photos have a broader gray distribution than that of the simulation results. Furthermore, there are some stripes in the photos. These stripes may come from the telescope lens.

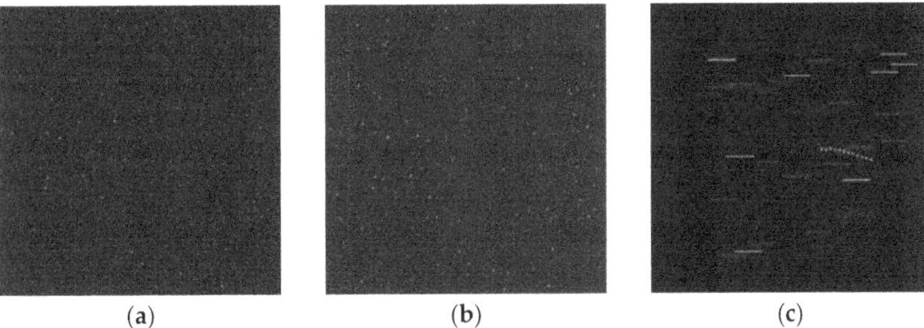

(a) (b) (c)

Figure 12. Simulation results using the proposed method: (**a**) simulation result for Figure 13a; (**b**) simulation result for Figure 13b; (**c**) simulation result for Figure 13c.

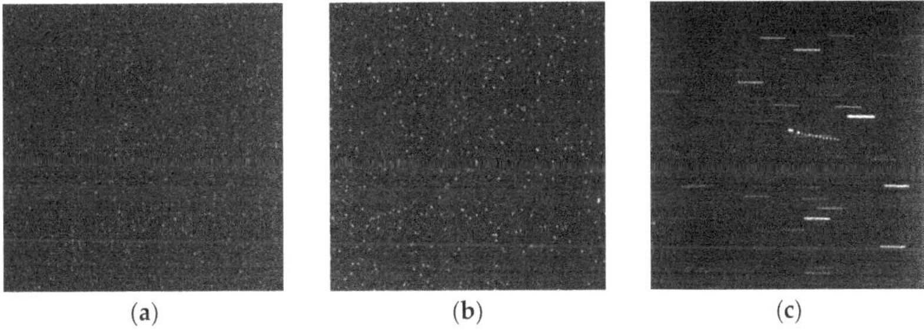

(a) (b) (c)

Figure 13. Real star map images: (**a**) taken by the Shanghai Sheshan Observatory; (**b**) taken by the Shanghai Sheshan Observatory; (**c**) taken by Canadian astronomer Michael A. Earl with a ground-based telescope ("Geostationary Satellite MSAT in Motion," http://www.castor2.ca/14_Images/Satellites/index.html accessed on 15 November 2021).

Here, we use two methods to evaluate the image similarity between the simulation results and actual photos. Figure 14 demonstrates the gray histograms of the simulated images in Figure 12. And Figure 15 demonstrates the gray histograms of the real images in Figure 13. First, the Bhattacharyya coefficient which is based on the gray histogram is used to indicate the color distribution of an image. The coefficient ranges from 0 to 1. The closer to 1, the more similar the two images are proved to be in terms of color distribution. The second standard is the cosine similarity, which changes an image into vectors by adding the divided gray level area numbers [19]. The calculation results are shown in Table 2, which point out that there is still a gap between the grayscale distribution of the simulated results and the actual photos, but the overall character of both is very close.

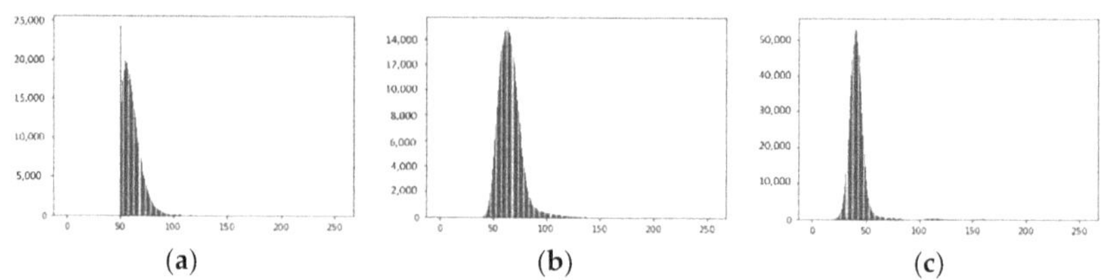

(a) (b) (c)

Figure 14. Gray histograms of the simulated images: (**a**) gray histogram of Figure 12a; (**b**) gray histogram of Figure 12b; (**c**) gray histogram of Figure 12c.

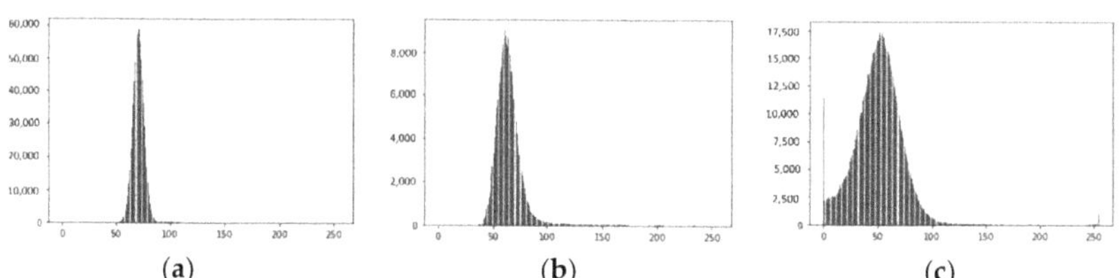

(a) (b) (c)

Figure 15. Gray histograms of the real images: (**a**) gray histogram of Figure 13a; (**b**) gray histogram of Figure 13b; (**c**) gray histogram of Figure 13c.

Table 2. Calculation results of image similarity.

Number	Bhattacharyya Coefficient in Gray Histogram	Cosine Similarity
Figure 12a vs. Figure 13a	0.778	0.977
Figure 12b vs. Figure 13b	0.639	0.992
Figure 12c vs. Figure 13c	0.639	0.893

Figure 12c shows the effect of a 2 s imaging time in a dynamic environment simulated using the proposed method's noise parameter $N_{img} = f(x, y) \cdot 1 \cdot \eta(0, 15), N_{\Delta x}, N_{\Delta y}, \sim \text{Gaussian}(0, 3)$. Figure 13c shows an actual image taken by Canadian astronomer Michael A. Earl with a ground-based telescope. These are some of the differences between Figures 12c and 13c and the main reasons behind them. Besides the direction of the background trajectories, which depends on the movement states of the simulated conditions, the trajectory style of the target is more noticeable in the actual photos. Because the satellite tumbles at a regular interval within an extended period, the satellite's orbit produces a periodic flickering phenomenon, which is not addressed by the method in this paper. Next, the noise distribution in Figure 13c is more complicated than in Figure 12c. Figure 13c becomes brighter due to the height of the image, which means the average of Gaussian noise becomes bigger. which means the average of Gaussian noise becomes bigger. In addition, some black dots and black areas exist in the figure in a more obvious way. It is necessary to determine different characteristics and modes of noise to improve the quality of the simulation results.

5.2. Star Centroid Extraction

The centroid of stars is the most important factor in attitude determination. We adopt the common centroid method to evaluate the accuracy of the simulation result of the proposed method. Figure 16a shows an actual image taken by Michael A. Earl. Four distinct targets are circled in the image, and the simulation result is shown in Figure 16b.

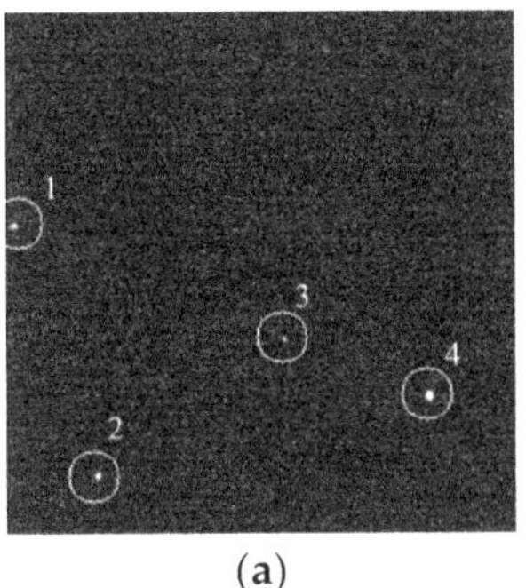
(a)

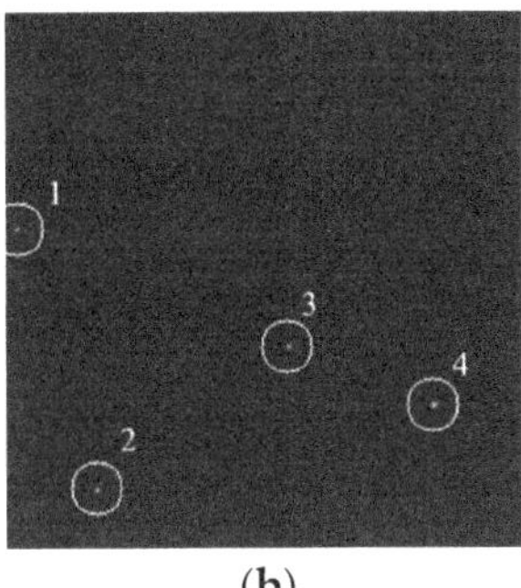
(b)

Figure 16. Star centroid extraction comparison: (**a**) courtesy of Mike Earl—CASTOR; (**b**) simulation result of the proposed method.

Next, before we perform the star centroid extraction process, we need to implement the Gaussian filter to reduce the interference from background noise in Figure 16a. The Gaussian kernel size is 5 × 5 and the Gaussian kernel standard deviation is computed with the formula provided by OpenCV (getGaussianKernel(), https://docs.opencv.org/4.0.0/d4/d86/group__imgproc__filter.html#gac05a120c1ae92a6060dd0db190a61afa accessed on 15 November 2021). Finally, we adopt the common centroid method to extract the "center of mass" as follows:

$$\begin{cases} x_0 = \dfrac{\sum_{i=1}^{m} \sum_{j=1}^{n} x_i f_{ij}}{\sum_{i=1}^{m} \sum_{j=1}^{n} f_{ij}} \\ y_0 = \dfrac{\sum_{i=1}^{m} \sum_{j=1}^{n} y_j f_{ij}}{\sum_{i=1}^{m} \sum_{j=1}^{n} f_{ij}} \end{cases} \tag{10}$$

where f_{ij} denotes the value of the image after background noise reduction.

The same steps will apply to the detection of Figure 16b. The results for the two images are shown in Table 3.

Table 3. Results of star centroid extraction in Figure 16a,b.

	Figure 16a		Figure 16b		Error	
No.	x	y	x	y	Δx	Δy
1.	7.329	121.973	7.498	121.983	0.169	0.01
2.	52.113	267.078	52.037	266.969	−0.076	−0.109
3.	163.037	187.762	162.989	187.964	−0.048	0.202
4.	248.171	220.978	247.978	220.970	−0.193	−0.008

Where the errors are calculated as the follow equation:

$$\begin{cases} \Delta x = x_\beta - x_\alpha \\ \Delta y = y_\beta - y_\alpha \end{cases} \tag{11}$$

where x_β and y_β denote the x and y coordinate values of the stars in Figure 16b. x_α and y_α denote the x and y coordinate values of the stars in Figure 16a.

As we can see, because of the noise of figures, the accuracy is not perfect, but it is acceptable for common use. If we perform the image processing in detail and apply a more specific centroid extraction method, the error will be smaller.

6. Conclusions

In summary, this paper carries out a simulation method for dynamic sequential star maps of space targets in multiple scenarios with complex star backgrounds. Numerical simulation provides a more convenient way to generate the abundant data of star maps to promote the study of star sensors. A static star point model and a dynamic sequence star map model are constructed, respectively. We take the random noises, such as detector noise and dark current noise, and uneven environment illumination into account. Meanwhile, the impact of the noise model analysis and platform motion disturbances on imaging will also provide a reference for performance testing and verification of star sensors. In the next stage of the work, we may consider the condition in which multiple interference target star points exist. In addition, it is essential to continue to study the noise sources and patterns in authentic images so that the parameters can be further optimized, making them more similar to those in the actual scenario.

Author Contributions: Conceptualization, H.Y.; methodology, Y.H.; software, Y.J.; validation, Y.H. and D.Z.; formal analysis, Y.Y. and J.L. (Jin Liu); investigation, J.L. (Jun Li); writing—original draft preparation, Y.J.; writing—review and editing, Y.J. and H.Y.; project administration, X.J. and X.Y. All authors have read and agreed to the published version of the manuscript.

Funding: This work was supported in part by the Joint Astronomical Fund of the National Natural Science Foundation of China (grant number U1831133), the Key Laboratory of Space Active Opto-electronics Technology of the Chinese Academy of Sciences (grant number 2021ZDKF4), and the Shanghai Science and Technology Innovation Action Plan (grant numbers 21S31904200,22S31903700).

Institutional Review Board Statement: Not applicable.

Informed Consent Statement: Not applicable.

Conflicts of Interest: The authors declare no conflict of interest.

References

1. Lu, R.; Wu, Y.P. An Approach of Star Image Simulation for Strapdown Star Sensor. *Aerosp. Control. Appl.* **2016**, *42*, 57–62. [CrossRef]
2. Wang, H.Y.; Song, Z.F.; Li, J.J.; Xu, E.S.; Qin, T.M. Simulating method study on stray light noise out of sunlight baffle of star tracker. In *SPIE 96750N, AOPC 2015: Image Processing and Analysis*; SPIE: Bellingham, WA, USA, 2015; pp. 166–172. [CrossRef]
3. Wei, X.G.; Zhang, G.J.; Fan, Q.Y.; Jiang, J. Ground function test method of star sensor using simulated sky image. *Infrared Laser Eng.* **2008**, *37*, 1087–1091.
4. Xing, F.; Wu, Y.P.; Dong, Y.; You, Z. Research of laboratory test system for micro star tracker. *Opt. Tech.* **2004**, *30*, 703–705+709.
5. Liu, H.B.; Su, D.Z.; Tan, J.; Yang, J.K.; Li, X.J. An approach to star image simulation for star sensor considering satellite orbit motion and effect of image shift. *J. Astronaut.* **2011**, *32*, 1190–1194. [CrossRef]
6. Zheng, X.J.; Shen, J.; Wei, Z.; Wang, H.Y. Star map simulation and platform influence of airborne star sensor based on J-band data of 2MASS catalog. *Infrared Phys. Technol.* **2020**, *111*, 103541. [CrossRef]
7. Wang, H.Y.; Yan, Z.Q.; Mao, X.N.; Wang, B.W.; Liu, X.; Kang, W. A new high-precision star map simulation model and experimental verification. *J. Mod. Opt.* **2021**, *68*, 856–867. [CrossRef]
8. Yang, J.; Liang, B.; Zhang, T.; Song, J.Y.; Song, L.L. Laboratory Test System Design for Star Sensor Performance Evaluation. *J. Comput.* **2012**, *7*, 1056–1063. [CrossRef]
9. Liu, F.C.; Liu, Z.H.; Liu, W.; Liang, D.S.; Yuan, H. Space target sequence image simulation based on STK/matlab. *Infrared Laser Eng.* **2014**, *43*, 3157–3161.
10. Yan, J.Y.; Liu, H.; Zhao, W.Q.; Jiang, J. Dynamic real-time star map simulation based on convolution surface. *J. Beijing Univ. Aeronaut. Astronaut.* **2019**, *45*, 681–686. [CrossRef]
11. Wang, Y.P.; Niu, Z.D.; Song, L. A Simulation Algorithm of The Space-based Optical Star Map with Any Length of Exposure Time. In Proceedings of the 2021 International Conference of Optical Imaging and Measurement (ICOIM), Xi'an, China, 2 September 2021; pp. 72–76. [CrossRef]
12. Zhu, P.; Jiang, Z.; Zhang, J.L.; Zhang, Y.; Wu, P. Remote sensing image watermarking based on motion blur degeneration and restoration model. *Optik* **2021**, *248*, 168018. [CrossRef]
13. Wang, H.Y.; Wang, T.F.; Zhu, H.Y.; Liu, T.; Ge, C.J. Method for Determining Optimal Radius Value of Defocused Image Spot of Star Sensor. *Laser Optoelectron. Prog.* **2019**, *56*, 101101. [CrossRef]
14. Zhang, G.J. *Star Identification*; Springer: Berlin/Heidelberg, Germany, 2017; pp. 53–55. [CrossRef]

15. Wang, H.Y.; Zhou, W.R.; Lin, H.Y.; Wang, X.L. Parameter Estimation of Gaussian Gray Diffusion Model of Static Image Spot. *Acta Opt. Sin.* **2012**, *32*, 0323004. [CrossRef]
16. Zhi, S.; Zhang, L.; Li, X.L. Realization of simulated star map with noise. *Chin. J. Opt. Appl. Opt.* **2014**, *7*, 581–587. [CrossRef]
17. Gao, Y.; Lin, Z.P.; Li, J.; An, W.; Xu, H. Imaging Simulation Algorithm for Star Field Based on CCD PSF and Space Targets Striation Characteristic. *Electron. Inf. Warf. Technol.* **2008**, *23*, 58–62.
18. He, Y.Y.; Wang, H.L.; Feng, L.; You, S.H.; Chen, Z.K. Star spot motion trajectory modeling and error evaluation under large angular velocity. *J. Beijing Univ. Aeronaut. Astronaut.* **2019**, *45*, 1653–1662. [CrossRef]
19. Wang, C.Q.; Shen, X.L.; Li, L. Analysis on image similarity calculation algorithm. *Mod. Electron. Tech.* **2019**, *42*, 31–38. [CrossRef]

MDPI AG
Grosspeteranlage 5
4052 Basel
Switzerland
Tel.: +41 61 683 77 34

Photonics Editorial Office
E-mail: photonics@mdpi.com
www.mdpi.com/journal/photonics

www.ingramcontent.com/pod-product-compliance
Lightning Source LLC
LaVergne TN
LVHW071518170726
843515LV00008B/2005